T0203053

IONIZED
GASES

AMERICAN VACUUM SOCIETY CLASSICS

H. F. Dylla, Series Editor-in-Chief

Basic Data of Plasma Physics:
The Fundamental Data on Electrical
Discharges in Gases
Sanborn C. Brown

Field Emission and Field Ionization
Robert Gomer

Vacuum Technology and Space Simulation
David H. Holkeboer, Donald W. Jones, Frank Pagano,
and Donald J. Santeler

The Physical Basis of Ultrahigh Vacuum
P. A. Redhead, J. P. Hobson, and E. V. Kornelsen

Handbook of Electron Tube and
Vacuum Techniques
Fred Rosebury

Vacuum Sealing Techniques
A. Roth

Ionized Gases
A. von Engel

IONIZED GASES

A. von Engel

Reprinted by arrangement with Oxford University Press.
©A. von Engel 1965.
Published by the Springer New York in 1994.

Library of Congress Cataloging-in-Publication Data

von Engel, A. (Alfred)
    Ionized gases/by A. von Engel.
        p. cm.—(American Vacuum Society classics)
    Includes bibliographical references and index.
    ISBN 978-1-56396-272-1
    1. Ionized gases. I. Title. II. Series.
QC702.V65   1993   93-27401
530.4'4–dc20   CIP

Series Preface

The science of producing and measuring high and ultrahigh vacuum environments has fundamental interest for basic research in addition to a wide variety of important practical applications. Basic research involving particle physics, atomic and molecular physics, plasma physics, physical chemistry, and surface science often involves careful production, control, and measurement of a vacuum environment in order to perform experiments. Practical applications of vacuum science and technology are found in many materials processing techniques used for microelectronic, photonic, and magnetic materials and in the simulation of space and rarefied gas environments.

As in most of modern science, the rapid development of the field has been accompanied by a parallel growth in the related technical literature including specialized journals, monographs, and textbooks. There exist early publications in vacuum science and technology which have attained the status of indispensable references among practitioners, lecturers, and students of the field. Many of these "classic" publications have gone out of print and are currently unavailable to newcomers to the field. The present series, commissioned by the American Vacuum Society, and published by AIP Press to celebrate the 40th anniversary of the Society in 1993, is entitled the "American Vacuum Society Classics."

The American Vacuum Society Classics will reprint important books from the last four decades that continue to have significant impact on the modern development of the field. It is the goal of the American Vacuum Society Classics to reprint these books in a high quality and affordable format to ensure wide availability to the technical community, individual researchers, and students.

H. F. Dylla
Continuous Electron Beam
Accelerator Facility
Newport News, VA
and
College of William and Mary
Williamsburg, VA

PREFACE TO SECOND EDITION

THE request for a second edition is of great satisfaction to every author. In this case, it seems that over the years the book has not only been widely used and approved by many of my senior colleagues but particularly by numerous young scholars at home and abroad. The first edition has also been translated into three eastern languages, all unfamiliar to me and spoken in countries I have not visited yet.

Though the main structure of the first edition has been maintained, quite a few changes have been made in this edition. In order to emphasize the textbook character a large number of questions and problems—some of them demanding essays—have been included at the end of each chapter. The answers to some of these, whenever necessary and possible, have been provided too and I am sure that the collection will stimulate some lively discussions.

A further innovation is the inclusion of empty pages in the reference section which will enable the reader to add new references to the bibliography presented. Also, it will be seen that the number of diagrams, tables, and references has been considerably increased and the appendixes enlarged. The index contains—as before—some page numbers with the letters F and T added, referring to figures and tables respectively. Finally, a short glossary has been added to help those interested in etymology. In spite of all that the new volume has been kept at a reasonable size and price.

Textual changes have been made foremost in Chapter 3 which has been extensively revised and amended. In particular certain sections dealing with the production and properties of excited atoms and molecules, photoionization including that by laser beams and kindred subjects have been added. Smaller changes will be found in nearly all chapters and appendixes. Although a number of errors contained in the first edition have been removed, I am sure that—in exchange for it—the reader will find new errors.

I have resisted any temptation to extend the field originally treated in this book, though I may have been induced to do so by my vested interests. However, such an extension has become now partly unnecessary in view of such lucid treatises as G. Francis's book *Ionization Phenomena in Gases*. However, another artifice was used to stem expansion : the treatment of some of the problems has been transferred to the questions at the end of each chapter, and hence the reader is strongly recommended to study the problems introduced there.

I should like to put on record the assistance I have received in my work by various authorities. They include the Governing Body of Keble College, the Clarendon Laboratory of Oxford University, the Atomic Energy Authority, the Department of Scientific and Industrial Research, the Central Electricity Generating Board, and the United States Air Force, Aeronautical Systems Division. Their support has enabled me to combine research in this field with undergraduate teaching in general physics, as a result of which I was able to keep a balanced mental diet throughout the years. For that and for the support in complex matters of administration and laboratory I am grateful to Professor B. Bleaney, F.R.S., and Mr. T. C. Keeley, C.B.E.

My many friends and colleagues here and in foreign lands have given me much help and have shown great patience, in particular Drs. R. N. Franklin and J. R. Cozens. On special problems I had the advice of Drs. S. J. B. Corrigan, C. C. Goodyear, and K. W. Arnold and editorial work in the final stages has been alleviated by Dr. J. M. Breare to whom I owe my thanks. I am indebted to Professor K. Yamamoto of Nagoya University and Dr. A. E. Robson of Harwell for sending me lists of mistakes and misprints contained in the first edition, to Mr. W. S. Barrett for correcting the glossary, and to Mr. P. H. Vidaud de Plaud for reading the proofs and spotting errors.

Finally, I wish to thank my publishers for their continuous help and for agreeing to my wishes.

A. v. E.

Oxford, Keble College
August 1964

PREFACE TO FIRST EDITION

IT is now more than twenty years since my first contribution to a text-book on electric discharges in gases. For reasons connected with the economics of publishing it was not possible to produce an English translation of these two volumes. However, during the last war the book was reproduced in the U.S.A. by the photo-offset process.

Since then the study of ionization phenomena has made great progress and various research schools in this country and abroad have been engaged in work on this subject. After the first swift advances in nuclear physics it appears that there are quite a number of important fundamental problems left to attract researchers to other fields of physics in spite of the oracles of our professional prophets. In fact, the understanding of a great many questions in atomic physics, astrophysics, spectroscopy, the physics of the atmosphere and of the solid state, to quote only a few, depend on the better grasp of the processes in the gaseous state. This is particularly felt now in the new field of semiconductors and their applications.

The text of this book is based on my lectures for undergraduates in the physics school at Oxford. It is a more modest enterprise than the book which I wrote with M. Steenbeck. Although it is intended as an introduction to this field, it should also be of value to researchers and scientific engineers. It puts emphasis on the fundamental concepts and the limitations of the various treatments. Consequently it deals less extensively with the details of the theories, their application to real discharges and their more practical uses. In order to bring out the salient features, a large number of simplifications have had to be made which will be disliked by some and hailed by others. Every teacher knows that these are often a necessary artifice, though they annoy the purists and are sometimes said to mislead beginners.

This book does not give such a complete account as one might expect from its title, but I have carefully considered what parts of the subject should be excluded in order to make the book more readable. However, I have included far more data and references than I originally intended, so as to help researchers in a quantitative analysis of their work. An extensive subject index allows discrimination between references to the text, figures, and tables. An author index has been omitted mainly to prevent the author from quoting his own work too often.

PREFACE TO FIRST EDITION

In Oxford, where I have the good fortune to teach and research, many of my early dreams have come true. I owe this largely to my friends who have given me advice and moral support. My thanks go to Sir John Townsend, whom I joined in 1940, and particularly to Lord Cherwell, whose help and encouragement I wish to record gratefully. To E. W. B. Gill, with whom I worked closely, go my special thanks for enriching my wisdom, physics, and fun. I wish to thank the Governing Body of Christ Church for giving me accommodation in the House and the Governing Body of Exeter College for encouraging post-graduate research. I also wish to thank the Royal Society, and in particular Sir George Thomson, for taking interest in this work and providing Royal grants.

I owe a great debt to T. C. Keeley, the Director of the Clarendon Laboratory, whose patience and assistance has been to me a constant source of strength. I wish to thank Dr. S. Whitehead, Sir Harold Hartley, Sir David Brunt, Sir George Nelson, Dr. J. S. Forrest, Dr. T. E. Allibone, and Mr. R. Davis, for the support our work has received through them. I have had in the past many stimulating discussions with Professors Llewellyn Jones, Emeleus, Loeb, Allis, Weissler, Meek, S. C. Brown, and Doctors Penning, de Groot, Prowse, Craggs, and Thonemann from whom I have drawn much inspiration. I have greatly benefited by my studies as a Fulbright Travelling Fellow in the U.S.A. for which I am most grateful to L. B. Loeb and W. P. Allis as well as to my numerous American colleagues.

Perhaps I have received most inspiration from my pupils who have worked and argued with me for many years, in particular Doctors G. Francis, W. L. Harries, P. F. Little, and R. J. Bickerton. I especially wish to acknowledge the help I have received from Mr. A. E. Robson, who has read and improved the manuscript, and from my wife, who has typed the larger part of it. Finally, I wish to thank the Superintendent and Staff of the Radcliffe Science Library for their constant helpfulness and the Clarendon Press for acceding to my numerous suggestions and the careful preparation of the printing.

A. v. E.

Oxford
January 1955

CONTENTS

CONTENTS

CONTENTS

1

HISTORICAL NOTES

THE electric gas discharge is a phenomenon which is observed when a gas or vapour becomes electrically conducting. Under these conditions free electric charges are present and can move through the gas, usually under the influence of an electric field: the gas is said to be ionized.

It appears that discharge physics originated in England, although it should not be too difficult to establish that basic ideas of this branch of knowledge sprang up at about the same time in other parts of the world, particularly in France, Italy, and Germany. One of the earliest sources seems to be W. Gilbert's *De magnete, magneticisque corporibus, et de magno magnete tellure*, published in London in 1600. Gilbert, Queen Elizabeth's physician, found that a charged conductor loses its charge when brought near to a flame, and that an electroscope becomes charged when it is connected to a flame. Coulomb (31) in 1785 was probably the first to deduce from his observations with the electrostatic torsion balance that a charged metal sphere loses the greater part of its charge through the air and not through imperfect insulation.

In the middle of the seventeenth century O. von Guericke (32) observed electric sparks with the first electrostatic machine, a large sphere of sulphur rubbing against a cloth. Sparks caused by harnessing atmospheric electricity were first produced in 1751 by B. Franklin (33) in his hazardous experiments with wires suspended from kites flown in thunder clouds. At about the same time (1777) Lichtenberg (34) studied dust patterns left by a spark discharge passing along the surface of an insulator.

About 1800 H. Davy (35) in England and Petroff (36) in Russia discovered the arc discharge. They observed that when the points of two pieces of charcoal connected to a battery were brought together and then drawn apart, a continuous discharge developed in air. It formed 'an ascending arc of light' of brilliancy unknown hitherto. The energy was supplied by a battery consisting of several thousand cells and the current must have been of the order of several amperes. The gas in the arc discharge was found to be at a very high temperature since platinum, lime, and magnesia fused readily in it, while diamond and plumbago (graphite) easily vaporized; the arc persisted even when the air pressure was somewhat reduced. The electric properties of the arc were not studied systematically until about a hundred years later when Mrs. Ayrton (37) started her researches. Her monograph, which deals with the short arc in air, contains a nearly complete history of its discovery.

Between the years 1831 and 1835 Faraday (38), working at the Royal Institution, took up research on gas discharges at low gas pressure. He discovered what he called a glow discharge which consisted of a series of alternate luminous and dark zones. They varied in length and colour; sometimes they were stationary and sometimes in motion (striations). All these phenomena were observed in tubes filled with air to a few millimetres of pressure while the discharge was maintained by a source of potential of about 1000 V. In addition, Faraday seems to have been the first to find that the current can pass through a discharge tube filled with a gas at low pressure without showing any luminosity at all. This he called a dark discharge. Particular attention should be paid to Faraday's printed works since they are masterpieces of simplicity, modesty, and conciseness.

It is convenient to regard the dark discharge, the glow discharge, and the arc discharge as the three fundamental types of continuous electric discharge. They are self-sustained in that they can be maintained without the support of an external ionizing agency.

During the twenty years following, not much spectacular progress was made in this field. Subsequently discharge physics was mainly concerned with research into the various kinds of rays produced by the electric discharge. In 1858 Plücker (39) discovered that a glow discharge working at a pressure of order 1/100 mm Hg emits cathode rays. The beam which originates from the cathode colours the gas along its path and when it impinges on the glass wall of the discharge tube a green fluorescent spot is produced. Hittorf (40) observed in 1869 that these cathode rays are deflected in a magnetic field, and Goldstein (41) in 1876 and Hertz (42) in 1883 showed that cathode rays are deflected by an electric field.

It was first thought that the transport of electricity through gases took place in a similar way to the transport of charges through conducting liquids. However, Crookes, who at that time investigated discharges at low pressure, visualized charge carriers quite different from those assumed to exist in electrolytes. It was probably at the Congress in Sheffield in 1874 that he put forward the first hypothesis that cathode rays are most likely to be fundamental particles of the atom. In 1879 Crookes published his well-known paper (43) in the *Philosophical*

Transactions, where he remarked: 'The phenomena in these exhausted tubes reveal to physical science a new world, a world where matter may exist in a fourth state. . . .'

There followed a period rich in discoveries. Righi (44) from 1876 onwards studied spark-discharges and detectors particularly in the centimetre-wave region. H. Hertz (45) observed in 1887 that light emitted by a spark caused a nearby spark gap to break down more easily. One year later Hallwachs (46) discovered that a zinc plate which is irradiated with ultraviolet light (from an arc lamp) becomes positively charged—as we know now through the emission of photo-electrons. Soon it became clear that the cathode rays must have a mass much smaller than that of the lightest gas atoms. Thus they were taken as being atoms of negative electricity (57), and Stoney (47) in 1891 proposed for these particles the name 'electron'. In 1874 at the Belfast Meeting of the British Association he said: 'Now the whole of the quantitative facts of electrolysis may be summed up in the statement that a definite quantity of electricity traverses the solution for each bond that is separated.' However, Stoney did not assume that the electrons could exist in a free state and independent of an atom.

There was a long-lasting controversy on whether cathode rays consisted of charged particles or whether they were disturbances propagated through the ether. It was not settled until Perrin (48) in 1895 led a beam of cathode rays into a Faraday cylinder and established that they carried a negative charge. Finally J. J. Thomson (49), Kaufmann (50), and Wiechert (51) derived from the fact that the charge to mass ratio of the cathode rays (reduced to zero speed) is independent of pressure, nature of the gas, and material of the cathode, that the particles are of a type common to all elements and thus distinct from electrolytic ions. Later Lenard (52) showed that cathode rays can be led from highly evacuated tubes into the atmosphere through metal windows of a few microns thickness.

The existence of rays of positive ions was first shown by Goldstein (53) in 1886, who passed these 'canal rays' into an adjoining chamber through a hole in the cathode of a glow discharge. Later investigations by J. J. Thomson (54), F. W. Aston (55), and particularly by W. Wien (56) furnished valuable information about the properties of beams of positive ions, a field which is by no means yet exhausted.

ESSAY QUESTIONS

(1) Explain how the experiments of Hallwachs, Lenard, and others led to Einstein's theory of photo-electric emission. Describe the various experiments which show the validity of the theory. Discuss the studies by Elster and Geitel, Richardson, and Millikan.

(2) Write an essay on the state of knowledge around the turn of the century when the inter-relation between the various branches of physics had become apparent. Show in particular how the relation between electromagnetic waves and light has been verified experimentally. Include works by Hertz, Righi, and others.

(3) Write notes on the fundamental discoveries and ideas of Michael Faraday with special reference to his work on gas discharges.

(4) Describe the origin of high-temperature chemistry: in particular Petroff's and Davy's investigations on the arc and its discovery. What sources of electricity were originally available and what restricted their use?

(5) Give an account of the steps leading to the present concept of electrons and ions. Emphasize the close relation between the transport of charges in electrolytes and ionized gases.

2

CONDUCTION IN FEEBLY IONIZED GASES

1. INTRODUCTION

A gas becomes a conductor of electricity if free charges such as ions, electrons, or heavy charged particles are present. Positive ions are atoms, molecules, or groups of molecules which have lost one or more electrons and are thus singly or multiply charged. Negative ions, however, are similar particles usually with one electron attached to them like H^-, O_2^-, I^-, OH^-, etc. In the majority of cases positive ions have a single charge like H^+, He^+, H_2^+, O_2^+, CO^+, etc.; an example of a doubly charged atomic ion is an α-particle, namely He^{++}. Rare gases form molecular ions like He_2^+, Ne_2^+, etc.; excited ionic species such as $(He^*)^-$ have been observed too.

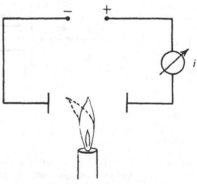

Ionization in gases and electrification of metallic or insulating surfaces can be produced by illumination with ultraviolet light or X-rays, by bombarding the substance with α-particles, and in many other ways which will be discussed in Chapter 3.

FIG. 1. Making visible the movement of ions in a hot gas and measuring the ion current flowing. Applied potential: several kV.

For demonstration purposes coal-gas or candle-flames (25, 61) are quite useful and respectable sources of gaseous ions, as can be shown by a simple experiment (Fig. 1). A lighted wax candle is arranged between two plane electrodes a few centimetres apart, the tip of the flame being at about the centre of the gap. When a potential difference of order 1 kV is applied to the gap, we observe that the flame is strongly deflected, particularly the tip of the flame which is bent towards the negative electrode. This experiment suggests that there are more positive ions than negative ions in the tip and the mantle. Further down the core of the flame exhibits a surplus negative charge. It follows that if more ions of one kind are present, more collisions between these ions and the neutral gas molecules take place, resulting in a displacement of the hot flame gas in the direction of the motion of the ions. A study of the properties of such hot flame gases reveals that the hottest part lies in the yellowish-white region where a temperature of order of 2000° C exists. There an intense thermal ionization of the gas occurs. The degree of deflexion increases with increasing electric field; it is determined by a balance between the electric force moving the ions horizontally and the 'elasticity' of the flame gas corresponding to a vertical force which is caused by convection currents in air. Details in (70f).

By inserting a galvanometer it can further be shown that a current flows in the circuit (Fig. 1) as a result of the motion of the ions through the atmospheric air. The current strongly increases when an alkali salt is vaporized in the flame. As has been indicated above, the ionization in the flame gas is not uniform. Therefore this experiment is not suitable for further quantitative deductions.

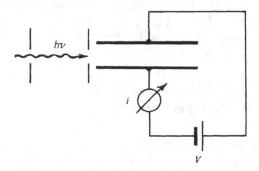

FIG. 2. Ionization of a gas by passing X-rays through it. These are equivalent to light quanta each of energy $h\nu$.

2. IONIZATION CURRENTS

In order to investigate quantitatively the dependence between the current flowing through an ionized gap and the applied field strength, a second experiment will be described (16, 27). Here a discharge gap with two large parallel plane electrodes in atmospheric air is used. The interspace is irradiated with a strong beam of X-rays which sets up a uniform ionization throughout the gas volume; the beam must be restricted laterally so as to avoid its falling on the electrodes (Fig. 2). If the applied voltage in this circuit is gradually increased, while the strength of the irradiation is kept constant, it is found that the current rises, at first relatively steeply and later on more slowly, until at relatively large voltages it becomes constant. This last effect indicates that in large fields all the ions which have been produced in the gap by the X-rays are moving to their respective electrodes, thereby setting up a current in the outer circuit. In weak fields, however, only a portion of the ions produced can reach the electrodes because a large number of ions of unlike sign will recombine, that is neutralize their charges in the gas, before having reached an electrode.

A negative ion which arrives at the positive electrode will deliver the electron which has been attached to it and in general the neutral molecule so formed returns to the gas. If, on the other hand, a positive ion arrives at the negative electrode it will pick up an electron from the huge pool of electrons in the metal and then return as a neutral molecule into the gas. It follows that the contribution of a newly-formed ion to the current will depend on the time during which the ion stays in the gas as a free entity.

Next we shall treat quantitatively what is sometimes called the characteristic of the discharge that is set up in a parallel gap according to Fig. 2. Let N be the number of ion pairs in a unit volume, that is the number concentration, and let an external source uniformly irradiate the gas volume enclosed by the two infinitely large plates. A constant rate of production dN/dt ion pairs per cm³ and sec is assumed, and hence equal numbers of positive and negative ions are present. Let a sufficient time pass so that equilibrium between production and disappearance is established. We then have at any time

$$N^+ = N^- = N. \tag{2.1}$$

If the ions produced disappear solely by recombination, the rate at which neutralization takes place in the volume is proportional to the product of their concentrations and because of (2.1) proportional to N^2. The constant of proportionality is called the coefficient of recombination ρ. (This relation corresponds to that for the rate of bimolecular chemical reactions.) For equilibrium we have

$$dN/dt = -\rho N^2. \tag{2.2}$$

This equation holds only if the applied field is sufficiently weak. However, it is possible to have a small current flowing through the gas without invalidating (2.2) if the number of ions neutralized by recombination in unit time is large compared with the number collected by the electrodes. The current flowing through the gap per unit area of cross-section, that is the current density j, is given by the equation of continuity (the j's add up because unlike charges move in opposite directions)

$$j = j^+ + j^- = e(N^+v^+ + N^-v^-) = eN(v^+ + v^-), \tag{2.3}$$

where e is the electronic charge and v^+, v^- the respective velocities of positive and negative ions in the field, singly charged ions being assumed. Because of the weak electric field the motion of an ion through the gas is governed by the accelerating field and the retarding 'viscous force'; the inverse of the latter is proportional to the mobility k or drift velocity in unit field; this corresponds to that of ions in electrolytes. The relation between the drift velocity v and the field strength X is

$$v = kX. \tag{2.4}$$

k is in general different for positive and negative ions and designated as k^+ and k^- respectively. Combining (2.2), (2.3), and (2.4) we obtain

$$j = \{(k^+ + k^-)e[(dN/dt)/\rho]^{\frac{1}{2}}\}X, \tag{2.5}$$

where j, k, e, X are in e.s.u. Remember 1 A = 3.10^9 e.s.u., 1 cm/sec per V/cm = 300 e.s.u., e = 4·8.10^{-10} e.s.u., 1 V/cm = 300 e.s.u.

Thus at low fields j is proportional to X. The term in curly brackets is usually called the 'conductivity' of the gas. However, the similarity between (2.5) and Ohm's law is purely superficial: for example, the conductivity depends here on the rate of production of ions. Turning to numerical example, let dN/dt be of the order 10^8 ion pairs per cm³ and sec and $X \approx 1$ V/cm. The mobilities at 1 atmosphere are of the order 1 cm/sec per V/cm, and $\rho \approx 10^{-6}$ cm³/sec, so a current of the order 10^{-12} A/cm² ($\approx 5.10^{-3}$ e.s.u.) will flow through the circuit. This means that $j/e = 10^7$ ions/cm² and sec are collected at the electrodes which is to be compared with the rate of loss by recombination. Assuming a gap of 10 cm width, thus 10^9 ions per sec and cm² recombine, which is large compared with the number reaching the electrodes, justifying our assumptions. Since in general $k^+ < k^-$, we deduce from (2.3) that the number of positive charges arriving at the cathode is smaller than that of the negative charges arriving at the anode (section 3), though equal numbers are produced and recombine.

Let us now apply a field large enough to draw all the ions to the electrodes before they have a chance to become neutralized in the gap. The number of ions which can be removed from a gap d cm wide per cm² of electrode surface in unit time is $d(dN/dt)$ and the current density is

$$j = j_s = ed(dN/dt). \tag{2.6}$$

The suffix s indicates that this is a saturation current and hence is independent of the field strength. Using the same data as in the example above we find for a gap width of 10 cm a current density of 10^{-10} amp/cm². Fig. 3 gives the relation between the current density and the applied field according to (2.5) and (2.6). (2.6) applies to ionization chambers

(62) used for the measurement of the intensity of X-ray sources and the rate of ionization in the atmosphere (Table 2.1).

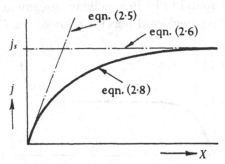

FIG. 3. Dependence of the current density j on the electric field X in a plane parallel condenser when the gas is uniformly irradiated with X-rays. j_s is the saturation current density.

TABLE 2.1

Ion concentration N and rate of ionization dN/dt in the atmosphere

Height	N^+ ions/cm³	N^-	dN/dt ions/cm³ sec
Ground	500–1100	400–850	4–10
2000 m ≈ 6000 ft .	650	550	4–10
4000 m ≈ 12 000 ft . .	1000	1000	12–20
5200 m ≈ 16 000 ft . .	2400	2000	≈ 30
16 000 m ≈ 50 000 ft .	..	..	≈ 200

The increase of N and dN/dt at greater heights is due to the reduced absorption of the soft components of the cosmic radiation (64). Over sea at zero level cosmic radiation provides about 2 ions/cm³ sec; over land 3·5 and 2 ions/cm³ sec are produced by radio-activity of the earth and air respectively, thus total 7·5 (63). Over sea $N^+ = N^-$; 500–700 small, 200 large ions. Over land 220 and 267 small, 1480 and 1510 large ions, always slightly more negatives. Small ions have mobilities of order 1 cm/sec per V/cm, large ions (on dust, water droplets, etc.) 10^{-2} to 10^{-4}.

In the range of medium field strength, some ions recombine in the gap while the others are neutralized at the electrodes. Assuming that the two corresponding rates are in the same ratio throughout the volume, then

$$(dN/dt)-(dN/dt)_{\text{rec}}-(dN/dt)_j = 0. \qquad (2.7)$$

Since the rate of recombination follows from (2.5) and the rate of loss to the electrodes from (2.6), the current density in the intermediate range is found from a quadratic in j to be

$$j = ey\left\{\left(1+\frac{2d(dN/dt)}{y}\right)^{\frac{1}{2}}-1\right\} \qquad (2.8)$$

with

$$y = (k^+ + k^-)^2 X^2/2\rho d.$$

which for small and large X leads to (2.5) and (2.6) respectively. (2.8) is called 'the characteristic' of the discharge. It has been investigated experimentally and found to be in excellent agreement with the theory when diffusion of ions (neglected here) is allowed for (16).

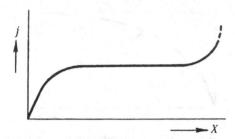

FIG. 4. Relation between current density j and field X in a gap with plane electrodes showing the rise of j at large X due to ionization by collision superimposed on the ionization in the gas by irradiation. (Plateau often shorter than shown.)

If the field strength is increased further the current which previously had reached saturation value will rise again (Fig. 4). This is because in very large fields some of the electrons originally formed by irradiation in the gas instead of being attached to gas molecules will be accelerated in the electric field and can reach velocities which enable them to ionize gas molecules. Thus the rate of production is no longer constant but depends now on the field, a phenomenon which will be dealt with in Chapter 4.

3. EFFECT OF SPACE CHARGES

So far it has been assumed that the charges move through the gas independently of each other and that the charge per unit volume is so small that the field at any point within the gap is given by the electric field $X = V/d$ of the plane parallel electrodes. If the current density is increased, the field will also depend on the distribution of the charges in the volume. The current density at which field distortion by space charges sets in can be estimated by comparing the number of field lines from the electrodes with that from charges in the volume (which are assumed to end on the electrodes). The charge per unit area on the plane parallel electrodes is $\sigma = X/4\pi$; from (2.3) the charge in the corresponding volume is $eNd = dj/(k^{+}+k^{-})X$, provided that $j_s \approx j$. Thus field distortion has to be considered when these two quantities become comparable. Assuming $X \approx 1$ e.s.u. $= 300$ V/cm, $d = 1$ cm, $(k^{+}+k^{-}) \approx 600$ e.s.u., $e \approx 5.10^{-10}$ e.s.u., then the field differs considerably from the static one when the current density becomes $j \geqslant 60$ e.s.u. $= 10^{-8}$ A/cm^2; this value rises with X^2.

Let us first consider qualitatively how a sufficiently powerful space charge, which is maintained by irradiating the gas, reacts on the electric field produced by a constant applied potential difference. Initially let equal numbers of ions of both signs be uniformly distributed throughout

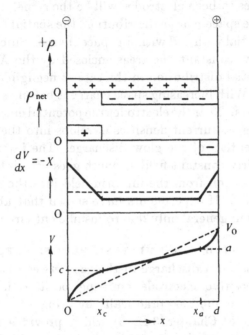

FIG 5. Spatial distribution of space charge ρ, field X, and potential V in a gas-filled condenser with plane electrodes and uniform ionization in the gas showing the influence of space charge. Oc and aV_0 correspond to cathode and anode fall in potential, respectively, Ox_c and $x_a d$ to the two fall spaces. V_0 is the applied potential. The dotted lines represent $X(x)$ and $V(x)$ for $\rho = 0$.

the volume. When the potential is applied the electric field will move positive and negative ions to their respective electrodes. Since their mobilities are in general different, the number of ions arriving per unit time at the electrodes and hence the positive and negative ion current will not be equal. (It should be mentioned that this was one of the inconsistencies connected with the problem treated in section 2. There we assumed that ions are produced in equal numbers but at the same time found that unequal numbers arrived at the electrodes. Though this will not have much effect on the numerical results, the treatment is, strictly speaking, not exact.) What happens is that at higher current densities the negative space charges are repelled and the positive attracted by the respective electrodes, so that in the vicinity of the cathode a surplus positive charge is present and correspondingly a negative charge is observed near the anode. However, because of their larger mobility the negative ions cause a smaller field at the anode than the positive ions at the cathode. Consequently when space-charge effects set in the rate of removal of charges to both electrodes will be the same.

Fig. 5 shows the space-charge distribution, the spatial (x) distribution of field, and potential with and without space charge. Since the potential difference V_0 is kept constant, the areas enclosed by the X–x curves are equal. The potential distribution in the case of negligible space charge is a straight line. With increasing space charge it becomes a curve which falls into three parts. Near the electrodes the potential changes relatively rapidly and at higher current densities develops into the 'cathode and anode fall in potential' of the glow discharge. The intermediate part which shows a fairly constant field strength goes over into the positive column. The transition from the uniform field into the distorted field takes place steadily. If experiments have shown that abrupt changes took place, it was in general only due to insufficient circuit stability.

4. DISCHARGES WITH ION-EMITTING ELECTRODES

Let us now consider a discharge in the gas between two large plane electrodes, the positive electrode emitting positive ions uniformly (Fig. 6). This can be obtained practically by using a thermionic anode (Kunsman electrode, Chapter 3 B, 6) which provides moderate ion currents. The nature of the gas and its pressure will determine the mobility of the emitted ions. The current density is assumed to be large enough to make it necessary to consider field distortion by space charges. Under these circumstances the current-voltage relation of the discharge can be derived in the following way: the equation of continuity is

$$j = \rho v, \tag{2.9}$$

where j is the current density, ρ the space-charge density, and v the velocity of the ions. From Poisson's equation we have

$$dX/dx = 4\pi\rho, \tag{2.10}$$

and the ion speed is given by the mobility k of the ions

$$v = kX. \tag{2.11}$$

From (2.9), (2.10), and (2.11) we find by integration

$$X^2 - X_e^2 = (8\pi j/k)x. \tag{2.12}$$

The constant X_e represents the field strength at the emitting electrode. Since ions are produced in nearly unlimited numbers, we may assume that the field at the anode becomes very small. This is the more true the smaller the initial velocity of the ion, and thus we may take the field X_e as approximately equal to zero. That is equivalent to saying that all the field lines starting at the cathode end at the surface of a very dense

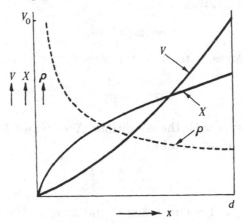

FIG. 6. Spatial distribution of space charge ρ, field X, and potential V in a gas when the ions are emitted from one electrode only. For exact cathode boundary conditions see (211).

space-charge layer which rests at the surface of the anode, but none of the lines reach the anode. The field distribution is thus

$$X^2 = (8\pi j/k)x \tag{2.13}$$

and the potential difference between the electrodes

$$-V_0 = \int_0^d X \, dx = \left(\frac{32\pi j}{9k}\right)^{\frac{1}{2}} d^{\frac{3}{2}}. \tag{2.14}$$

The current-voltage relation is therefore

$$j = \left(\frac{9k}{32\pi}\right)\frac{V^2}{d^3}. \tag{2.15}$$

This relation shows the parabolic rise of the current as a function of the voltage. Here the current is no longer a function of the (undistorted) field. Fig. 6 shows the spatial distribution of the field and the potential, as calculated from (2.13) and (2.14).

If we were to take a similar electrode system where the cathode emits electrons and reduce the gas pressure p until the mean free electron path

$\lambda_e \gg d$ (say $p < 10^{-4}$ mm Hg), then the relation between j, V, and d becomes different. In this case we have from (2.9) and (2.10)

$$j = \rho v = \frac{v}{4\pi} \frac{dX}{dx}. \tag{2.16}$$

An electron starting with zero speed falls freely through the distance d and hence $mv^2/2 = eV$ or

$$v = (2e/m)^{\frac{1}{2}} V^{\frac{1}{2}}, \tag{2.17}$$

so that (2.16) becomes, with $A = 4\pi j/(2e/m)^{\frac{1}{2}}$,

$$\frac{d^2 V}{dx^2} = \frac{A}{V^{\frac{1}{2}}} \tag{2.18}$$

and integrating twice with the conditions $X = 0$ and $V = 0$ at $x = 0$ and $V = V_0$ at $x = d$, we obtain

$$j = \frac{(2e/m)^{\frac{1}{2}}}{9\pi} \frac{V_0^{\frac{3}{2}}}{d^2}. \tag{2.19}$$

With $d = 1$ cm, $V_0 = 1$ e.s.u. $= 300$ V and $2e/m = 10^{18}$ e.s.u., we find $j \approx 3.10^7$ e.s.u. $= 10$ mA/cm². Comparison of (2.15) and (2.19) shows that at sufficiently low pressure j is less sensitive to variations of either V_0 and d. If both electrodes emit charges, the constant in (2.19) is changed (1).

There is, however, a serious objection which can be raised against these derivations. From (2.11) it follows that when $X_e = 0$, $v = 0$ at $x = 0$, and thus because of (2.9) for any finite value of j the space-charge density ρ at the emitting surface becomes infinite. This difficulty can be overcome by allowing for the finite initial velocity of the emitted charges. The calculations for the general case are rather involved (1). However, it is easy to show the sense in which the above results must be corrected with a finite field at $x = 0$, giving a potential trough.

For the purpose of obtaining a simple mathematical answer we shall treat now a similar case but with a space-charge distribution (66) which cannot be physically realized. A potential difference V_0 is applied to a plane parallel gap of width $2d$ which contains a constant uniformly distributed positive space-charge ρ_1 between the cathode and a plane d cm from it, while the remaining space is entirely free of space charge (Fig. 7). In the space 1, therefore, the field distribution is given by integration of Poisson's equation (2.10):

$$X_1 = 4\pi\rho_1 x - c_1 \tag{2.20}$$

with the constant $c_1 > 0$ because for $\rho_1 = 0$, $X_1 < 0$. The potential distribution is given by

$$V_1 = -2\pi\rho_1 x^2 + c_1 x + c_2. \qquad (2.21)$$

$c_1 x > 0$ because $X = -dV/dx$. Fixing $V = 0$ at $x = 0$ makes $c_2 = 0$.

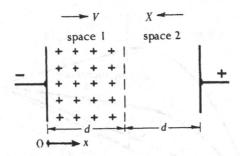

FIG. 7. Gap bounded by plane parallel electrodes with positive ions uniformly distributed in one-half of the interspace.

In the space 2 we have $\rho_1 = 0$ and hence

$$X_2 = -c_3, \qquad (2.22)$$

$$V_2 = c_3 x + c_4. \qquad (2.23)$$

At the boundary d, $X_1 = X_2$ and $V_1 = V_2$. Also at $x = 2d$ from (2.23)

$$V_2 = c_3 2d + c_4 = V_0. \qquad (2.24)$$

Furthermore at $x = 0$ and $x = 2d$ the potentials are 0 and V_0 respectively. With these boundary conditions we obtain three equations for c_1, c_3, c_4 which give the following solutions:

$$X_1 = \pi\rho_1(4x - 3d) - X_0, \qquad (2.25)$$

$$X_2 = \pi\rho_1 d - X_0, \qquad (2.26)$$

$$V_1 = \pi\rho_1 x(3d - 2x) + X_0 x, \qquad (2.27)$$

$$V_2 = \pi\rho_1 d(2d - x) + X_0 x, \qquad (2.28)$$

where $X_0 = V_0/2d$ is the field when space charges are absent. By (2.25) the field at the cathode is negative (a positive charge moves in negative x direction, the more so the larger ρ_1. Fig. 8 shows the field and potential distribution). The field changes linearly in space 1 and is constant in space 2. For moderate values of X_0, $X_2 < 0$ unless ρ_1 is very strong, say 10^7 to 10^9 charges/cm³. The potential rises first steeply, the steeper the larger ρ_1, and then increases linearly in space 2 except for very large values of ρ_1, when it passes a maximum and then decreases. Thus when very intense positive space charges are present in front of the cathode, the potential difference between a point in space and the cathode can exceed that between cathode and anode. For certain values of ρ_1 the field in space 2 can become zero because the number of field lines originating in the positive space charge and going towards the anode

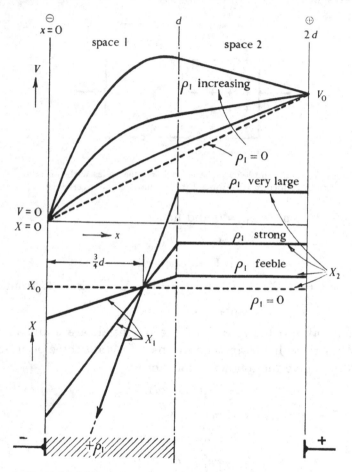

FIG. 8. Spatial field and potential distribution for the gap in Fig. 7 when a constant potential difference V_0 is applied and ρ_1 is raised.

is just equal to the number of lines originating at the anode and going towards the cathode.

When the space charge in space 1 is negative and sufficiently strong, then the field will be positive at $x = 0$, which means that the negative charges are driven towards the cathode. X_1 becomes negative at larger values of x. $X_1 = 0$ at

$$x_m = \tfrac{3}{4}d - X_0/4\pi\rho_1. \tag{2.29}$$

At this point V is negative and also passes through a minimum. $V = 0$ at $x = 0$ and also at

$$x_0 = \tfrac{3}{4}d - X_0/2\pi\rho_1. \qquad (2.30)$$

Fig. 9 shows how such a potential trough can develop near the cathode. It has been found to exist in the neighbourhood of thermionic cathodes

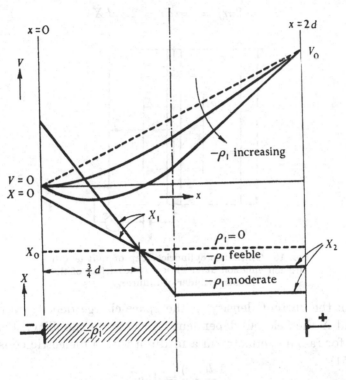

FIG. 9. Spatial distribution in X and V when a negative space charge fills the left half space and ρ_1 is varied. When ρ_1 is sufficiently raised a potential trough forms near the cathode.

where the field has been investigated by the deflexion of a narrow pencil of electrons. Similar conditions would exist when a positive space charge is situated near the anode. Thus when space charges have the same sign as the nearby electrode, the emitting surface can be thought to lie at the potential minimum, that is a virtual anode or cathode can be assumed to exist a certain distance away from the real one.

Much attention has been paid to the discharge between two concentric cylinders (7, 8, 11, 13, 21), particularly when the inner cylinder emits ions of the same sign as its polarity. These are the conditions in an ideal corona discharge which is maintained in a gas between a thin wire and a large cylinder. Ionization of higher pressure takes place in

an extremely narrow region around the inner wire. Ions of one sign are
attracted by it and neutralized, ions of the other are repelled and travel
in the radial symmetrical field to the outer cylinder. Let the radii be
designated according to Fig. 10, then the current i per unit length
follows from the equation of continuity

$$i = 2\pi r j = 2\pi r \rho v = 2\pi r \rho k X, \tag{2.31}$$

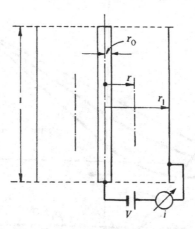

FIG. 10. Concentric cylindrical gap of unit length
with ion emission at or close to the surface of the
inner cylinder of radius r_0.

where j is the current density, ρ the space-charge density, v the ion
speed, and X the field, all dependent on r; k is the mobility. Poisson's
equation for radial symmetry in a medium with a dielectric constant 1
is (see 8.51)

$$\frac{1}{r}\frac{d(Xr)}{dr} = 4\pi\rho. \tag{2.32}$$

Substituting from (2.31) in (2.32) we obtain

$$(Xr)\frac{d(Xr)}{dr} = \frac{2i}{k}r. \tag{2.33}$$

By integration we find with $X = X_0$ at $r = r_0$:

$$(Xr)^2 - (X_0 r_0)^2 = (2i/k)(r^2 - r_0^2). \tag{2.34}$$

The product Xr is a measure of the number of field lines at r when a
space charge is distributed in a manner controlled by k and i. At radii
far away from the inner wire $r \gg r_0$, we have

$$X' = (2i/k)^{\frac{1}{2}}. \tag{2.35}$$

This equation shows that for large r the field strength is constant and independent of position; it rises with increasing current and decreasing mobility. Fig. 11 shows the field distribution which has been confirmed by measurements. The dotted line represents the distribution if no space charges were present. Since the applied voltage is assumed to be the same in each case, the two areas are equal.

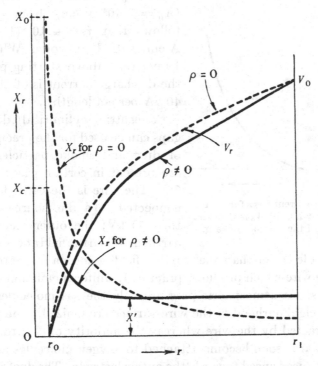

FIG. 11. Radial field and potential distribution in gap, Fig. 10, when the space charge is zero or not zero. X' is the value of X_r at large r when $\rho \neq 0$. X_c and X_0 are the fields at the surface of the inner cylinder when $\rho \neq 0$ and $\rho = 0$ respectively. V is the applied potential.

This radial discharge is an example of particular interest: it shows that a uniform field can be produced over the major portion of a cylindrical gap in the presence of strong space charges which distort the originally hyperbolic field distribution. The potential distribution (Fig. 11) can be found by integrating (2.34). We shall only discuss the case $X_c < X_0$, meaning that the current is small, though the assumption $X_c > X_0$ does not present any more mathematical difficulties. With given X_0 and $i \ll k(X_0 r_0/r_1)^2$, we find for i as a function of the applied potential V (see Chapter 8)

$$i = \frac{k}{r_1^2 \ln (r_1/r_0)} V_0(V - V_0), \qquad (2.36)$$

where $V_0 = X_0 r_0 \ln (r_1/r_0)$ and is the starting potential of the corona discharge since for $V < V_0$, $i = 0$. The current rises with V as shown in Fig. 12.

A numerical example should illustrate these points: in atmospheric air ($k = 2$ cm^2/V sec) let $V = 6.10^2$ e.s.u., $r_0 = 0\cdot1$ cm, and $r_1 = 10$ cm. With $V_0 = 3.10^4$ V $= 100$ e.s.u. ($X_0 = 7.10^4$ V/cm), from (2.36) it follows that $i = 4.10^{-4}(V-3.10^4)$ A/cm, with V in volts. When V is 10 per cent above starting potential the discharge current is of the order 40 μA per cm length.

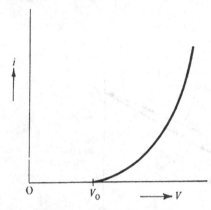

FIG. 12. Corona current i as a function of the voltage V for Fig. 10. V_0 is the starting potential of the corona discharge.

Concentric cylindrical discharge gaps can be used for the precipitation of dust and smoke particles which are present in certain gases (17, 68, 70). The wire is the cathode and is connected to a d.c. source of -20 to -50 kV; the outer electrode is earthed. When the field at the cathode surface is made larger than the starting field, a corona discharge is set up along the wire which produces practically uniform emission of negative particles. At atmospheric pressure only the gas molecules in the immediate neighbourhood of the wire surface are ionized. The positive ions are attracted by the wire whereas the majority of electrons while moving outwards soon become attached to oxygen molecules and then move with reduced speed towards the outer electrode. The dust particles are gradually charged up by these negative ions until the charge on them reaches an equilibrium value. The field drives these heavy charged particles slowly towards the outer electrode while at the same time the streaming gas moves them upwards. When the heavy clusters reach the outer electrode they give up their charge and stick to the electrode surface. By introducing dust particles 'heavy ions' are formed which reduce the average mobility. Therefore the current through the discharge will fall when particles are added to the gas; this agrees with observations.

Discharges between a positive wire and a concentric negative cylinder in rare gases or gas mixtures are also used in the Geiger–Müller counter, now a conventional instrument for measuring or detecting α, β, γ, X, and cosmic rays, beams of neutrons, protons, etc. (62, 69).

5. Effect of Back-Scattering

We return now to a system with two parallel plane electrodes and shall try to explain the variation of the current in the gas with the applied electric field when photo-electrons are released from the cathode by irradiation with ultraviolet light. Fig. 13 shows the results of

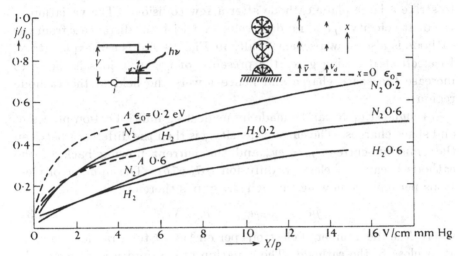

Fig. 13. Effect of a uniform electric field on the back-diffusion of photo-electrons in various gases and for different initial energies ϵ_0 in eV (67 a). Curves for A are dashed.

measurements in A, N_2, and H_2, i.e. the dependence of the current density of electrons j per unit photo-electric emission current density j_0 on X/p, the electric field per unit pressure, being a measure of the average energy an electron acquires along a mean free path λ_e in the field direction (note that $X/p \propto X.\lambda_e$). Another parameter involved here is the mean energy ϵ_0 of the photo-electrons at the cathode which can be varied, e.g. by using filtered and unfiltered ultraviolet radiation from a Hg lamp.

The first striking result is that j is found always considerably smaller than j_0, the photo-electric saturation current *in vacuo*. We conclude that in the presence of a gas photo-electrons partly return to the cathode in spite of the electric field driving them away from it. Secondly, as p is raised, while X/p is kept constant, j/j_0 remains constant showing that the mean energy of electrons in the gas is a proper variable. Thirdly, the larger the mean initial energy ϵ_0 of the photo-electrons the smaller is j. This is because a higher initial energy means a larger initial random velocity $\bar{v}$ since the electrons are emitted into an angle 2π (and not a larger drift velocity since this is an average velocity with which an electron swarm moves, see Chapter 4). When $\bar{v}$ is large a large number of electrons after being scattered elastically by gas molecules will return to the cathode. A 0·6 V electron in N_2 at $p = 100$ mm Hg with $\lambda_e = 4.10^{-4}$ cm (see Fig. 21) at $X/p = 10$ will only acquire $X\lambda_e = 0·4$ V if the first free path λ is parallel to the field and thus has a good chance to strike a large plane cathode after a few collisions. The variation of the drift velocity v_d and random velocity $\bar{v}$ with the distance x from the cathode is also shown schematically in Fig. 13 and is self-explanatory. In electron-attaching gases the presence of negative ions is likely to increase the space charge and hence lowers the field in the cathode region.

A rough estimate can be made by neglecting multiplication processes and space charges. The current density j is then the difference between the emission current $j_0 = en_0$ and the current flowing back to the cathode because of electron diffusion (67, 67 b). The number of electrons per cm^2/sec flowing through the gap is therefore

$$j/e = n_0 - n_{\text{diff}} = n_0 - N\bar{v}/4, \qquad (2.37)$$

where N is the number of electrons per cm^3 and $\bar{v}$ their random velocity at or close to the cathode. The equation of continuity in terms of the drift velocity v_d is

$$j/e = Nv_d. \qquad (2.38)$$

From (2.37) and (2.38) we have

$$j/j_0 = n/n_0 = 1/(1 + \bar{v}/4v_d), \qquad (2.39)$$

where both $\bar{v}$ and v_d depend on X/p only. The rise of j/j_0 with increasing X/p is essentially the dependence of v_d on X/p (see Chapter 4) since $\bar{v}$ corresponds to the initial energy rather than the random energy of the swarm in the gas. This is because close to the cathode the field contributes too little to $\bar{v}$ to be effective so that back-diffusion is prominent, whereas at larger distances (say $10\lambda_e$) the field acts on fewer electrons. The effect of inelastic losses is to lower $\bar{v}$ in the zone of back-diffusion, a few λ_e in width. The values of j/j_0 in A are higher than the corresponding values in molecular gases probably because of the large electron mean free path at low energies and the forward scattering.

6. FLOW OF CURRENT THROUGH A GAS

Let the gas be bounded by two large plane electrodes and let free charges of the same sign be present in the gas (Fig. 14). The current which flows through the circuit when a potential difference is applied can be determined in two ways: either by counting the number of charges which arrive at the electrodes in unit time, or by determining the rate of change of charge which is induced by the moving charges on one of the electrodes (17).

If the first treatment were applicable the current due to a single charge q moving from the gas to one electrode would be recorded by

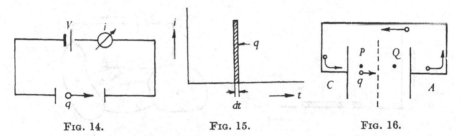

FIG. 14.   FIG. 15.   FIG. 16.

FIG. 14. Circuit in which a charge q moves from the gas to an electrode.

FIG. 15. Hypothetical oscillogram of the current pulse if the flow of current were associated with the charge q entering an electrode.

FIG. 16. A current pulse accompanies the motion of q from P to Q.

an oscillograph as a single pulse (Fig. 15). The area of the current-time diagram corresponds to the amount of charge. According to this view no current would be recorded while the charge moves through the gas and the whole charge is delivered to the electrode at the instant when the charged particle strikes it.

This picture cannot be correct for the following reason: When the charge q moves from P to Q (Fig. 16), two points which lie on different sides of the median plane, the number of field lines or lines of displacement ending on C and A will vary continuously. In particular, when the charge passes the median plane half of the lines end at C, the other half at A. A similar argument holds for P and Q lying on the same side of the median plane. A calculation for a given system can be made by the method of electric images.

Thus according to the second treatment when a charge moves through a gas the induced charge on the two electrodes varies with time as long as the motion persists. Assuming, for example, a negative charge which is moved by some means from P to Q, we find that the rate of increase of the induced positive charge at A is equal to the rate of decrease of the negative charge at C. In the external circuit a certain number of electrons are driven through the wire from A to C and the amount of charge passing that way is equal to the corresponding change of charge induced on the electrodes. An oscillogram of a current is given in Fig. 17, showing that the current flows while the charge is in motion (70 a).

This approach is, of course, only applicable if the acceleration of the charge is so small that the corresponding changes of the field accompanying the charge do not induce radiation. It means that the wavelength of the pulse has to be large compared with the dimension of the gap or that

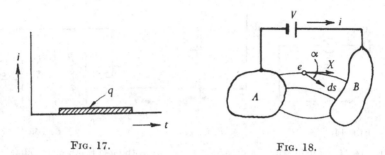

Fig. 17.   Fig. 18.

FIG. 17. Actual oscillogram of the current pulse due to a motion of a charge. Experimental confirmation for example in (70 a).

FIG. 18. To derive the instantaneous value of current of a negative charge moving between electrodes A and B of general shape.

its dimension divided by the time necessary for displacing the charge must be small compared with the velocity of light. If, for example, P and Q are 1 cm apart and a charge travels between them in 10^{-8} sec, then the speed of the disturbance is 10^8 cm/sec, which is small compared with the speed of light.

Though the first of these two views (Fig. 15) is in principle incorrect, it yields the same numerical result as the correct treatment when applied to a large number of charges which flow into an electrode continuously and uniformly. Under non-steady conditions, however, when there is a growth or decay of current in a discharge only the second treatment will give the right answer.

We shall now show by means of a simple theorem that it is possible to determine the instantaneous value of current which flows in the circuit when a charge e is moving between two electrodes of general shape (Fig. 18). The work done by the electric field X on e when e moves along ds in the time dt is equal to the energy delivered by the source V. Then if α is the angle between the field and path vectors,

$$eX\widehat{\ ds} = eX\,ds\cos\alpha = Vi_t\,dt \tag{2.40}$$

or
$$i_t = e(X/V)v\cos\alpha. \tag{2.41}$$

Thus for a given velocity v the current at any instant depends on the ratio X/V which is independent of V and depends only on position, that is the field configuration. In the special case of a uniform field between planes d cm apart when e moves parallel to the field, we obtain from (2.41)

$$i_t = ev/d, \qquad (2.42)$$

and similarly one finds for a charge moving through the interspace between two concentric cylinders of radius r_1 and r_2, when $r_2 \gg r_1$,

$$i_t = ev/(r \ln r_2/r_1). \qquad (2.43)$$

If e is moved from A to B (Fig. 18) we have from (2.41) or (2.43) a charge q flowing through the circuit of an amount

$$q = \int i\, dt = \frac{e}{V} \int_{V_A}^{V_B} \widehat{X\, ds} = \frac{e}{\ln r_2/r_1} \int_{r_1}^{r_2} \frac{dr}{r} = e.$$

An electron of an energy equivalent to 1 eV moving 1 cm in the field direction produces a current of $4 \cdot 8 . 10^{-10} . 6 . 10^7 / 3 . 10^9 \approx 10^{-11}$ A which lasts for $d/v = 1/6 . 10^7 \approx 1 \cdot 6 . 10^{-8}$ sec.

7. Current fluctuations

A steady electric current *in vacuo* or through a gas is in general associated with a large number of charges moving between two electrodes. Since the passage of each single charge produces a current pulse (Fig. 17), it follows that the actual current consists of a steady component i, plus a randomly fluctuating part $\Delta i = (i_t - i_1)$—the noise—which manifests the presence of discrete charges (Fig. 18 a). The value i is so chosen that $\overline{\Delta i} = \int_0^\tau \frac{\Delta i\, dt}{\tau} = 0$, where τ is a convenient time interval comprising a very large number of fluctuations. However, note that $\overline{\Delta i^2} \neq 0$. Also because of the randomness of the variations all frequencies occur.

Each pulse is proportional to the charge e and $\Delta i \propto i_1$. The mean-square fluctuation within a frequency interval f and $f+df$ (measured by an electric filter which passes only df) is $\overline{di^2} \propto ei_1\, df$, provided $f^{-1} \gg d/v$, the transit time of e. If $\overline{di^2}$ is integrated over the whole frequency range, a factor (< 1) must be included which makes $\int \overline{di^2}$ finite. This is because at high frequencies, i.e. when $d/v \geqslant 1/f$, the length of the pulse increases relative to the periodic time chosen and the noise must become weaker. For the extreme case of the pulse length $\to \infty$, we have a pure continuous current without fluctuations.

The result of an exact calculation for zero net space charge (70 b) is

$$\overline{di^2} = 2F(f, d/v)ei_1\, df. \tag{2.44}$$

A numerical example illustrates the foregoing. A current $i_1 = 1$ A consisting of single charges ($e = 1{\cdot}6{.}10^{-19}$ coulomb) passes between

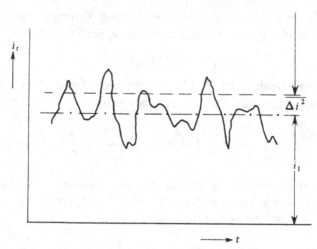

Fig. 18 a. Electric noise or random fluctuations of current. The net area under the curve and $\overline{\Delta i}$ are both zero, the average of the square of Δi is finite.

two electrodes; the transit time of a charge is 10^{-8} sec. Assuming that F is a step function which changes from 1 to 0 at $f = 10^9$ sec^{-1}, we obtain $\int\limits_{f=0}^{f=\infty} \overline{di^2} = 2{.}1{.}1{\cdot}6{.}10^{-19}{.}10^9 = 3{\cdot}2{.}10^{-10}$ A^2 and the noise current $\sqrt{(\overline{di^2})} \approx 10^{-5}$ A which is small compared with 1 A. It might appear here as if the law of conservation of energy would be violated. However, it has to be remembered that voltage fluctuations have been omitted, the inclusion of which would reaffirm its validity.

QUESTIONS

(Answers at the end of the book)

(1) The gas contained in a long concentric cylindrical condenser ($r_1 = 1$ cm, $r_2 = 4$ cm) is uniformly irradiated with X-rays of intensity (power) of 1 röntgen/sec, the radiation being constricted so as to avoid reaching the electrodes. A potential difference V is applied to the electrodes which is of such value that multiplication of charges is negligible. Find the relation between V and the current i per cm axial length flowing through atmospheric air for low and moderate values of V. Assume recombination with $\rho = 1{.}10^{-6}$ cm^3/sec, the mobilities to be $k^+ \approx k^- \approx 2$ cm/sec per V/cm and neglect diffusion and space charge effects.

Show that for small values of V the current per unit axial length is independent of the radius r if $r_1 < r < r_2$.

(1 röntgen = 1 r is the quantity of X-ray radiation, i.e. an energy, which produces an electric charge of 1 e.s.u. ($\approx 2.10^9$ elementary charges) in 1 cm³ of dry air at s.t.p. The amount of energy which corresponds to 1 r is smaller for long than for short wavelength radiation because the length of an ionization chamber for complete absorption would be impracticable or the gas pressure too high even if gas atoms with high atomic numbers were used.)

(2) A concentric spherical ionization chamber contains a gas at pressure p. The radius of the outer electrode is 1 cm; the surface of the inner electrode of radius $r \ll 1$ cm is uniformly covered with a radioactive source of 100 millicurie strength. If each particle, emitted from the source produces 50 ion pairs per cm of path and per mm Hg pressure in the gas, find the relation between the saturation current i_s and the gas pressure p. Assume that the applied voltage between source and shell produces negligible gas ionization by collision.

In practice radiation may be absorbed by the electrodes thereby increasing the total current. Indicate how to discriminate from observations between gas and wall effects.

$$(1 \text{ curie } = 3 \cdot 7 \times 10^{10} \text{ disintegrations/sec.})$$

(3) Given are two infinite plane parallel electrodes in a gas at atmospheric pressure. The electrode separation is $d \gg \lambda$, λ being the mean free path of ions. One electrode emits an unlimited number of positive ions, the other emits electrons at the same rate. If the mobility ratio $k^+:k^- = 10^{-3}$ and if emission velocities and back scattering are neglected, find

(a) the spatial distribution of potential, electric field, and net space charge;

(b) the relation between the current densities j^+, j^- and the applied voltage, V_0. Also show that for zero electron emission the solution to (b) is

$$j^+ = (9k^+/32\pi)V_0^2/d^3.$$

(4) Set up the differential equation for dV/dx for a plane parallel electrode system when one electrode emits an unlimited number of positive and the other of negative charges at zero gas pressure. (These conditions prevail when the mean free path of charges is much larger than the electrode separation.)

(5) Given a concentric cylindrical system of unit axial length where the inner cylinder of radius R_1 emits positive ions and the outer of radius R_2 emits electrons. Set up the differential equation for $V = f(r)$ when

(a) $(R_2 - R_1)$ is much less than the mean free path;

(b) the mean free path is small compared with $(R_2 - R_1)$.

For (b) find the current-voltage relation when electron emission is zero, and

(c) the positive emission so intense as to cause zero electric field at the anode.

(6) Calculate the fraction f of electrons scattered back to the cathode of a gas-filled photo-electric cell with large plane parallel electrodes, 1 cm apart. A potential difference of 100 V is applied to the electrodes and the cell is filled with argon at a pressure of 25 mm Hg. For drift and random velocities use Figs. 61 and 93.

(7) Derive an expression for the variation with time of the instantaneous current flowing in a circuit containing an evacuated concentric spherical shell condenser assuming the inner shell emits a burst of n electrons of energy ϵ which

start simultaneously and move radially outwards. Calculate and plot the absolute current in amperes as a function of time and the duration of the current pulse in seconds when $n = 10^3$ electrons, $r_1 = 1$ cm, $r_2 = 4$ cm, and $\epsilon = 2$ eV.

(8) Write an essay on the origin of the electric noise in ionized gases. Compare the ionic noise with that produced by ohmic resistances. Indicate how space charges in the gas are likely to affect the intensity of noise (70 b).

(9) A cylindrical vessel of length l, containing a mixture of positive ions $(Z.e)$ and electrons and two electrodes at the ends connected to a ballistic galvano-meter of resistance R, is dropped end on from a certain height h on to the ground. Describe and calculate what you expect to occur on impact, linear momentum being conserved. Discuss this problem in juxtaposition to Tolman's experiment with a spinning coil of metal wire. (The first free fall experiment was performed by R. Colley (*Ann. d. Phys.* **17**, 55 (1882)) with an electrolyte of CdI.)

(10) Explain why an electron or singly charged positive ion moving from A to B in Fig. 18 brings to B a charge e, whereas a positive ion and an electron starting from a point P in space but moving to A and B respectively convey a total charge e and not $2e$.

Describe the current pulse you would observe in a wire connected to A and B if starting from P first an electron and, after its arrival at A, a positive ion moved to A.

(11) Derive from first principles the increase in current which is observed when electrons, emitted from a cathode and accelerated, ionize the gas so that a positive ion current flows to the cathode. Apply this to the theory of space charge detectors (H. F. Ivey, *Adv. Electronics,* **6** (1954), 135.

3

PRODUCTION OF CHARGED PARTICLES

A. EXCITATION AND IONIZATION IN GASES

1. INTRODUCTION

Ions and electrons can be produced in a gas by a variety of processes, the most important of which we are going to treat in this chapter. Because of its complex character ionization by collision in an electric field will be left until Chapter 7. Before dealing with the main topics it seems to be desirable to give a short account of the fundamental parameters which are used in kinetic theory; this should help one to recognize that such parameters, though convenient, are only applicable within a restricted range of experimental conditions (72 a).

(a) *Mean free path and velocity distribution*. Molecules or atoms of a gas or vapour move in general at random. In thermal equilibrium their velocity distribution is Maxwellian provided no beams of particles are present. If a small disturbance causes a departure from the equilibrium, this is restored after a time of the order of 1 collision period. At room temperature the restoration time is about 1 sec at 10^{-6} mm Hg.

The velocities of molecules irrespective of their direction are distributed according to

$$dN/N = (4/\sqrt{\pi})x^2 e^{-x^2}\, dx \quad \text{with} \quad x = v/v_m, \tag{3.1}$$

where dN/N is the fraction of molecules with velocities in the range v to $v+dv$ as shown in Fig. 19. The most probable velocity v_m is the maximum of the distribution curve, and it can be shown that

$$mv_m^2/2 = kT, \tag{3.2}$$

where m is the mass of a molecule ($m_H = 1.66 \cdot 10^{-24}$g) and $k = 1.37 \cdot 10^{-16}$ ergs/°K. v_m is of the order of the speed of sound at room temperature (for N_2: $v_m \approx 420$ m/sec). Also $v_m \propto (T/M)^{\frac{1}{2}}$, where M is the molecular weight. The experimental verification of the distribution law is given in (71) and (72).

The linearly averaged velocity

$$\bar{v} = (2/\sqrt{\pi})v_m = (8kT/\pi m)^{\frac{1}{2}}. \tag{3.3}$$

The root mean square velocity

$$(\overline{v^2})^{\frac{1}{2}} = (3kT/m)^{\frac{1}{2}}. \tag{3.4}$$

The three velocity values are in the ratio $1:1.13:1.22$.

The gas molecules collide incessantly with one another. The distance covered between two successive collisions is called the free path l, the time interval the collision period. Magnitude and direction of l are distributed at random. In the absence of any external field the free path is a straight line; the deviation produced by gravity is negligible.

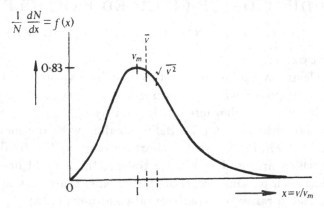

$$\frac{1}{N}\frac{dN}{dx}=f(x)$$

FIG. 19. Maxwellian velocity distribution. The velocities are expressed in units of v_m.

If λ is the mean free path (in cm per collision), the ratio of the number of molecules z which have not collided after travelling a distance x' from an origin to the total number z_0 initially present at the origin is

$$z/z_0 = e^{-x'/\lambda}. \tag{3.5}$$

Hence out of 100 molecules about 63 have suffered a collision at $x' \geqslant \lambda$. About 37 have a free path larger than or equal to λ, about 14 a free path longer than 2λ, and only 2 longer than 4λ. All directions of λ are equally likely. These are the results of the classical treatment which regards molecules as solid elastic spheres. λ is inversely proportional to the gas density, and if T, the absolute gas temperature, is not too high,

$$\lambda \propto T/p. \tag{3.6}$$

In general λ increases less than linearly with T, see (3.7).

The values for λ given in Table 3.1 are obtained from viscosity measurements. The total target area per cm³ is the cross-section Q in cm²/cm³ ($Q = 1/\lambda$). (3.5) is not strictly correct since $\lambda = f(v)$ (73 a). The absolute values for λ vary according to the method of measurement (molecular beams, heat conductivity, diffusion, critical data, dielectric constant, etc.). It is for this reason that figures found, for example, in German tables are often different from those given in English tables. At higher gas temperatures a correction has to be made which takes into consideration that with increasing molecular speed the 'time of contact' in a collision decreases. The ratio of two mean free paths at different temperatures follows roughly a relation of the form $(72a)$

$$\lambda_1/\lambda_2 = T_1(T_2+C)/T_2(T_1+C), \tag{3.7}$$

where C is the Sutherland-constant in °K; it means that at a temperature C the mean free path is half of that at infinitely large temperature. C increases in general with the size of the molecule as shown in Table 3.1 and is a measure of the strength of the mutual molecular attraction.

TABLE 3.1

Mean free path λ of molecules at 1 mm Hg and 0° C in 10^{-3} cm in their own gas $(6,76a)$; cross-section q from (3.12)

Gas	He	Ne	A	Kr	Xe	H$_2$	N$_2$	O$_2$	Cl$_2$	Hg	Na	Cs
λ	17·6	12	8·1	6·6	5·6	14·2	6·7	7·0	4·9	≈ 3	< 12	0·12

For larger molecules like C_2H_5OH $\lambda \approx 2$–3.10^{-3} cm.

Sutherland-constant C in °K (6) for $T \leqslant 600$ °K

Gas	He	Ne	A	Kr	Xe	H$_2$	N$_2$	O$_2$	Cl$_2$	I$_2$	H$_2$O
C	80	61	170	190	250	80	110	136	325	590	550

The classical mean free path λ_i of an ion is given by

$$\lambda \leqslant \lambda_i \leqslant \sqrt{2}\lambda. \tag{3.8}$$

The factor $\sqrt{2}$ applies only when $v_{\text{ion}} \gg v_{\text{mol}}$.

The mean free path of an electron λ_e has been found by classical reasoning to be $4\sqrt{2}\lambda$, where the factor 4 is the result of treating the electron as a molecule of radius zero. The figures so obtained often do not even give the right order of magnitude. In fact λ_e can be larger or smaller than λ, and this depends on the energy of the electron, an effect found by Ramsauer and Townsend (10, 18); $\lambda_e \propto 1/p$ always.

The dependence on the energy cannot be explained by the laws of classical mechanics and its treatment is one of the great successes of wave mechanics. Bohr suggested quite early that the extremely small cross-sections q of atoms for electron collision $(Q_e = qN_e = 1/\lambda_e$, see section (c)) in rare gases are associated with a diffraction effect which manifests itself in the angular distribution as well as in the cross-section of the scattered electrons. In wave mechanics an electron of energy V corresponds to a (de Broglie) wave of matter of wavelength $\Lambda = h/mv = (154/V)^{\frac{1}{2}}$ Ångström units (19). This wave is scattered by an atom which represents a centre of force whose potential falls off sharply with the distance, thus forming a 'well'. As in the theory of

light, the amplitude of the waves at a distant point—here the inverse
of the total cross-section—is found by taking the sum of the ampli-
tudes of the wavelets allowing for the phase changes which occur at the
scattering obstacle. The phases have to be corrected for the polarization
caused by the displacement of the atomic electrons with respect to the
nucleus. The result so found is that the elastic cross-section (in cm²)
for electron collision is (18)

$$q_{el} = (\Lambda^2/\pi) \sum_{0}^{\infty} (2l+1)\sin^2\eta_l, \qquad (3.9)$$

where l is the azimuthal quantum number describing the distance from
the scattering point to the centre of the atom, and η_l the partial phase

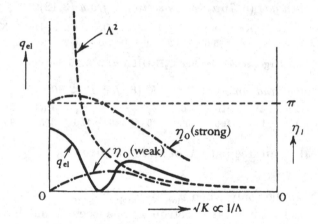

Fig. 20. Atomic cross-section q_e for elastic collisions between electrons and
atoms as a function of the electron energy K or their wavelength Λ. Illustrating
the Ramsauer–Townsend effect (3.9) for $l = 0$ and small values of K (18).
η_l is the difference of phase angles between the scattered and the incident wave.

shift between the scattered and the incident wave, describing the probability of a deviation, which is a function of the electron energy K and
$\Lambda^2 \propto 1/K$.

Fig. 20 explains how zero points in q_{el} can arise. It is seen that as
$K \to 0$, $\eta_0 \to 0$ when the atomic field is weak (He, Ne) and $\eta_0 \to \pi$ or $n\pi$
when it is strong (A, etc.). In this case η_0 becomes π at $K \neq 0$ and
$q_{el} = 0$. Hence a head-on collision ($l = 0$) between an electron and an
atom cannot be observed. On the classical picture no explanation can
be offered of why the atomic target should be without effect on the
electron (18).

At higher values of K inelastic scattering is to be included. This leads
to an expression similar to (3.9), that is $q_{inel} \propto 1/K$ approximately.

Figs. 21 and 22 show the total cross-section $Q_e = Q_{el} + Q_{inel}$ for electrons of uniform energy in various atomic and molecular gases. For energies > 50 eV, Q_e decreases with the energy K in accordance with (3.9), roughly as $Q_e K = $ const. At small energies of a few eV, that is when Λ is larger or of the order of the 'effective' radius of the atom, maxima and minima in Q_e occur. He and Ne have few electrons

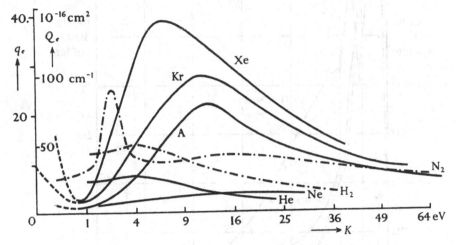

FIG. 21. Cross-section Q_e for momentum transfer collisions between electrons and molecules as a function of the electron energy K (6). $\lambda_e = 1/Q_e$.

and hence potential wells with less steep walls than the heavier rare gases; thus Q_e varies more strongly for A, Kr, and Xe. The alkalis have the largest values of Q_e; a single electron fills their outermost shell and so the range of the potential is likely to extend far out. From (3.9) $q_{e\,max}$ is of order Λ^2, assuming unity for the phase terms. When $K = 1$ eV, $\Lambda^2 \approx 1 \cdot 5 \cdot 10^{-14}$ cm^2 and $Q_e = 3 \cdot 6 \cdot 10^{16} \cdot 1 \cdot 5 \cdot 10^{-14} \approx 500$ cm^2/cm^3, which is the same order as the values given in Fig. 22.

There is another way of predicting the occurrence of the Ramsauer–Townsend effect. The uncertainty principle can be applied to show that the classical collision between an electron and a molecule becomes indefinite when their relative velocity becomes too small. The uncertainty in linear momentum times the uncertainty in coordinate (the root mean square expectation of the variation) must be large compared with $h/2\pi$ for the classical concept to hold, that is

$$\Delta(mv)\,\Delta x \geqslant h/2\pi.$$

Taking the uncertainty in position to be equal to the diameter of the atom, $\Delta x \approx 10^{-8}$ cm, and with $m \approx 10^{-27}$ g, we find from the above

$\Delta v \approx 10^8$ cm/sec or about 2·8 eV. Thus electrons of order 1 eV moving in a gas are indeterminate to such a degree that $\Delta v \approx v$ as far as atomic dimensions are concerned. It follows that for speeds of that order classical collision theory fails but holds when v is sufficiently large.

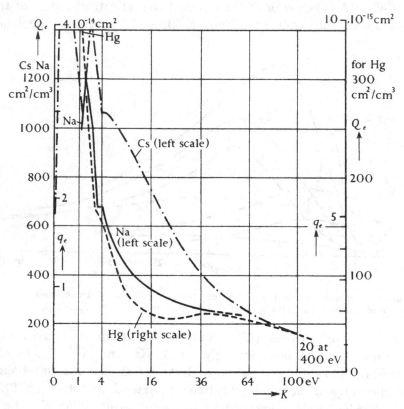

FIG. 22. Cross-section Q_e for collisions between electrons and atoms of vapours as a function of the electron energy K (6). ($Q_{e\,max}$ of Hg $\approx$ 400–500 cm⁻¹.)

The values of Q_e for energies < 1 eV are not so well known. The following table gives some results at $p = 1$ mm Hg for energies of $\approx 3.10^{-2}$ eV (molecular energies).

TABLE 3.2

Collision cross-sections Q_e of electrons of low energy in cm²/cm³ at 1 mm Hg and 0° C (73, 73b, 73c)

Gas	He	Ne	A	Kr	Xe	H₂	N₂	Hg	Cs
Q_e	5	7	8	40	130	4	1·6	37	530
	(23)	(3·5)	(2·5)	(36)	(240)	..	..	..	..

The mean free path of an excited particle moving in its own gas is in general shorter than that of the same particle in its ground state. The concept of a mean free path is more seldom applied to an excited atom or molecule (see collisions of the second kind). The mean free path of ions varies with their energy too (Fig. 25).

(b) *Mean distance.* In a gas with a uniform distribution of molecules the mean distance between them when their concentration is N molecules/cm³ is of the order

$$d = 1/N^{\frac{1}{3}}. \qquad (3.10)$$

Let us compare d with λ. At 1 mm Hg in any gas $d = 1{\cdot}7 \times 10^{-6}$ cm, whereas λ is between 10^{-2} and 10^{-3} cm. Since $d \propto 1/p^{\frac{1}{3}}$ and $\lambda \propto 1/p$, d approaches λ at large p; however, at large p, $\lambda < d$ when the size of the molecules becomes apparent.

(c) *Cross-sections for collision processes.* The total target area for an elastic collision between equal molecules (the inverse of the mean free path) is

$$Q = 1/\lambda = (4\sqrt{2})r^2\pi N, \qquad (3.11)$$

where r is the 'radius' of the molecules in cm, N the concentration in number/cm³, and Q the elastic collision cross-section in cm²/cm³. The cross-sections and mean free paths for electrons and ions have been treated above. For gas molecules, for example, at 1 mm Hg and 0° C with $N \approx 3{\cdot}6 \times 10^{16}$ molecules/cm³ and with r of order 10^{-8} cm, Q is of order 10 cm²/cm³.

The use of collision cross-sections instead of mean free paths has often proved to be an advantage: the total effective cross-section for simultaneously occurring collision processes (excitation, ionization, chemical change, charge transfer, etc.) is simply obtained by adding the cross-sections for the single processes. Of course this is only permissible so long as the state of the gas remains substantially unchanged, that is the total number of collision products is small compared with the number of molecules present, and so long as the processes are independent of one another.

If Q is the collision cross-section per unit volume for a certain type of collision and q the corresponding collision cross-section of the atom in cm², we have

$$Q = qN. \qquad (3.12)$$

Q is usually referred to 1 mm Hg and 0° C, and the corresponding value of $N = 3{\cdot}56 \times 10^{16}$ molecules/cm³. Another unit of q is $\pi a_0^2 = 0{\cdot}88 . 10^{-16}$ cm², where a is the first Bohr radius.

Let a molecule of cross-section q_m for electron encounters be hit by an electron of energy $\geqslant eV_i$; the effective ionization cross-section of an atom is then defined by

$$q_e = q_m f_e, \qquad (3.13)$$

where f_e is the probability of ionization by electrons or the 'ionization function', that is the fraction of collisions resulting in ionization; note that f_e is density-independent.

Another term used is the ionization efficiency s. It is equal to the number of ions and electrons produced by one primary particle per cm of path usually reduced to 1 mm Hg and 0° C. If the primary particles are electrons, s_e is related to f_e by

$$s_e = f_e/\lambda_e. \qquad (3.14)$$

It follows that s_e is numerically equal to the cross-section Q for ionization of molecules by electrons. The ionization efficiency for electrons is obtained directly from experiment (12, 19). The corresponding relations for ions and atoms for any type of collision are

$$s_i = f_i/\lambda_i \qquad (3.15)$$

and

$$s_a = f_a/\lambda. \qquad (3.16)$$

When a beam of electrons passes through a gas, the number of electrons z in the beam at a distance x from the origin diminishes with x because of the collisions between electron and gas molecules which scatter the electrons out of the beam. Assuming uniformity and symmetry and z_0 electrons entering at the origin, we have

$$z/z_0 = e^{-\alpha x}, \qquad (3.17)$$

where α is the absorption coefficient ($\alpha \propto 1/p$) which is related to the cross-section (3.11) by

$$\alpha = \sum_{1}^{n} Q_n. \qquad (3.18)$$

In the case of absorption of light the corresponding coefficient is $\mu = aN$, where a is the absorption cross-section for light quanta of energy $h\nu$.

The cross-section for the various collision processes can be derived theoretically (18, 93). A classical treatment is possible in certain cases (11), but only tedious wave-mechanical calculations give exact solutions.

(d) *Collision frequency.* The number of collisions which one particle makes with other particles in unit time is called the collision frequency f_c. Its inverse is the collision period τ.

$$f_c = 1/\tau = v/\lambda, \qquad (3.19)$$

where v is the random velocity and λ the mean free path. For example, the collision frequency of a 1-eV electron moving at random in Ne at 1 mm Hg is $f_c = v_e/\lambda_c \approx 6.10^7.5 = 3.10^8$ sec^{-1}, $1/\lambda_e$ being taken from Fig. 21. The collision frequency of electrons is in general a complex function of p because both λ_e and v_e depend on p; for atoms $f_c \propto p/\sqrt{T}$.

2. SCATTERING

(a) *Scattering of electrons.* Assume that an electron of given energy moves through a field-free box filled with a gas at such low pressure that the electron collides only once on its way from the entrance to a wall. The result of the collision is twofold in that the electron changes its energy and direction of motion: it can either collide elastically and thus lose a very small amount of energy which, according to the laws of classical mechanics, is on the average a fraction $2m/M$ of its energy, or it can lose an appreciable amount of its energy in an inelastic collision. The direction of its path in general is changed in the collision in the case of both elastic and inelastic scattering.

Let us discuss the 'single' scattering for electrons by He atoms. In Fig. 23 is shown the number of electrons which are observed in forward direction $\theta = 0$ and scattered at, say, 16°. It is seen that the majority of the primary electrons of energy E_p go straight on while a smaller number is scattered into all angles (Fig. 23 b and Fig. 24). The number of scattered electrons is found (Fig. 23 c) by measuring (with electrometer) the current received in a Faraday cage F, and their energy E_s by varying the retarding potential of the cage with respect to the box B which contains the gas. Fig. 23 also shows that at all angles there is a group of scattered electrons whose energy is about 22 eV smaller than E_p (75 eV). Since the excitation potentials of He lie between 19·7 and 24·5 eV, the small peak represents those electrons which have lost about that energy and have been inelastically scattered, whereas the large peak is due to the elastic scattering with negligible loss of energy. If electrons of larger energy are scattered the result is that the elastic and inelastic peaks become more equal: for example at $E_p = 150$ eV and $\theta = 16°$ the areas under the peaks are practically equal.

According to the collision theory based on solid elastic spheres I_θ, the scattered intensity per unit solid angle (Fig. 24), should be the same for all angles, that is if this were applied to electrons they would be scattered isotropically. If a charged particle of small mass collides with a heavy charged particle whose field is of the Coulomb type (force $\propto 1/r^2$, potential $\propto 1/r$), then the intensity of scattering varies with $1/\sin^4(\frac{1}{2}\theta)$. This is Rutherford's law which has been confirmed, for example, for α-particles singly scattered in gold foils, and for fast electrons. The scattered intensity should then approach infinity when $\theta \to 0$, which is not in accordance with experience. The quantum theory also shows

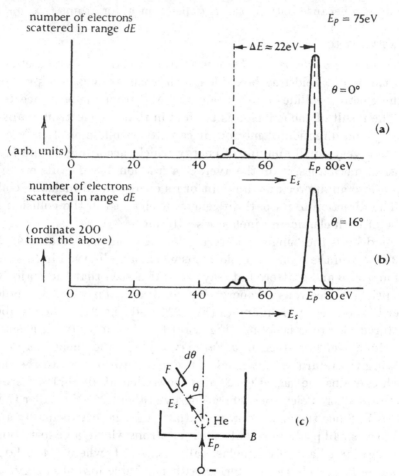

FIG. 23. Elastic and inelastic scattering of 75 eV electrons in He in the forward direction ($\theta = 0$) and into $\theta = 16°$. E_p, E_s energy of primary and scattered electrons respectively. ΔE energy loss (18).

that more electrons are scattered forward than sideways or backwards, but asserts that $I_\theta \, 2\pi \sin \theta \, d\theta$, the total number of electrons scattered between θ and $\theta + d\theta$, increases with decreasing θ and after passing a maximum lying between $\theta = 2°$ and $10°$ becomes zero for $\theta = 0$ in agreement with experiment.

For gases like A or Hg the general shape of the scattering curves is the same as for He except that for larger values of θ one or more maxima

and minima appear. These support the evidence from cross-section
curves regarding the wave nature of the electrons in this phenomenon.
Since molecular gases also have a pronounced dependence of the cross-
sections on the electron energy it can be expected that the scattering

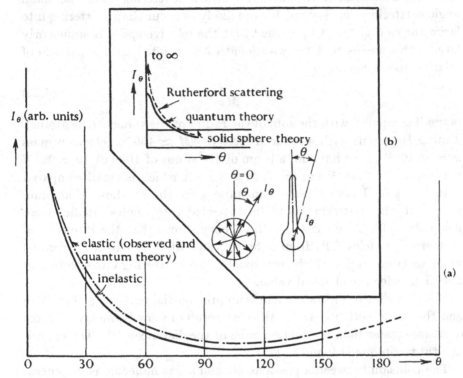

Fig. 24. (a) Intensity I_θ of the scattered beam of 80 eV in He (per unit solid angle)
as a function of the angle θ. Ordinate of inelastic curve $< 10^{-2}$ of elastic curve.
24 (b) Isotropic and forward scattering. Rutherford, solid sphere, and quantum
theory scattering (18).

has all the characteristic features we have found in atomic gases. The
observations confirm these conclusions.

Details of the measurement of scattering can be found in (18).

(b) *Scattering of ions.* When a beam of ions passes through its own gas
and the energy of the beam is low, say less than 10^3 times the thermal
energy, then the beam is scattered elastically in the gas. According to
the classical theory the scattering of neutral particles is isotropic and
thus on the average the number of particles found in an elementary solid
angle $d\theta$ is the same for any θ. Also the energy transferred in a collision
is on the average equal to half of the energy which would have been

transferred in a head-on collision, or simply half the beam energy. It has been found, however, that this description is far from the truth. Whereas classically the direction of the relative velocity of two hard spheres is turned through any angle θ with equal probability, wave-mechanical treatment shows that forward scattering—that is, small angle scattering—is very much more likely to occur than scattering into large angles (say $> 10°$) provided that the relative speed is sufficiently large. This means that the wavelength λ associated with a particle of relative speed v given by

$$\lambda = \frac{h}{Mv} \tag{3.20}$$

is small compared with the sum of the radii r of the spheres (the atoms). Taking He atoms with a reduced mass of $\frac{1}{2}M \approx 4000m$ (electron mass $m = 9.10^{-28}$ g), we have for a beam of He atoms of 1000 eV ($v \approx 2.10^7$ cm/sec, $h = 6·6.10^{-27}$ erg sec) $\lambda \approx 8.10^{-3}$ Å which is small compared with $r = \frac{1}{2}$ Å. Under these circumstances the theory shows that prac-tically all the scattering is to be expected into angles smaller than $\theta = \lambda/2r = 1°$. Also for $\lambda \ll r$ the theory shows that the intensity of scattered particles falls from a finite value at $\theta = 0$ to an (average) value at large angles which corresponds to a scattering cross-section of $2\pi r^2$, i.e. twice the classical value.

If $\lambda \gg r$, that is for low v, there is no preferential scattering (Fig. 24 b) and the cross-section is $4\pi r^2$. All these results follow if the law of force decreases faster than the inverse cube of the distance (74). The experi-mental technique is found in (75).

The collision between a positive ion and a gas molecule is in general not very different from that between two molecules of the same relative speed. The effect of the charge is to induce dipoles on neighbouring molecules causing only small deviations of the ions and of the molecules because of the large masses involved. The ion is thus scattered essen-tially like an uncharged particle.

For low energies the ions are partly scattered elastically. Fig. 25 shows the cross-sections of hydrogen molecular and atomic ions (pro-tons) scattered through angles $> 4°$ in their own gas. At energies above about 10^3 eV the curve rises and then represents the total cross-section for scattering which includes charge transfer, etc. The variation of Q_{el} with ion energy K at low K is entirely instrumental; the true value of Q_{el} may be constant though larger than that at 100 eV. The later rise is due to charge transfer associated with loss of energy (see Fig. 66). At very large K ($> 10^3$ eV) inelastic scattering by excitation and ionization occurs. The conversion of kinetic into internal energy, the angles of scattering, and the velocities before and after collision are controlled by the conservation of energy and momentum (section 4a (i)).

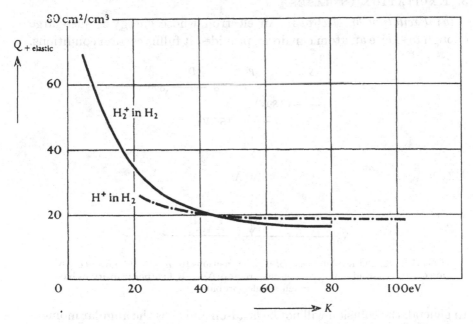

FIG. 25. Observed elastic scattering cross-section $Q_{+\text{elastic}}$ for positive molecular hydrogen ions and protons in molecular hydrogen as a function of their kinetic energy K (76).

(c) *Scattering of excited atoms.* Large cross-sections have been observed and calculated for metastable He $2\,^3S$ and Ne ^3P atoms in collision with Ne and A gas. Table 3.2 a gives the results:

TABLE 3.2 a

Collision cross-sections Q of He_{met} and Ne_{met} of thermal energy, in cm^2/cm^3 at 1 mm Hg and $0°$ C, and q in 10^{-16} cm^2 (76 b)

	He		Ne		A	
	Q	q	Q	q	Q	q
He_{met} .	552	145	460	128	720	200
Ne_{met} .	740	206	550	153	1000	278

For He, Ne, A in resonance states $q \approx 10^{-17}$–10^{-16} cm^2 (76 d).

This result can be understood by assuming that metastable atoms and rare gas atoms in the ground state interact strongly to form temporarily a molecule which dissociates spontaneously into its original constituents.

3. EXCITATION IN GASES

(a) *Excitation by electrons.* An electron whose energy is just large enough to excite an atom can do so, provided it fulfils certain conditions.

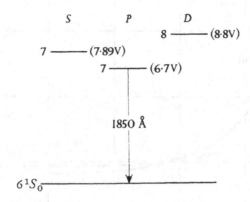

FIG. 25 a. Singlet level diagram of Hg. Transitions from $7S \to 6S$ and $8D \to 6S$ are optically forbidden. Selection rules require $\Delta j = \pm 1$; other transitions have a much smaller probability.

In general, the collision will not be head-on and thus the angular momentum of electron and atom with respect to the common centre of mass of the system has to be conserved. This means that the change in angular momentum of the system before and after collision $p_1 - p_2$ has to be equal to Δp, the change in internal angular momentum of the atom. However, Δp is quantized, that is it can only have values which are a multiple of $h/2\pi$ and hence

$$\Delta \mathbf{p} = \Delta \mathbf{j}(h/2\pi), \qquad (3.21)$$

where $\Delta \mathbf{j}$ is the change in the inner quantum number ($\mathbf{j} = \mathbf{l} + \mathbf{s}$ = azimuthal + spin quantum number).

Taking Hg (Fig. 25 a), a transition of one of the electrons in the sixth shell from the ground state $6 \ ^1S_0$ into $7 \ ^1S_0$ ($n = 6$ into $n = 7$) is associated with $\Delta j = 0$ as shown by the suffixes. Hence from (3.21) $\Delta p = 0$ and thus only head-on collisions are effective. Since their number is negligible, zero excitation probability is to be expected for electrons of energy 7·89 eV, the critical energy corresponding to that transition.

A transition from $6 \ ^1S_0$ into $7 \ ^1P_1$ (or $8 \ ^1D_2$) is associated with $\Delta j = 1$ (or 2) and 6·67 (or 8·8) eV electron energy. Excitation will result only if the atom is hit in such a direction that Δp has the required value.

Only a negligible number of electrons fulfil this condition, and thus the excitation probability is again zero at the critical potential. However, the excitation probability becomes positive when the electron energy is larger than the minimum energy necessary for the particular transition because the primary electron can then carry away the excess energy and also balance the angular momentum. The excitation probability therefore rises from zero when the colliding electrons exceed the corresponding critical energy. Values of the energy required to raise an electron from the ground state of an atom into the lowest excited state from which transitions back are optically allowed—the resonance state—are given in Table 3.3, also the energy needed to raise electrons to metastable states from which transitions into the ground state are not allowed. Actually the 'degree of metastability' determines the likelihood of such an event to occur (see end of this section).

TABLE 3.3
Resonance and metastable potentials (volts) (23, 24, 84)

	V_r		V_{met}	
He . . .	21·2		19·8	20·7
Ne . . .	16·7	16·8	16·6	16·7
A . . .	11·6	11·8	11·5	11·7
Kr . . .	10·0	10·6	9·9	10·5
Xe . . .	8·5	9·6	8·3	9·4
Rn . . .	7·0	8·5	..	
H . . .		10·2	..	
H$_2$ . . .	11·2	12·2	..	
N . . .		10·2	2·4	3·6
N$_2$ . . .		6·1	6·2	
O . . .		9·1	2·0	4·2
O$_2$ . . .	$\sim$ 5·0		1·0	1·8
Cl . . .		9·2	0·1	8·9
Cl$_2$ . . .		3·6	..	
Na . . .		2·1	..	
K . . .		1·6	..	
Cs . . .		1·4	..	
Hg . . .	4·9	6·7	4·7	5·4
Cd . . .	3·7	3·8	3·9	
C . . .		7·5	1·2	2·7

The probability of excitation to a certain energy level is often called the 'excitation function', defined as the number of electron collisions leading to the required transition, divided by the total number of collisions per electron. This fraction P is, in general, of the order of 10^{-2} or smaller, which means that every 100th collision or more will lead to excitation (Fig. 26 and Table 3.4). The excitation probability

P corresponds to the ionization probability f_e (3.13). $P = Q_{exc}/Q_{tot}$ becomes a maximum for an electron energy corresponding to V_{max} volts.

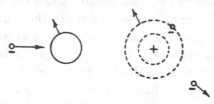

FIG. 26. Collision between an electron and gas atom leading to excitation.

TABLE 3.4

Maximum probability P_{max} of excitation of atoms by electron collision (23, 30, 80, 82, 82 a); $q = Q/3 \cdot 6 \cdot 10^{16}$

	Transition	λ_λ	Crit. pot. (V)	P_{max}	Q_{max} cm²/cm³	V_{max}
Na	$3\,^2S_{\frac{1}{2}} - 3\,^2P_{\frac{1}{2},\frac{3}{2}}$	$\begin{cases}5896\\5890\end{cases}$	2·1	~ 0·2	110	9
He	$1\,^1S_0 - 2\,^1P_1$	584	21·2	1.10⁻²	0·15	50
Ne	$2\,^1S_0 - 3\,^1P_1$	736	16·7	6.10⁻³	7.10⁻²	..
A	$3\,^1S_0 - 4\,^3P_1$	1066	11·6	2.10⁻²	1·6	..
H	$1\,^2S_{\frac{1}{2}} - 2\,^2P_{\frac{1}{2}}$	1216	10	< 0·2	5·8	16
He	$4\,^1P_1 - 4\,^1D_2$	4921	~ 24	8.10⁻⁴	6.10⁻³	50
Hg	$6\,^1S_0 - 6\,^1P_1$	1850	6·7	~ 0·7	15	15
Hg	$6\,^3P_2 - 7\,^3S_1$	5461	5·4	3.10⁻²	11	~ 10
Hg	$6\,^1S_0 - 6\,^3P_1$	2537	4·9	8.10⁻²	6·3	5·6
Hg	$6\,^1S_0 - 6\,^3P_2$	..	5·4	0·1	11·5	6·8
Hg	$6\,^1S_0 - 6\,^3P_0$	..	4·7	1·4.10⁻²	2·1	5·5
K	$4\,^2S_{\frac{1}{2}} - 4\,^2P_{\frac{1}{2},\frac{3}{2}}$	$\begin{cases}7645\\7699\end{cases}$	1·5	..	≈ 700	7

There is a marked difference in the slope and shape of the curves of the excitation function plotted against the electron energy for different transitions. For example, for atoms with two valency electrons the function for singlet–singlet transitions rises relatively slowly, and has a broad maximum at an energy which is a few times the excitation energy. On the other hand, the function for singlet–triplet transitions reaches a sharp maximum at an energy just above the critical energy and then falls quickly. Figs. 27–27 d illustrate these points. Compare, for example, the excitation curves for the ^3S and ^3P state in He with those for the ^1S and ^1P state or the excitation to any of three ^3P states in Hg with that to the ^1P state. The excitation cross-sections in H have been calculated (Fig. 27); the agreement with experiment, using a beam of H atoms crossed by an electron beam of energy K, is not entirely satisfactory. The excitation of He is now quantitatively known owing to advances in resolving the contributions to levels which have slightly different onset potentials. Fig. 27 b shows q_{exc} to various states as well as the total excitation cross-section. The excitation of Na (Fig. 27 c) comprises both levels of the doublet. The curves for Hg (Fig. 27 d)

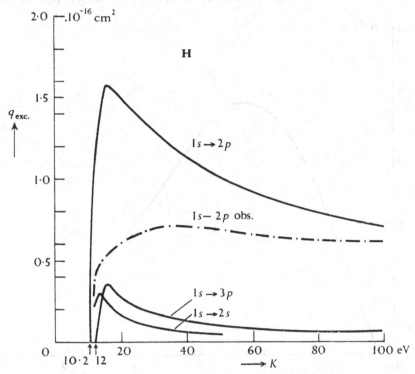

FIG. 27. Excitation cross-section q_{exc} for electrons in H as a function of the energy K.

———— calculated (18)
—·——·— observed (86 c)

shows the total inelastic and the sum of the excitation cross-sections as well as the resonance cross-sections. The latter are good examples for the closeness of the maximum to the critical potential for singlet–triplet transitions, while the maximum for a singlet–singlet transition lies at an energy which is usually several times larger than the critical potential.

The reason for this difference is that for singlet–singlet transitions the total spin quantum number is unchanged and thus before and after excitation the spin vectors of the two valency electrons must be anti-parallel (Fig. 28). For singlet–triplet transitions s goes from 0 to 1, that is the spin vectors are reoriented to be parallel (Fig. 28). For elements with strong spin-orbit coupling like Hg this can be brought about by electron impact, but for elements with weak coupling like He, it can only occur when the impinging electron itself is captured by the atom and one valency electron is expelled. This is called an electron exchange and is likely to occur within a narrow energy range only. Hence the

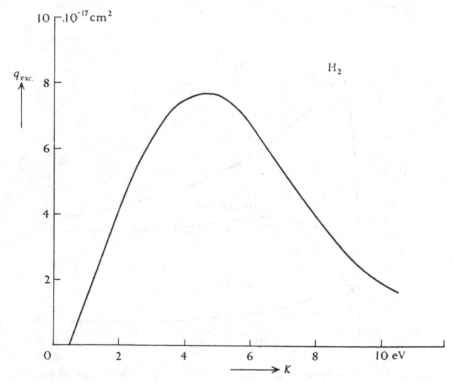

FIG. 27 a. Excitation cross-section q_{exc} for electrons in H_2 for excitation to all rotational and vibrational states as a function of the energy K (86 a).

corresponding excitation function has a sharper maximum for He than for Hg. Such stringent conditions are not attached to singlet–singlet transitions and hence the maximum is broad (18, 19).

The excitation from the ground to a metastable state is probably a rare event. Weak discontinuities have been observed at metastable potentials in the curves for the determination of critical potentials. Thus metastable levels are mainly filled from higher levels.

An electron which collides with an atom can sometimes excite the atom to a state which can only be reached by raising two electrons from the ground state. This double-excitation produces the emission of lines which correspond to 'anomalous terms'; that is the energy of the term is often larger than the ionization energy (Fig. 28 a). High-current arcs emitting Ca, Be, Cu lines, etc., show this effect. Further, since anomalous terms lie in the continuum, 'auto-ionization' can take place (77, 78).

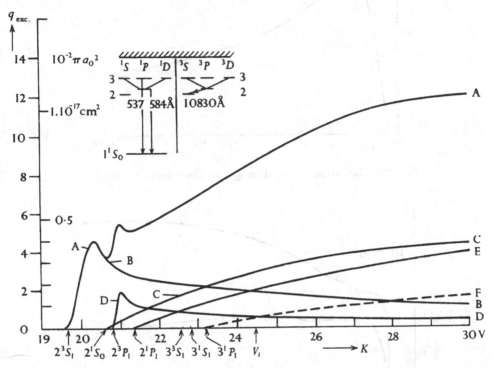

FIG. 27 b. Excitation cross-section q_{exc} for electrons in He. (145 a). A, total; B to F, partial inelastic cross-sections to various states.

We can think of this process as occurring in three steps (Fig. 28 a): first, by electron impact or absorption of a quantum, an electron is lifted in the atom from the ground to a higher state a. This excited atom now receives further energy, from the light source or from a particle which lifts another electron to a higher state. (The energy-level diagram of an excited atom (23) is not always known, but it can be assumed to a first approximation to be similar to that of the atom in the ground state but lifted up by an amount corresponding to the first excitation energy. It lies below the spark-line spectrum, which is that of the excited ion.) The lowest excitation level b of such a doubly excited atom is thus about or usually more than twice the first excitation energy which—as already stated—is often higher than V_i. Finally the electron can leave the atom with a kinetic energy which is equal to the difference between anomalous

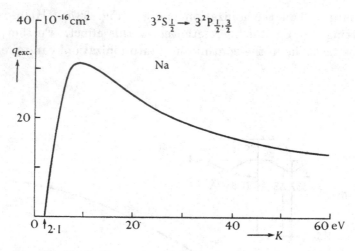

FIG. 27 c. Excitation cross-section q_{exc} for electrons in Na. (98 d)

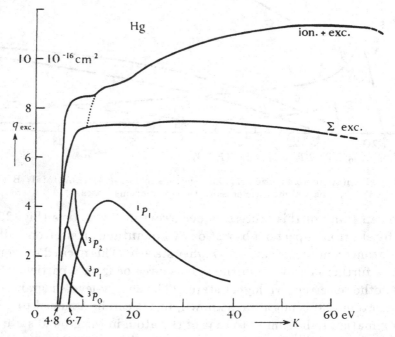

FIG. 27 d. Approximate inelastic cross-sections q_{exc} for electrons in Hg (30), the sum of all excitation cross-sections and the sum of the latter plus the ionization cross-section; dotted line: onset of ionization.

and ionization energy and no radiation is emitted; the other electron returns to its former level and thus a positive unexcited ion results. This process is called auto-ionization in the field of atomic spectra and Auger effect in the field of X-ray spectra (81). It can also be regarded as a kind of absorption of excitation energy by the atom itself, energy which would otherwise be emitted as a photon. The transition probabilities become larger as the energy difference between the anomalous term and ionization energy decreases (83).

FIG. 28. Antiparallel and parallel electron spins.

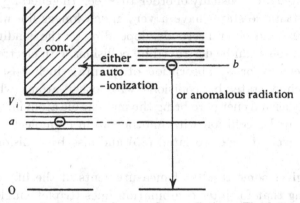

FIG. 28 a. Auto-ionization and anomalous radiation. (The latter is a rare process of probability $\approx 10^{13}$/sec compared with 10^8/sec for radiation.)

When two excited atoms collide, one of them can become ionized provided the sum of the excitation energies exceeds the ionization energy. This process is particularly effective when the excitation energy is about half of the ionization energy and when the atoms are of the same kind and present in large concentration; that is when their life is long. Any such collision in which excitation energy is exchanged is termed a collision of the second kind (see Appendix 2). Ionization of rare gas molecules occurs via excited atoms. The 'appearance' potential of He_2^+ from $He+e \rightarrow He^*+e$ followed by $He^*+He \rightarrow He_2^+$ is 23·2 V; for the corresponding reactions in Ne, A, Kr it is 20·9, 15·1, 13·2 V ($V_{i\,at} > V_{i\,mol}$).

The result of an exciting collision is an atom in a higher state. The number of excited atoms in a certain state depends on the rate of excitation and the energy loss by radiation, assuming a single excited state combining with a lower state, and equilibrium. The emission rate—the number of quanta emitted per unit time—is controlled by 'spontaneous' and 'induced' emission. The excitation rate depends on the electron collision rate and the rate of absorption of quanta in the lower state.

If $h\nu$ is the quantum energy associated with a certain transition, the emission rate E in ergs/sec per atom is given by

$$E = Ah\nu, \qquad (3.22)$$

where A is Einstein's coefficient of spontaneous emission. It can be found from classical radiation theory or wave-mechanics ($A \propto \nu^3 M^2$, where ν is the emitted frequency and M the electric dipole moment of the oscillator), by measuring the intensity profile of spectral lines or the dispersion of light in the gas. Also often used is $1/A = \tau$, the mean life of the excited atom, usually of order 10^{-8} sec. In general $\sum A = 1/\tau$. However, metastable states have a very much longer life which may be between 10^{-4} and several seconds, depending on the conditions.

Metastable atoms can be destroyed by collisions with electrons which gain kinetic energy or by absorption of quanta which raise them to a normal excited state, by collisions with other atoms which may transfer their thermal energy to bring the metastable atom into a nearby excited state, or by collision with another atom or molecule which is then ionized, excited, or dissociated (85) and also by collisions with a wall.

Table 3.5 gives some results of measurements of the life of excited states, showing that for inter-combination lines (triplet–singlet transitions) τ is large.

Collisions between electrons and molecules will in general raise the latter into higher states of rotations and vibrations. If the electron energy is sufficiently large 'electronic excitations' will occur; this excitation can lead to emission of light or to dissociation of the molecule but transfer of the excitation energy to other particles is also possible within the lifetime of the excited molecule (Fig. 27 a) (103 d).

However, the cross-sections for the various processes are widely different. An ordinary molecule of H_2 is usually in the ground state ($1 \, ^1\Sigma_g$) and in the lowest vibrational level. An electron colliding with it can excite it electronically to a higher state forming H_2^*; the 'transition', i.e. the time the outer electrons require to rearrange themselves, occurs, according to Frank–Condon's hypothesis, in a time interval which is small compared with the vibrational period (order 10^{-12} sec) which depends on the nuclear mass. This means that in the diagram of potential energy E as a function of the interatomic distance r (Fig. 29)

TABLE 3.5

Mean radiative life τ of excited atoms and molecules in sec (23, 30a, et al.)

H	.	.	$2P-1S$ (res)	$\lambda = 1216$ Å	$\tau = 2\cdot1.10^{-9}$
			$3P-1S$	1026	$1\cdot8.10^{-8}$
			$-\alpha$	6563	$2\cdot3.10^{-8}$
			$4F-3D$	18 751	$1\cdot1.10^{-7}$
			$5G-4F$	40 512	$3\cdot7.10^{-7}$
He	.	.	$2\,^1P-1\,^1S$ (res)	584	$4\cdot3.10^{-10}$
			$3\,^1P-1\,^1S$ (res)	537	$2.10^{\,0}$
			$3\,^1D-2\,^1P$	6678	$1\cdot5.10^{-8}$
			$2\,^3P-2\,^3S$	10 830	$9\cdot7.10^{-10}$
			$3\,^3P-2\,^3S$	3889	$1\cdot2.10^{-7}$
			$4\,^3P-2\,^3S$	3188	$1\cdot5.10^{-7}$
			$3\,^3D-2\,^3P$	5876	$1\cdot6.10^{-8}$
			$4\,^3S-2\,^3P$	4713	$6\cdot8.10^{-8}$
Li	.	.	$^2P-^2S$ (res)	6708	$2\cdot7.10^{-8}$
Na	.	.	$^2P-^2S$ (res)	5890/6	$1\cdot6.10^{-8}$
K	.	.	$^2P-^2S$ (res)	7665/99	$2\cdot7.10^{-8}$
Rb	.	.	$^2P-^2S$	7800	$2\cdot8.10^{-8}$
Mg	.	.	$^3P-^1S$	2852	$2\cdot2.10^{-9}$
Cd	.	.	$^3P-^1S$	3261	$2\cdot5.10^{-6}$
			$^1P-^1S$	2288	2.10^{-9}
Tl	.	.	$^2P_{\frac{3}{2}}-^2S_{\frac{1}{2}}$	5350	$1\cdot4.10^{-8}$
Hg	.	.	$^3P_1-^1S$	2537	$1\cdot1.10^{-7}$
			$^1P-^1S$	1849	1.10^{-9}
N_2	.	.	$A\,^3\Sigma_u$		$< 10^{-2}$
N_2	.	.	$a\,^1\Pi_g$		$\approx 2.10^{-4}$
NH	.	.	$A\,^2\Pi$ 0, 1		$4\cdot3.10^{-7}$
OH	.	.	$^2\Pi-^2\Sigma^+$		1.10^{-6}
CN	.	.	$^2\Sigma^+-^2\Sigma^+$		1.10^{-8}
CH	.	.	$A\,^2\Delta$ $v=0$		$5\cdot6.10^{-7}$
CH	.	.	$B\,^2\Sigma$ $v=0$		1.10^{-6}
H_2	.	.	$^1\Pi_u$ (C state) $\rightarrow\,^1\Sigma_g$		$3\cdot5.10^{-8}$
H_2	.	.	$^3\Sigma_g$ (b state) $\rightarrow\,^1\Sigma_g$		$1\cdot1.10^{-8}$

τ for H_α to H_ζ is $2\cdot3, 12, 40, 100, 230, 450 . 10^{-8}$ sec respectively.

a transition will take place at constant r, i.e. along vertical lines within the shaded area, the total amplitude of the first vibrational level. Thus the difference between a thermal and an electron collision leading to dissociation is essentially this: by heating the gas absorption of vibrational+rotational quanta increases gradually the 'amplitude' of vibration of the diatomic molecules until at $E = 4\cdot5$ eV the two atoms are not bound together any longer. When electrons of energy above $8\cdot8$ eV hit the molecule, it can be excited into the $1\,^3\Sigma_u$ state which has no potential minimum and is repulsive. As a result two H atoms are formed, each with kinetic energy $\frac{1}{2}(8\cdot8-4\cdot5) \approx 2\cdot1$ eV, which fly apart. For thermal and collision dissociation energies see Table 3.5a.

The cross-sections for dissociation of H_2 by electron collision have been calculated as a function of the electron energy (Fig. 29a). In

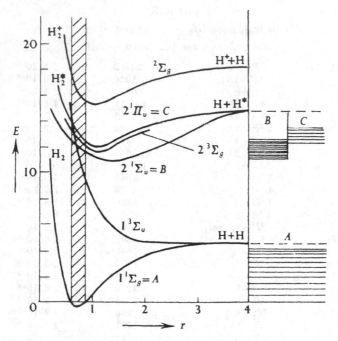

FIG. 29. Potential energy E as a function of the interatomic distance r (in Å) of a hydrogen molecule in various states (24). H_2^* and H^* are electronically excited molecules and atoms respectively. Werner bands $(C \to A)$ and Lyman bands $(B \to A)$ originate from the vibrational levels. The shaded area is the Frank–Condon range. $^3\Sigma_g$ and $^3\Sigma_u$ are termed b and a states.

TABLE 3.5 a

Thermal dissociation energy of diatomic molecules in eV (1 eV = 23·1 kcal/ g mole) and, in square brackets, onset energy for dissociation by electron collisions (76 c)

Li_2	. . 1·1		N_2^+	. 8·7
Na_2	. . 0·75		O_2	. 5·1 [$\approx$ 8]
K_2	. . 0·51		O_2^-	. 6·5
Rb_2	. . 0·47		Hg_2	. 0·15
Cs_2	. . 0·45		Cd_2	. 0·2
F_2	. . 1·6 [$\approx$ 2·5]		Zn_2	. 0·25
Cl_2	. . 2·5 [3·7]		Se_2	. 2·8
Cl_2^-	. . 4·2		Sb_2	. 3·0
Br_2	. . 1·9 [2·5]		Bi_2	. 1·7
I_2	. . 1·5 [2·7]		B_2	. 3·6
H_2	. . 4·5 [8·8]		As_2	. 3·9
H_2^+	. . 2·65		Pb_2	. 0·6
$He_2^*(^3\Sigma_u^+)$	. 2·6		NO	. 6·5 [> 10]
N_2	. . 9·8 [24·3]		OH$^+$, OH	. 4·5
Cu_2	. . 0·17 (?)		C_2	. $\begin{cases} 3\cdot6 \ ? \ [\approx 7] \\ 4\cdot9 \ ? \end{cases}$
CN	. $\begin{cases} 6\cdot5 \ ? \ [> 10] \\ 8\cdot1 \ ? \end{cases}$			

analogy to atomic singlet–triplet transitions (Fig. 27) there is a sharp peak at ≈ 12 eV, where electrons are most effective dissociation agents. The maximum, $q = 0.62 . 10^{-16}$ cm^2, is fairly large and higher than other excitation cross-sections which are associated with light emission. The latter occur e.g. through triplet–triplet transitions ($2\ ^3\Sigma_g \to 1\ ^3\Sigma_u$) giving rise to the uv-continuum which is probably weaker than the

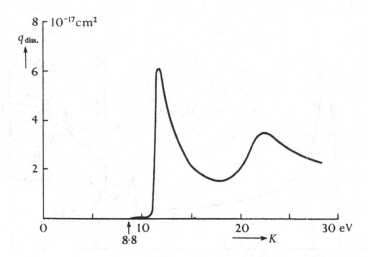

FIG. 29 a. Calculated dissociation cross-section q_{diss} for electrons in H$_2$ as a function of the energy K (93 b).

Lyman ($C \to A$) and Werner ($B \to A$) bands, of which the latter—the $\Pi \to \Sigma$ transition—is the most likely one. The maximum cross-section of $C \to A$ can only be estimated ($\leqslant 0.2 . 10^{-16}$ cm^2). The vibrational excitation cross-section has been calculated for $v = 0 \to 1$. It is zero up to 0.54 eV and has a maximum of $\approx 8 . 10^{-17}$ cm^2 at 4.5 eV (86a).

At higher electron energies ionization combined with dissociation of the molecule is observed. In the low energy range the process

$$\mathrm{H_2} + e \to \mathrm{H^+} + \mathrm{H} + 2e$$

sets in at ≈ 18 V of which 13.5 V are for ionization and 4.5 V for dissociation. The cross-sections are 0.4, 1, $3.2 . 10^{-18}$ cm^2 for electron energies of 22, 30, 50 V respectively.

An important and frequent reaction is between a slow molecular ion, produced e.g. by photo-ionization, and an H$_2$ molecule in the ground state

$$\mathrm{H_2^+} + \mathrm{H_2} \to \mathrm{H_3^+} + \mathrm{H}$$

which has a cross-section $q \approx 1.7 . 10^{-14}$ cm^2 at room temperature. q should decrease with increasing relative velocity of the colliding particles. It seems that most of the molecular ions, which were thought to be H_2^+, are in fact H_3^+. This ion is very stable: note that two protons in H_2^+ are 'held together' by one electron whereas in H_3^+ three protons are held together by two electrons. The cross-section for the corresponding formation of D_3^+ has been observed to be $1 \cdot 2 . 10^{-14}$ cm^2. At low p, reactions between H_2 adsorbed on the wall and H_2^+ have to be considered.

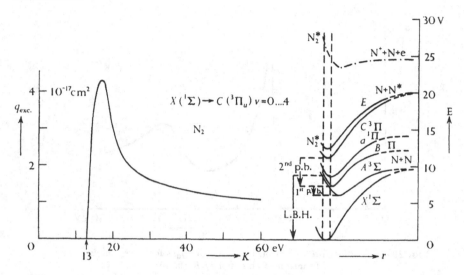

FIG. 30. Measured excitation cross-section for electrons q_{exc} (ground to C state) as a function of the energy K (209) and potential energy diagram of N_2 (98 c). p.b. = positive bands; L.B.H. = Lyman—Birge—Hopkins bands.

Our information on other molecular gases is more scanty. Fig. 30 shows the excitation cross-section for electron collisions leading to the C-state of N_2. Again it is seen that this singlet–triplet transition is characterized by a sharp maximum just above threshold. This state decays into the B-state by emission of light, known as the second positive band (since it was first observed in the positive column of a glow discharge). Dissociation by electron collision is a rare process since the Frank–Condon range in N_2 lies too near to the minima of the curves. Only for larger electron energies of ≈ 24 eV is dissociation into $N^+ + N$ to be expected (98 c).

(b) *Excitation by ions and atoms.* Critical potentials for excitation of atoms by collision with ions or atoms have not yet been measured because of lack of sufficiently sensitive apparatus. The excitation cross-sections for this process at energies sufficiently far above the critical energy have, however, been determined as a function of the ion and atom energy K. Fig. 31 shows that the excitation efficiencies for collisions between He atoms rise slowly with the energy and have broad maxima starting at several keV and extending probably beyond 10 keV. The corresponding curves for collision with ions have a similar shape (for example, Hg ions in Hg); the cross-sections may, however, be larger.

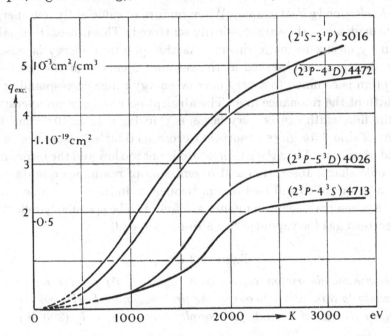

FIG. 31. Excitation cross-section q_{exc} for He atoms in He as a function of their energy K (18). In brackets the two quantum states of the He atom are given which describe the transition caused by a collision between a fast and a slow He atom. The following figure is the wavelength in Å of the radiation which is emitted when the excited He atom returns to the ground state 2 ^1S or to the lower excited state 2 ^3P.

Modern collision theory predicts that for large ion energy the excitation cross-section for allowed (and some forbidden) transitions decreases with increasing energy as $(1/E)\ln E$ or $1/E$ approximately (18). The maximum cross-section is of the same order as the corresponding maximum for electron collisions and lies at about $E = (M/m)eV_{exc}$, where M/m is the ratio of the ion and electron mass. This result is, for example, applicable to excitation of gases by α-particles.

On the other hand, where the relative speed of the ion with respect to the atom is small compared with the speed of the electrons in their orbits, the atomic electrons, and in particular the outer ones, will have sufficient time to redistribute themselves and return the excitation energy to the ion. It is to be expected that this is the more likely the slower the ion. Hence the excitation cross-section increases as the ion velocity rises, reaching a maximum where the velocities of bound electrons and of the ion are about equal. However, the theoretical predictions do not always agree satisfactorily with observations.

(c) *Excitation by light quanta.* When quanta of sufficiently low energy pass through a gas, they are elastically scattered. This process is usually treated by electromagnetic theory. As the quantum energy is raised, strong absorption is observed at the resonance energy of the gas (85). Absorption is confined to a very narrow energy range corresponding to the width of the resonance line. The absorption cross-section averaged over the line width can be very large, e.g. in Hg it is $\approx 10^{-13}$ cm² for 2537 Å. Table 3.5 b gives absorption cross-sections of resonance lines derived from (85). When the gas pressure is not too low and the resonance line profile sharp, absorption and re-emission of resonance quanta will occur, a state called diffused or imprisoned radiation. At sufficiently high gas pressures resonance quanta are found to be specularly reflected by their own gas (or vapour). See also section 4 (d).

TABLE 3.5 b

Maximum absorption cross-section q_a (in cm²) for resonance quanta to raise atoms to one of the first resonance levels and the life τ (in sec) of the state for Doppler broadening only (300° K)

Atom	State	Line	q_a	τ
H . .	^2P	1216 Å	$\approx 1.10^{-13}$	$1 \cdot 6 . 10^{-9}$
He .	^1P	584	$\approx 5.10^{-14}$	$4 \cdot 3 . 10^{-10}$
Cd . .	$^1P_1$	2288	$1 \cdot 6 . 10^{-11}$	$2 . 10^{-9}$
	$^3P_1$	3261	$\approx 2 \cdot 5 . 10^{-14}$	$2 \cdot 5 . 10^{-6}$
Hg . .	$^3P_1$	2536	$1 \cdot 4 . 10^{-13}$	$1 . 10^{-7}$
	$^1P_1$	1850	$\approx 4 \cdot 6 . 10^{-12}$	$1 \cdot 3 – 1 \cdot 6 . 10^{-9}$
Na . .	$^2P_{1\frac{3}{2},\frac{1}{2}}$	5896/90	$1 \cdot 4 . 10^{-12}$	$1 \cdot 6 . 10^{-8}$
K . .	$^2P_{1\frac{3}{2},\frac{1}{2}}$	7699/65	$\approx 1 \cdot 7 . 10^{-13}$	$2 \cdot 7 . 10^{-8}$
Cs . .	$^2P_{\frac{1}{2}}$	8944	$\approx 1 \cdot 5 . 10^{-12}$	$3 \cdot 8 . 10^{-8}$
	$^2P_{\frac{3}{2}}$	8521	$\approx 3 . 10^{-12}$	$3 \cdot 3 . 10^{-8}$

4. IONIZATION IN GASES

(a) *Ionization by electrons.* Consider an electron whose kinetic energy is above that required for ionization which collides with gas molecules or atoms which are in the ground state. The classical view is to regard the electron and the atom as hard elastic spheres of constant mass which in a collision exchange energy and momentum subject to the laws of conservation. As a result of the impact the swift primary electron usually produces a singly charged positive ion and another electron (Fig. 32). In general the energy of the two emergent electrons will not be the same but will depend on the angle into which they are scattered. It is possible to treat the collision process on classical lines, but before doing so we shall discuss qualitatively the mechanism of the liberation of an electron from an atom.

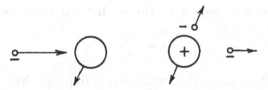

FIG. 32. Collision between an electron and a gas atom leading to ionization.

When a primary electron approaches an atom it may by virtue of its electric field interact with one of the bound electrons nearest to its path. If the force exerted by the 'primary' electron on the atomic one is sufficiently large and acts for a sufficient time, a 'secondary' electron may be ejected from the atom. To what degree the primary electron will penetrate into the atom depends on its initial direction and its speed, and also on the number and configuration of the atomic electrons. The larger the speed and the smaller the atomic number the smaller is the deviation of the primary electron from its initial path. The deviation or scattering is caused by the repulsion of the primary by the atomic electrons. However, the atom cannot be regarded as a rigid distribution of charges. Thus the approaching primary electron will displace the atomic electrons with respect to the nucleus and the atom becomes an induced electric dipole. The degree of polarization is larger for the heavier atoms, that is it increases with the atomic number. It follows that the exchange of momentum, the ionization potential, and the probability of ionization will depend on the polarizability of the atom. To develop an induced dipole requires a certain time. On the other hand, the time interval during which the primary electron is close to the atom and its nearest distance of approach determines the amplitude of the dipole moment. Consequently a very fast electron can only act upon the atomic electrons for a very short time and polarization cannot fully develop. Other effects must also be considered (18).

The ionization energy of an atom can be easily derived only in certain cases. The energy required to detach the electron from a hydrogen atom, which has one positive nuclear charge ($Z = 1$) is given by the sum of potential and kinetic energy:

$$eV_i = -\frac{Ze^2}{r} + \frac{mv^2}{2}, \qquad (3.23)$$

where v is the velocity of the electron (e, m) in the orbit. On Bohr's theory the electron being in the lowest level ($n = 1$) moves in a circular orbit whose radius r is given by equating the electrostatic attraction between electron and proton to the centrifugal force acting on the electron. Hence

$$\frac{Ze^2}{r^2} = \frac{mv^2}{r}. \qquad (3.24)$$

The quantization of the angular momentum requires that

$$mvr = \frac{nh}{2\pi}. \qquad (3.25)$$

By eliminating v from (3.23), (3.24), and (3.25) an expression for r is obtained. Substituting it in (3.23), we find

$$|V_i| = \frac{2\pi^2 me^3}{h^2}\left(\frac{Z}{n}\right)^2. \qquad (3.26)$$

Inserting the values of the constants, we find for hydrogen $V_i = 13 \cdot 5$ V. The value for He cannot be found in this way because only a fraction of the attraction from the two positive charges acts on one electron. What can be found exactly, however, is the energy required to remove the second electron from singly ionized He. Here we have $Z = 2, n = 1$, and thus from (3.26) V_i for He$^+ \to$ He^{++} is $4 \cdot 13 \cdot 5 = 54$ V, in agreement with observations. From this result it appears to be possible to give an approximate value of the ionization potential of He, viz. twice that of H (27 V). This conclusion is not correct since the shielding of the nuclear electrostatic field by the remaining electron cannot be calculated in an equally simple manner. The ionization energy to obtain 'stripped' atoms like Li^{+++} from Li^{++} etc. has also been found to agree with (3.26).

From the foregoing it follows that in general V_i is expected to be the larger the more completely the outermost shell is filled with electrons. Thus V_i is highest for the rare gases and lowest for the alkali metals. Also as Z increases V_i decreases, apart from periodic variations (see Appendix 8, Table A). The influence of screening or shielding on the ionization potential is treated in (90 b).

For the measurement of critical potentials see (12, 19).

TABLE 3.6

Ionization potentials V_i of atoms and molecules in V (6, 77 a, 83, 84)

H	. 13·6		H$_2$	. 15·4
He	. 24·5 (54·2)		C$_2$.	. 12
Li	. 5·4 (75·3, 122)		N$_2$	. 15·5
C	. 11·3 (24·4, 48, 65, 390)		O$_2$	. 12·1
N	. 14·5 (29·5, 47, 73, 97)		F$_2$	. 16·5
O	. 13·5 (35, 55, 77,...)		Cl$_2$	. 11·5
F	. 17·4 (35, 63, 87, 114)		Br$_2$	. 10·7
Ne	. 21·5 (41, 63, 97, 126)		I$_2$.	. 9·4
Na	. 5·1 (47·5, 72)		CO	. 14·0
Cl	. 13 (22·5, 40, 47, 68)		NO	. 9·25
A	. 15·7 (28, 41)		OH	. 13·8
K	. 4·3 (32, 47)		H$_2$O	. 13
Ca	. 6·1 (12, 51, 67)		CO$_2$	. 13·7
Fe	. 7·9 (16, 30)		NO$_2$	. 11
Ni	. 7·6 (18)		BF$_3$	. 17
Cu	. 7·7 (20·3)		BCl$_3$	. 11
Br	. 11·8 (19, 36)		Paraffins	. 10–12
Kr	. 14 (27, 37, 68)		NH$_3$	. 11·2
Rb	. 4·2 (28, 47, 80)		Hg$_2$	. 9·7
Mo	. 7·4		C$_6$H$_6$	. 9·6
I	. 10·4 (19)			
Xe	. 12·1 (21, 32, 46, 76)			
Cs	. 3·9 (33, 35, 51, 58)			
W	. 8·0			
Hg	. 10·4 (19, 35, 72)			
Rn	. 10·7			
Cd	. 9			

The figures in the brackets are the potentials necessary to remove a second, third, etc., electron. For instance, the ionization potential of He leading to doubly ionized He is thus 78·7 V.

The figures in column 2 are the lowest ionization potentials; e.g. H_2O into H_2O^+ requires 13 V but into $OH^+ + H$ 18·7 V.

(i) *Classical treatment of ionization by collision.* We shall determine now the amount of energy which is exchanged in a collision between a free charged particle and an atom and derive a condition for ionization. For simplicity let us assume that the atom of mass m is at rest and an electron of constant mass m_e collides head-on with it. By the principle of conservation of energy we have

$$\tfrac{1}{2}m_e v_{e0}^2 = \tfrac{1}{2}mv_1^2 + \tfrac{1}{2}m_e v_{e1}^2 + \Delta. \tag{3.27}$$

The subscripts e, 0, and 1 stand for electron, before and after impact respectively. Δ is the energy which can be transferred from the primary electron to the atom to increase its internal potential energy. Ionization can set in when
$$\Delta \geqslant eV_i. \tag{3.28}$$

Equating the linear momentum before and after impact we have

$$m_e v_{e0} = mv_1 + m_e v_{e1}. \tag{3.29}$$

By substituting v_1 in (3.27) from (3.29), we obtain an expression containing only electron speeds and Δ, viz.

$$v_{e0}^2 = v_{e1}^2 + (m_e/m)(v_{e0} - v_{e1})^2 + 2\Delta/m_e. \tag{3.30}$$

Since v_{e0} is given, Δ_{max} is found; from $d\Delta/dv_{e1} = 0$, viz.

$$v_{e1}/v_{e0} = m_e/(m + m_e). \tag{3.31}$$

It follows that the primary electron leaves the atom with less than $1/2000$ of its initial speed. Substituting v_{e1} in (3.30) and putting $K_0 = m_e v_{e0}^2/2$, we have

$$\Delta_{max}/K_0 = m/(m + m_e). \tag{3.32}$$

Since $m_e \ll m$ the mass ratio is unity and $\Delta_{max} = K_0$, but from (3.28), for ionization to occur

$$K_0 \geqslant eV_i. \tag{3.33}$$

The initial kinetic energy of an electron can thus be converted into potential energy of an atom in a single inelastic collision, and an electron can be released from an atom when the kinetic energy of the hitting electron exceeds the ionization energy of the atom.

From (3.32) we see that if m_e were the mass of an ion or a fast atom moving through its own gas, we obtain

$$\Delta_{max} = K_0/2. \tag{3.34}$$

This result has been confirmed by experiments with He atoms (see section 8).

It might appear that the convertible part Δ of the kinetic energy would depend on whether m is at rest and m_e moving or vice versa; this is not so. If v is the relative velocity of the two particles, $K_0 = m_e v^2/2$ or $K_0' = m v^2/2$. From (3.32) it follows that, with m_e moving,

$$\Delta = \frac{m K_0}{m + m_e} = \frac{m m_e v^2}{2(m + m_e)} = m_r v^2/2, \tag{3.35}$$

and in the case of m moving we have to interchange m and m_e and obtain the same result, viz.

$$\Delta = \frac{m_e K_0'}{m + m_e} = m_r v^2/2, \tag{3.36}$$

m_r being the so-called reduced mass.

Suppose a fast electron or ion of mass m_2 makes a head-on collision with an atom of mass m_1. Here its kinetic energy is large compared with the binding energy of the atomic electron,

$$K_0 \gg eV_i, \tag{3.37}$$

and so the energy term Δ (not Δ_{max}) is negligible:

$$\Delta \approx 0. \tag{3.38}$$

The collision is equivalent to one between m_2 and a free atomic electron. Hence from (3.27) and (3.29) it follows that the ratio of velocity of m_1 after collision to that of m_2 before impact is

$$v_1/v_{20} = 2m_2/(m_1+m_2). \tag{3.39}$$

Thus in an elastic impact with $m_2 \gg m_1$ the energy transfer from the fast particle on to the one at rest, in units of the initial energy of the hitting particle, equals $4m_1/m_2$; hence only a small fraction of the energy is transferable. If we average over the oblique impacts we find that the mean energy which can be transferred is about $2m_1/m_2$. If we assume that a fast electron collides essentially with an atomic electron we find from (3.39) by putting $m_1 = m_2$ that the atomic electron acquires the speed of the primary. The general expression for a head-on collision between two masses m_1 and m_2 follows from (3.27) and (3.29) with $\Delta = 0$:

$$v_{e1}/v_{20} = (m_2-m_1)/(m_2+m_1). \tag{3.40}$$

The case $m_2 \gg m_1$ is, for example, realized when a direct collision between a fast ion and an atomic electron occurs. According to (3.40) an H^+ ion of 50 keV can produce an electron of energy up to $(4/2000)5.10^4 = 100$ eV in agreement with observations.

If the electron collides obliquely with the atom the relations become rather complex because the degree of eccentricity of collisions has to be included. However, one result is noteworthy: when $\Delta \ll K_0$, that is for fast electrons, the two emerging electrons escape so that their paths enclose an angle of 90°. This has been confirmed by cloud-chamber experiments. Quantum rules do not have to be applied to the conservation of angular momentum since the final states of the free electrons lie in the continuum.

The expressions (3.39) and (3.40) have been derived as if the collisions were between elastic spheres but in addition inelastic processes were assumed to occur. This is not contradictory. For slow electrons the ionization process and the ejection of an electron have to be treated separately; for fast electrons $eV_i \ll K_0$ and thus can be neglected in any calculation.

Atoms can also be ionized by application of strong electric fields (82 b). The order of magnitude of the field strength required to remove a bound electron from the atom must exceed that of the field with which the electron is held by the nuclear charge. In the case of a hydrogen atom in the ground state this field is less than

$$X \approx e/r^2 \approx 5.10^{-10}/0.25.10^{-16} \text{ e.s.u.} = 6.10^9 \text{ V/cm.}$$

Wave-mechanical calculations give lower values, of the order 10^8 V/cm. From (3.24) and (3.25) $r \propto n^2$ and $X \propto 1/n^4$. It follows, for example, that classically H atoms in the first excited state $n = 2$ should be ionized by a field smaller by a factor 16 than atoms with $n = 1$, that is 4.10^8 V/cm. Experiments have shown that a component of H_γ ($n = 5 \rightarrow 2$) disappears at $\approx 10^6$ V/cm but already at slightly lower fields the line intensity begins to decrease. This occurs when the life of the excited state becomes comparable with the average time necessary for ionization by the electric field. Thus the observed disappearance of spectral lines in an external electric field indicates the ionization of excited atoms.

Calculations have been made for molecules in strong electric fields and it is found that relatively low fields (10^5 V/cm) can dissociate H_2 if it is in a high vibrational state ($v = 18$), i.e. close to the dissociation limit. A similar effect is observed in large magnetic fields.

(ii) *Efficiency of ionization.* From the foregoing we know that an electron which collides with a gas atom has to have an energy $K_0 \geqslant eV_i$ (Table 3.6) in order to be able to release one atomic electron. An electron with a smaller energy cannot ionize the atom and so the probability of ionization is zero for $V < V_i$. Since on the basis of Bohr's theory the ionization energy is a sharply defined quantity, we must expect the ionization probability curve to rise sharply from zero at $V = V_i$.

A measure of this probability is the ionization efficiency, usually defined as the number of ion pairs which one incident electron produces per cm of path at 1 mm Hg and 0° C. By an ion pair is understood a singly charged positive ion and one electron. The ionization efficiency is numerically equal to the ionization cross-section Q_i. Fig. 33 shows measured values of the ionization efficiency s_e as a function of the primary electron energy K for various gases and vapours. The observations made with vapours have been standardized in the same way as those obtained with gases. For the technique of measurement see (12, 18, 19); for the influence of impurities see (81 a).

The curves in Fig. 33 are plotted on a double-logarithmic scale in order to cover a large range of energy. A linear graph would show that the ionization efficiency curve has the anticipated shape: it rises first steeply and approximately linearly (masked in Fig. 33 by the use of log scales), then passes a maximum, and finally decreases at large values of K_0 roughly hyperbolically. Thus the ionization efficiency for electron energies below the maximum (say $V \leqslant 2V_i$) can be approximated by the relation

$$s_e = ap(V - V_i) = \frac{dN}{dx}, \qquad (3.41)$$

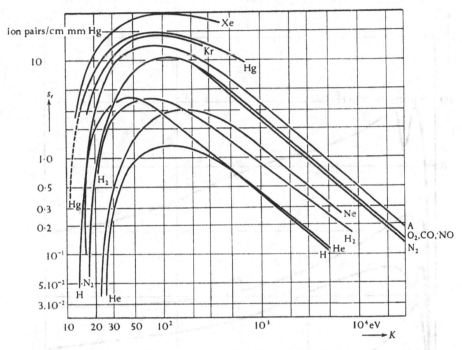

Fig. 33. Ionization efficiency s_e as a function of the electron energy K for various gases at 1 mm Hg and 0° C (17, 18, 81 a, 97 c, 97 d). The curves for CO and NO are near O_2. H measured and calculated (100 g).

where a is a constant expressed in ion pairs/cm per mm Hg per volt per primary electron; Table 3.7 shows that a is large for molecular gases and atoms with many electrons.

TABLE 3.7

Initial slope a of the efficiency of ionization curve in units of 10^{-2}

Gas .	.	.	He	Ne	A	Hg	H_2	N_2	O_2	Air	Na	Cs
a .	.	.	4·6	5·6	71	83	21	26	24	26	45	280

The maximum of s_e (Fig. 33) lies for the majority of gases so far investigated between about 80 and 120 eV with the exception of the alkali vapours whose maximum is between about 15 and 30 eV (Fig. 34). The corresponding values $s_{e\,max}$ lie between 1 and 20 ions/cm. The rule that the ionization efficiency is larger for lower ionization potentials is not generally applicable.

Attempts have been made to obtain analytical expressions for s_e by means of a classical treatment. If the electrons in the molecule are assumed to be nearly free so that the time of revolution is larger than

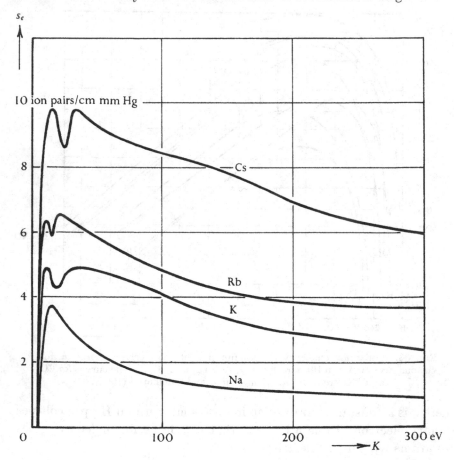

FIG. 34. Ionization efficiency s_e as a function of the electron energy K for the alkalis (the curve for Li is expected to lie below that for Na) (92, 100h).

the time the hitting electron requires to cross the atom, it can be shown that the fraction of energy transferred is $K_t/K_0 = 1/[1+\{K_0/(e^2/p)\}^2]$. K_t and K_0 are the transferred and the kinetic energy respectively, and e^2/p the potential energy, with p, the collision parameter—the perpendicular distance from the path to the scattering electron. Then (11)

$$s_e = \pi e^2 \frac{N_e}{V}\left\{\frac{1}{V_i} - \frac{1}{V}\right\}, \tag{3.42}$$

where N_e is the total number of 'optical' electrons per cm^3. (3.42) gives the right order of magnitude for s_e, a maximum at $2V_i$, and a decrease of s_e at large V, which agrees well with experiments. For electron energies of more than several hundred eV wave-mechanics and other classical treatments lead to an equation of the type (93)

$$s_e = (c_1/VV_i)\ln(c_2 V/V_i),\qquad (3.43)$$

where c_1 and c_2 are constants (85 a, b). Although the shape of this function, $(\ln V)/V$, shows an early maximum and subsequently decreases in a way roughly similar to Fig. 33, it must be emphasized that it only applies accurately to the region of high electron energies. However, in the case of H and He only, the entire ionization efficiency curve has been worked out from long and laborious wave-mechanical calculations.

Since s_e is of order 1 to 10 and numerically equal to Q_e, it follows from (3.12) that the cross-section of an atom for ionizing collisions with electrons of about 100 eV is approximately equal to or a little larger than the cross-section for atomic collisions. This applies only for the removal of an outer electron (see Appendix 8, Table A).

The experimental curves of Figs. 33 and 34 show the number of ions produced per cm per electron and do not disclose whether these ions are singly or multiply charged. For example, an electron whose speed exceeds 29 eV is capable of producing doubly charged Hg ions and 2 electrons; above 71 eV it can produce triply charged ones, etc. The occurrence of multiply charged ions has two effects: the maximum of s_e is increased above the value for singly charged ions, and secondly the maximum—which for the singly charged ions lies at about 50 eV— moves to larger energy values, here about 80 eV. Similar observations have been made with rare gases. For energy distribution of 'secondary' electrons after ionization, see (78 a) and Fig. 34 a.

An interesting observation has been made on the efficiency of ionization curve for Hg. The initial rise appears to be smooth when readings are taken at intervals of 0·1 V. If, however, the resolution is increased by, say, one order of magnitude, it is found that small but definite maxima and minima are superimposed upon the continuous curve. There is as yet no really satisfactory explanation available of what has been called ultra-ionization potentials.

The ionization cross-sections for removing more than the first electron decrease as the multiplicity of the ion is increased. Fig. 34 b shows this for $He^+ + e \rightarrow He^{++} + 2e$ as compared with the first ionization process (84 a). $q_{e\max} \propto 1/V_i^2$, V_i being the corresponding potential.

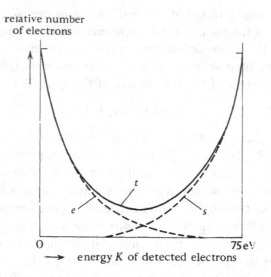

FIG. 34 *a*. Energy distribution of secondary electrons in He after ionization; total (*t*), ejected (*e*), and scattered (*s*) electrons (78 *a*).

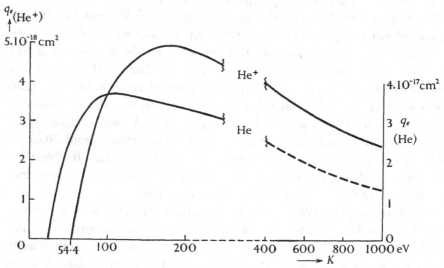

FIG. 34 *b*. Ionization cross-section q_e to produce doubly from singly charged He ions as a function of the electron energy K (84 *a*). For comparison the curve for He atoms has been added.

When a beam of electrons of energy K is shot into a gas it produces Z electrons along its path—the range. If $K > 4.10^3$ eV it is found that

$$Z = K/\epsilon, \qquad (3.44)$$

where ϵ is the average energy required to form an ion pair. ϵ is larger than V_i because it includes inelastic and elastic energy losses. For lower values of K, ϵ rises as can be seen from the following table.

For fast electrons $\epsilon \approx 30$ eV/ion pair and since $V_i \approx 15$ about one-half of the electron energy is used in exciting collisions. This view seems to be confirmed by measurements of ϵ in He with small amounts of other rare gases (of lower V_i). ϵ was found to decrease because of the ionization of the traces by metastable He atoms (Penning effect).

TABLE 3.8

Total ionization Z (ion pairs) and ionization loss ϵ (eV/ion pair)
(1, 17, 81 a)

ϵ for	$K < 4\,keV$	$K \approx 1\,keV$	Z for $K = 30\,eV$	$50\,eV$	$75\,eV$	$100\,eV$	$200\,eV$
He .	42·7	> 43	..	1·2	..	2·9	..
Ne .	36·8	..	..	..	1·2	2·0	..
A .	25·8	33	0·45	0·9	..	1·6	..
Kr .	24·7	..	..	..	..	..	..
Xe .	22·0	..	..	..	..	..	..
Hg .	..	..	1·1	1·4	..	2·7	5·3
H_2 .	36·3	> 36	..	..	..	1·4	2·8
N_2 .	36	45	..	..	1·3	1·6	..
O_2 .	31	..	..	..	..	..	..
Air .	33·9	45	..	..	..	..	..

So far it has been assumed that electrons collide with atoms in the ground state. In discharges, particularly at high current densities, the ionized gas often contains a considerable number of excited atoms and their ionization cross-section is larger than that of the unexcited atoms. If the cross-section is taken to be equal to the area of the circular Bohr orbit, we expect the cross-section of an excited state to be $\propto n^4/Z^2$; it follows that for the hydrogen atom and $n = 2$ the cross-section would be 16 times that of the ground state. Also in general there is always a larger number of slower electrons present which are now able to ionize since the least energy required to ionize an excited atom is $e(V_i - V_{exc})$. Data in this field are scarce. Fig. 35 shows the calculated ionization cross-sections for H in the ground and excited state (2p). It is seen that electrons with energies above 3·4 eV can ionize H* and that 8 eV electrons have a more than 10 times greater chance to ionize H* than have 50-eV electrons to ionize H atoms (88). Ionization cross-sections for excited Hg and He are also included in Fig. 35.

When fast electrons of order 10^4 eV traverse a gas, the electrons liberated from the atoms partly originate from inner shells. (At the same time the open places are filled by electrons coming from higher levels in the same atom and X-rays are emitted in the course of this

rearrangement.) The probability of removing an inner electron to infinity is very much lower than that of an outer one. Again the efficiency
of ionization has a maximum which lies at a few times the ionization
energy for that particular level but the absolute values for the K electrons are perhaps 10^{-4} times that for the outer level.

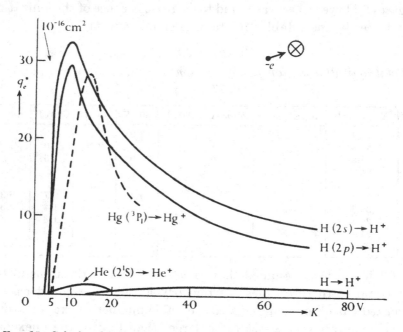

FIG. 35. Calculated ionization cross-section q_e^* for electrons colliding with excited atoms as a function of the energy K (88 a). Ionization of He and H in the
ground state added for comparison. For Ne, A, Hg, see (86 d).

(b) *Ionization of a gas by positive ions.* If positive ions of sufficient
speed collide with gas molecules ionization can take place as well as
excitation. This was shown over forty years ago with canal rays, that
is fast positive ions of 10^3–10^4 eV (91). As before, the onset potential
and the efficiency of ionization will be discussed separately.

Let us assume that the ion which produces ionization originates from
the gas. According to (3.34) we should expect then that an ion or atom
of energy equal to twice the ionization energy should just be able to
ionize an atom it strikes. This has not been confirmed yet. However,
neutral He atoms shot into He gas produced positive ions and electrons
at the expected value of energy (section (c)). We can therefore assume
that for all ions moving in their own gas the critical potential for ionization will be $2V_i$.

Measurements of the efficiency of ions of higher energy reveal that the absolute values of efficiency as well as the slope of the curve of ionization efficiency against energy near the onset is very much smaller than for electron collisions. Ions as well as atoms of low energy are very inefficient in ionizing the gas because of their large mass and low relative speeds. The ion simply spends a relatively long time in the neighbourhood of the atom with which it collides, so that there is a good chance that the electrons adjust their positions but remain in the atom while momentum between the ion and the atom is exchanged. The time during which the two particles actually collide is very much larger than the classical period of revolution of an electron in its orbit. This—rather vaguely—explains why a large number of collisions do not lead to ionization. In addition there is charge transfer (see Chapter 4) which for not too fast ions occurs with a probability of the same order as that of an elastic collision.

There are only few data on the efficiency of ionization of ions moving in their own gas and the accuracy of the measurements is probably not very high. Results are shown in Fig. 36. All curves seem to tend to a maximum which is probably somewhat higher than 10^4 eV. The figure shows also that, for example, potassium ions in rare gases have a lower efficiency than ions moving in their own gas. In contrast, the ionization efficiency of fast α-particles in air shows a maximum at about 1·5 MeV corresponding to 90 ion pairs/cm. Each α-particle produces about 2.10^5 ion pairs along its range in air. It is only possible to give a rough estimate of the initial slope of the efficiency function. For example, for protons in H_2 by extrapolating the measurements linearly to about 30 eV, a gradient of about 5.10^{-4} ions per cm per eV at $p = 1$ mm Hg is found, which is about 1/50 of the corresponding slope of the ionization efficiency for electrons.

The data of s_i do not give the real contribution to ionization by fast ions. In fact the primary ionization by ions is much smaller. The greater part of the ionization comes from the secondary electrons whose energy is often above 100 eV and thus just in the energy range in which the ionization by electron collisions is a maximum. Charge exchange is another source of error.

When the ionization efficiency for ions is compared with that for electrons it is found that the respective maxima lie approximately at the same speed. Thus an α-particle with a mass of about 8000 times that of the electron should be most efficient at about 10^6 eV and the proton at about 2.10^5 eV, which is roughly correct.

The ionization efficiency can be found from classical theory, which gives good results at large ion energies. The cross-section for ionization Q_i for ions with a charge e_2 passing through a gas is (11)

$$Q_i = s_i = N_e \frac{m_2 \pi e_1 e_2^2}{m_1 V_i} \frac{1}{K_2},$$
(3.45)

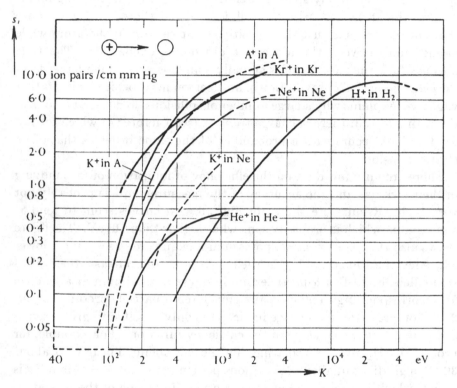

FIG. 36. Efficiency of ionization s_i of ions as a function of the energy K at 1 mm Hg and 0° C (89, 91). The curve for A^+ is uncertain.

where N_e is the total number of electrons in 1 cm³ of gas, m_1 the mass of the electron and e_1 its charge, K_2 the ion energy, and V_i the ionization potential of the gas molecule. For α-particles of 2 MeV in N_2 at $p = 1$ mm Hg this gives about 30 ion pairs/cm. In fact about 80 ion pairs/cm are observed but only a quarter to a third are produced by α-particles; the larger fraction is due to ionization by secondary electrons accelerated by α-particles.

(c) *Ionization by fast neutral atoms.* If a beam of atoms passes through the gas and its energy is increased it is observed that, besides elastic scattering, excitation and ionization occur. Older measurements with rare gas atoms give values for the critical potential at which ionization sets in which are of the order 30 to 70 eV. These values are probably too high, the errors being mainly due to the low sensitivity of the detectors used. Modern investigations of the ionization of He by He atoms where

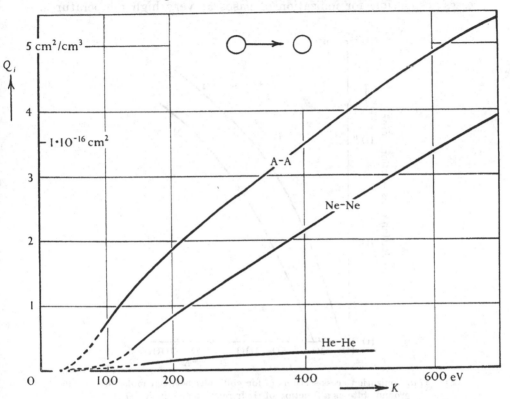

FIG. 37. Efficiency of ionization s_i (number of ion pairs/cm) of atoms colliding in their own gas as a function of their energy K at $p = 1$ mm Hg and $0°$ C (89).

currents of the order 10^{-15} A could be measured show that ionization sets in at 49·5 eV, which is exactly twice the ionization energy (87 a) and thus confirms (3.34). The corresponding critical potential for ionization is larger if the atoms of the beam have a larger mass than those of the gas because the kinetic energy relative to the mass of the system must be twice the ionization energy and not the energy of the particle relative to the laboratory system.

Fig. 37 gives a few curves of the ionization efficiency for atom–atom collisions in rare gases. A comparison with the curves in Fig. 36 shows that this process is at least as efficient as that for collisions between ions and atoms. The slope of the curves $s_i = f(K)$ for N_2–N_2 and H_2–H_2 is of order 5.10^{-4} and 10^{-4} cm²/cm³ per eV respectively (87); $q_i = f(K)$ for O_2–O_2 and N_2–N_2 is shown in Fig. 37 a; for He q_i is $< \frac{1}{10}$ of N_2.

Ionization by collision between neutral atoms is of great interest in electric discharges as well as in astrophysics. It is one of the rate processes responsible for ionization of gases at very high temperatures.

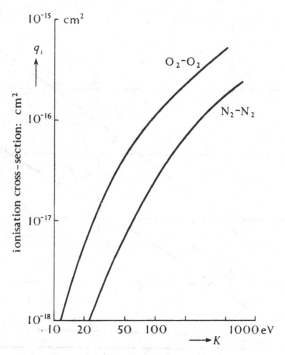

FIG. 37 a. Ionization cross-sections q_i for collisions between molecules in the ground state as a function of their relative energy K (97 e).

Here the gas molecules have velocities which are distributed at random, with a Maxwellian distribution. The number dz of collisions between molecules per cm³ and sec of mass M in a range of relative velocity v and $v+dv$ is

$$dz = \frac{\sqrt{2}\pi N}{\lambda} \left(\frac{M}{4\pi kT}\right)^{\frac{3}{2}} e^{-Mv^2/2kT} v^3 \, dv. \qquad (3.46)$$

N is the concentration in molecules/cm³ and T the gas temperature. Here the relative velocity v_i at which ionization occurs is given by $\frac{1}{2}Mv_i^2 = 2eV_i$. Assuming a constant average ionization probability f, we obtain by integrating (3.46) between v_i and ∞ (98 b)

$$\frac{dN}{dt} = fz = \frac{2fN}{\lambda} \left(\frac{kT}{\pi M}\right)^{\frac{1}{2}} e^{-2eV_i/kT}. \qquad (3.47)$$

f depends on T and the gas; it is probably of order 10^{-4} or less. (3.47) gives the number of ionizing collisions per cm^3 and sec by molecules in the ground state. Approximate values of z are found below.

TABLE 3.9

Rate z of ionizing collisions by molecules in cm^{-3} sec^{-1} for unit ionization probability and $p = 1$ mm Hg

$T =$	1000° K	3000° K	10 000° K	30 000° K
He . . .	. .	10^{-61}	1.10^{-1}	1.10^{16}
Cs . . .	10^{-15}	$1.6.10^{12}$	$2.8.10^{21}$	$2.4.10^{24}$

(d) *Ionization of gases and vapours by light quanta.* Light quanta passing through a gas may under certain circumstances ionize and excite the atoms. It is convenient to discriminate between two cases of photo-ionization when the energy $h\nu$ of the quanta (photons) absorbed is of the order of the ionization energy eV_i or when it is large compared with it.

(i) $h\nu \geqslant eV_i$: The alkali vapours are readily ionized by photons whose energy is larger, but still of the order eV_i. The following table gives the ionization potentials and the corresponding wavelengths λ_i. The radiation required lies in the ultraviolet region.

Since $eV = h\nu = hc/\lambda$ we have with $h = 6.6.10^{-27}$ erg sec, $c = 3.10^{10}$ cm/sec, $e = 4.8.10^{-10}$ e.s.u. and V (e.s.u.) $= V$ (volts)/300:

$$\lambda(\text{Å}) = 12,400/V \text{ (volts)}. \tag{3.48}$$

Rare gases, metal vapours, and molecular gases require quanta of large energy from the short ultraviolet or soft X-ray range.

TABLE 3.10

Ionization V_i and lowest excitation potentials V_1 and wavelengths λ_i and λ_1

Li	Na	K	Rb	Cs
$V_i = 5.39$	5.14	4.34	4.18	3.89 V
$\lambda_i = 2298$	2410	2858	2969	3184 Å
$V_1 = 1.85$	2.1	1.61	1.97	1.4 V
$\lambda_1 = 6707$	5896	7700	6299	8944 Å

He	Ne	A	Kr	X	Hg	H_2	N_2	Cu
$V_i = 24.6$	21.6	15.8	14	12.1	10.4	15.4	15.8	7.7 V
$\lambda_i = 505$	575	785	885	1022	1190	805	785	1610 Å
$V_1 = 19.8$	16.6	11.6	10	8.4	4.86	7.0	6.3	1.4 V
$\lambda_1 = 626$	746	1070	1240	1475	2537	1770	1970	8850 Å

Table 3.10 also gives the lowest values of energy of quanta which can produce an excitation from the ground state of an atom or molecule. V_1 refers to the lowest excitation potential and λ_1 is the corresponding wavelength of the 'resonance radiation'.

Suppose we irradiate a gas with monochromatic light of controllable wavelength, precautions being taken that the walls of the vessel are not illuminated and thus no secondary electrons are emitted from them. A gas atom which absorbs a sufficiently powerful quantum may release an electron whose energy of emission is

$$\tfrac{1}{2}mv^2 = h\nu - eV_i. \tag{3.49}$$

For example, a rubidium atom colliding with a photon of energy $h\nu = 4\cdot4$ eV, that is $\lambda = 2800$ Å, will release an electron (at the cost of $V_i = 4\cdot18$ eV) of initial energy $0\cdot22$ eV.

It should be noted that collisions between photons and atoms are different in character from those between electrons and atoms. A photon ionizes (or excites) the atom with a maximum probability at a certain critical wavelength or energy which is of order $0\cdot1$ to 1 eV above the minimum energy. An electron of that energy, however, has nearly zero probability of transferring energy irreversibly to the atom and requires perhaps 5 to 10 times the ionization energy to reach maximum ionization probability. The general reason seems to be that after an electron collision there are three bodies (the ion and two electrons) to carry away the energy and momentum, whereas after a photon collision there are only two bodies (ion and electron) and hence more stringent conditions apply.

Ionization by photons will occur with a certain probability depending on the wavelength and the nature and density of the gas; the probability can be expressed by the coefficient μ of absorption of light (corrected for scattering). Fig. 38 shows the absorption coefficient μ of metal vapours at 1 mm Hg and 0° C as a function of λ. We note that the principal maximum lies at a value of λ which agrees with that for λ_i given in Table 3.10. Absorption and ionization are a maximum when the quantum can just ionize. As λ decreases μ passes a minimum and then rises continuously. Moreover we find that at $\lambda > \lambda_i$ absorption (ionization) still occurs. This can be explained thus.

A quantum of energy less than eV_i cannot ionize unless the atom has already been excited (the atom may have collided a little while ago with an electron, an excited atom, or a photon). The absorption and ionization at wavelengths longer than the series limit is thus due to cumulative processes. This view is supported by the fact that certain lines in the absorption spectrum, e.g. of K coincide with certain relative maxima between 3300 and 3200 Å, not shown in the curve, Fig. 38.

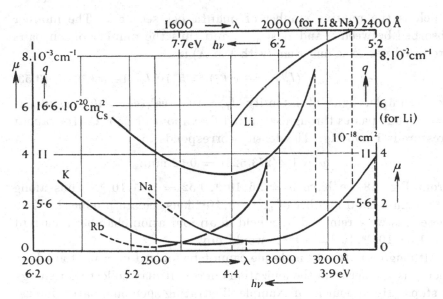

FIG. 38. Absorption coefficient μ in cm^{-1} at 1 mm Hg and 0° C and absorption cross-sections q in cm^2 of alkali vapours as a function of the wavelength λ and the quantum energy $h\nu$ (94). Dashed lines: ionization limit.

At wavelengths lower than the one corresponding to the first maximum, the theory of detailed balancing (Appendix 2) using a recombination cross-section which varies inversely with the square of the electron energy predicts (94 a) that

$$\mu \propto \lambda^4/(\lambda_i - \lambda), \qquad (3.50)$$

which is in good agreement with experiments if the range is not extended too far. The steady rise of μ as λ decreases is due to the absorption of photons by K_2 molecules. (Earlier experiments gave a large second maximum in this region which was due to the use of space-charge detectors which have often a tendency to oscillate.)

We shall now show how the ionization in a gas can be found numerically. Assume that a beam of monochromatic light of intensity I_0 enters a gas-filled chamber. I_x is the intensity (in ergs/sec) at a distance x. Since I_0 is the 'power' carried by the beam, $I_0/h\nu = z_0$ is the number of quanta entering per sec. The number of quanta absorbed along dx is

$$dz = -\mu z \, dx, \qquad (3.51)$$

μ being the absorption coefficient in cm^{-1} for a given wavelength and gas density. With $z = z_0$ at $x = 0$ we have

$$z/z_0 = e^{-\mu x}, \tag{3.52}$$

which is the relative number of quanta per sec at x. The number absorbed between 0 and x is $z_0 - z$, and thus the number of ion pairs produced per sec along x cm (with λ in Å) is

$$z_0 - z = {}_0 z_x = (I_0/h\nu)(1 - e^{-\mu x}) \doteq 5.10^7 I_0 \lambda (1 - e^{-\mu x}). \tag{3.53}$$

For example, light of intensity $I_0 = 10^{-6}$ cal/sec $\doteq$ 40 ergs/sec at $\lambda = 3000$ Å passes through saturated Cs vapour at 250° C. Its vapour pressure is 1 mm Hg. The density corresponds to

$$p = 1 \times (273/523) = 0.52 \text{ mm}.$$

From Fig. 38 we have $\mu = 2.6.10^{-3}.0.52 = 1.35.10^{-3}$. Thus along $x = 3$ cm a total number of ${}_0 z_x \approx 1.10^{10}$ ions/sec are produced. If all these ions were removed by a field in an ionization chamber a current $i = 1.59.10^{-19}.10^{10} \approx 10^{-9}$ A would flow.

Although an atom or molecule cannot be ionized by a photon whose energy is smaller than the ionization energy, it may collect that energy in steps. Hg vapour is an example illustrating such cumulative ionization. If irradiated with the intercombination line 2537 (the $6\,{}^1S_0 - 6\,{}^3P_1$ resonance line) a Hg atom can become ionized, though the transition is associated with only 4·86 eV, as compared with $V_i = 10.4$ eV. The ionization occurs only if resonance radiation without reversal is used, i.e. the intensity profile of the line must have one sharp maximum. This is readily obtained by using a cooled mercury lamp or by making sure that the beam does not cross a thick dense stratum of vapour. The ionization process is most probably (100 f)

$$\text{Hg} + h\nu \rightarrow \text{Hg} \,(6\,{}^3P_1),$$

$$\text{Hg} \,(6\,{}^3P_1) + \text{Hg} \,(6\,{}^3P_0) \rightarrow \text{Hg}_2^+ + e.$$

The second step is a collision between a metastable and a resonance state. The cross-section is likely to be very large, probably of order 10^{-13} cm^2 or so and the available energy is about $4.9 + 4.7 = 9.6$ eV. This together with the heat of dissociation of the Hg molecule of about 0·15 eV (Table 3.5 a) provides sufficient energy to ionize the Hg molecule ($V_i \doteq 9.7$ eV) with a reasonable probability (19). The ionization (dN/dt) so produced has been found to be $\propto p^2$ at very low p rising to p^3 at higher p. This shows that a two- (or even three-) stage process is operative (100 a). A $6\,{}^3P_0$ Hg atom can also be ionized by absorption of a quantum; the photo-ionization cross-section has been calculated to be about 0.24 cm^2/cm^3 = $6.6 . 10^{-18}$ cm^2 per atom (100b). Similarly Cs vapour is ionized by the Cs resonance line 8521 Å = 1·45 eV. Two excited Cs atoms collide forming $Cs_2^+ + e$. The total energy available is $2.9 + 0.45$ eV (Table 3·5a), which is probably the ionization potential of Cs_2 and like that of Hg_2 about 0·5–0·6 V lower than the ionization

FIG. 39. Collision of the second kind involving an excited atom and an atom in its ground state.

potential of the atom. Stepwise ionization occurs also in Zn and Cd. Collisions between metastable atoms can also produce ionization (93a).

A similar 'collision of the second kind' (Fig. 39) leading to ionization is one between a metastable Ne atom (V_{exc} = 16·5) and a normal A atom (V_i = 15·7). The excitation energy is used to ionize A and to accelerate the electron (Penning effect). The probability of ionization is of order unity. It is thought to be smaller when $V_i - V_{exc}$ is large (see Appendix 2).

(ii) $hv \gg eV_i$. The absorption of a quantum of large energy is quite distinct from the process discussed above. An ultraviolet photon hands over its energy to the least firmly bound electron whereas a powerful X-ray photon acts preferably on an electron of the inner shell. The fraction of the photon energy which is not used for ionization can either appear as kinetic energy of the ejected electron or can be used to increase temporarily the potential energy of the atom (98).

It has become apparent that absorption of light in the far ultraviolet and soft X-ray range has great significance in assessing the photo-electric processes in the gas and at the electrodes and walls. Recent work in the vacuum ultraviolet gives the dependence of the absorption coefficient μ as a function of the wavelength λ for continuous absorption in various gases (Table 3.11 and Fig. 39a). As λ decreases μ shows usually a series of peaks belonging to various bands or lines and then rises steeply at the ionization limit (Table 3.10), and after passing a maximum decreases slowly. At still shorter λ discontinuous changes—the X-ray absorption edges of the shells—are observed (Fig. 40): see (100j).

TABLE 3.11

Absorption coefficient μ at $p = 760\ mm$ Hg and $0°$ C as a function of the wavelength λ (90, 90 a)

Gas	$\lambda = 150$	300	500	600	700	800	1000 Å
CH_4	..	≈ 270	600	800	950	1100	970
NH_3	..	≈ 320	700	810	890	860	210
CO_2	..	≈ 400	780	700	660	340	band
CO	..	≈ 300	460	440	450	420	0
NO	..	≈ 350	550	800	500	520	450
H_2O	..	..	≈ 500	570	620	420	..

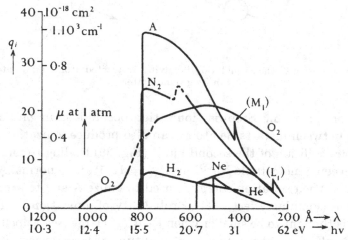

FIG. 39 a. Absorption coefficient μ at 1 atm and $0°$ C and cross-section q_i for ionization as a function of wavelength λ and quantum energy $h\nu$ (97 b).

Let He gas be irradiated by radiation of $\lambda = 1$ Å, corresponding (3.48) to a quantum of 12 400 eV. This may be absorbed by a He atom and one of the two electrons ejected at the expense of only 25 eV. The free electron thus obtains a kinetic energy practically equal to the quantum energy and is thus capable of producing further ionization in the gas. However, we know that fast electrons produce 1 ion pair per 30 eV of their energy; the 12 keV electrons will therefore give about 400 ion pairs. We see that almost all the ionization is due to the action of the ejected electron rather than the quantum itself.

Next let us irradiate argon gas. To remove an electron from the M, L, or K shell of an argon atom requires about 16, 250, and 3200 eV respectively (the first figure is the ionization energy). A 1-Å quantum is absorbed preferably by a K electron which escapes with an initial energy of about 9 keV which in turn produces about 300 ion pairs. If we now increase the wavelength to 4 Å, that is reduce the quantum energy to a value below 3200 eV, the available energy is too small to eject a K electron and hence instead an L electron is liberated. At the same time the probability of this ionization process is reduced by nearly one order of magnitude—a fact which finds its expression in a discontinuous change of the absorption coefficient (Fig. 40). Such jumps in μ always occur at values of μ at which the corresponding energy is just insufficient to liberate an electron from the next inner shell.

It has been assumed that the powerful X-ray quantum gives up the larger part of its energy to the most firmly bound electron whereas the excess energy goes into kinetic energy of the electron. This is not the complete story. We can assume that most of the quantum energy is used in removing electrons starting with the most firmly bound and gradually stripping the remaining electrons in higher orbits. As regards the total ionization in the gas, however, the numerical answer is hardly different.

We have now to consider the case when the excess energy $(h\nu - eV)$ of the photon is relatively small. How do the electrons rearrange in an atom after an inner electron has been ejected? It is to be expected that the vacancy in an inner shell will soon be filled by an electron from a higher level. This transition is accompanied by the emission of a quantum. The life of these excited states has been estimated to be of the order 10^{-15} sec. In the case of the argon atom, the transition of an electron from the L to the K shell produces the emission of a 2950-eV quantum whose energy can be absorbed by the same atom; the result is that one or more electrons can be emitted from the L shell. This process of ionization is known as 'Auger effect'. Up to 4 or 5 'Auger' electrons per atom have been observed in cloud-chamber photographs. An analogous effect (auto-ionization, section 3 a) occurs in the range of atomic spectra. For Auger cross-sections see (81).

We now turn to the quantitative side of the problem. The total number of ions produced per second along a certain path can be found very easily by assuming that the secondary electrons produced by the fast primary one represents the main bulk of ionization. If ϵ is the energy loss per ion pair produced (which is of order 30 eV), the total number of electrons and ions along a path of x cm is readily obtained. If ΔL is the intensity (power) of monochromatic radiation absorbed along the path and L_0 its value at $x = 0$, we have

$$_0z_x = \Delta L/\epsilon = (L_0/\epsilon)(1 - e^{-\mu x}) = [6 \cdot 3 \cdot 10^{11} L_0/\epsilon](1 - e^{-\mu x}), \qquad (3.54)$$

where $_0z_x$ is the number of ion pairs per sec along x cm, μ is the absorp-

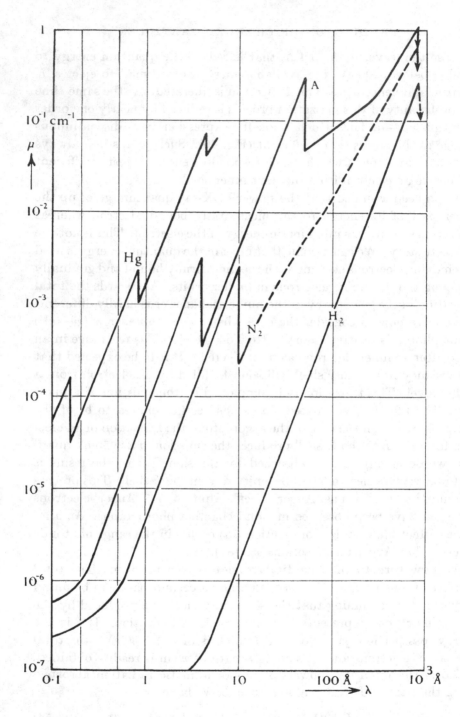

FIG. 40. Absorption coefficient μ at 1 mm Hg and 0° C as a function of wavelength λ and quantum energy $h\nu$ of X-ray radiation. Discontinuities: removal of $K, L,...$ electron (fine structure neglected, see Fig. 51). Non-linearity: Compton scattering. 124 Å correspond to 100 eV. $\mu \propto Z^4\lambda^3$ approximately for inner shell absorption. $\mu = 0.1$ cm^{-1} corresponds to $q = 2.8 . 10^{-18}$ cm^2.

tion coefficient; the numerical factor in (3.54) is due to expressing L_0 in ergs/sec and ϵ in eV (Table 3.8).

As to L_0 the intensity of radiation, an empirical relation has been found for X-ray tubes, viz.

$$L_0 \doteq 10^{-9}ZV^2i, \qquad (3.55)$$

where L_0, V, and i are in W, V, and A respectively and Z is the atomic number of the anode material.

The total absorption coefficient μ at 0° C and 1 mm Hg as a function of the wavelength λ is shown in Fig. 40 for various gases. The dotted parts of the curve indicate the regions where scattering becomes predominant. X-ray scattering produces ionization by absorption of scattered radiation, by absorption of light scattered by the Compton effect, and by the Compton electrons. The value of μ is proportional to the density and depends on λ and Z, the atomic number ($\mu \propto \lambda^3 Z^4$). For small λ, Compton-scattering, elastic scattering and pair production ($e^+ + e^-$) exceed absorption.

(e) *Thermal ionization and excitation.* In the preceding sections it has been pointed out that electrons colliding with atoms can produce ionization; so can atoms or photons colliding with atoms if the energy of the hitting particle exceeds a minimum value. There are, however, a variety of other processes which lead to ionization such as atoms and electrons colliding with excited atoms, etc. All these processes operate simultaneously if a gas is maintained at a sufficiently high temperature.

Let us assume a large amount of gas enclosed in a box which is heated from the outside; it is then isolated and finally a thermal equilibrium is attained. The object is to find the relative number of electrons (and ions) and excited atoms for a given gas as a function of its absolute temperature and pressure.

The answer can be obtained in two different ways. Since the gas is in thermal equilibrium with the walls—which we assume do not take part in ionization or excitation processes—we could obtain in principle the equilibrium concentration of electrons by summing up the rates of ionization and excitation for all elementary processes and equating the result to the sum of the rates of all neutralization—and 'de-excitation'—processes. The second way is the thermodynamic treatment. The gas is regarded as being in a state of dynamic equilibrium with 'chemical' changes taking place in two different directions. Instead of the rate constants for the single processes as above, the equilibrium constant for the reaction enters into the thermodynamic equations; thus besides the temperature the ionization potential of the substance is used to characterize the nature of the gas which 'dissociates' into electrons and ions. The thermodynamic treatment is an example of the application of Nernst's heat theorem to a dissociation equilibrium, an idea first applied by Lindemann (95) and later developed by Saha (96). The results have not only an important bearing on discharge problems but also on the physics of hot stars and astrophysics in general (97). No direct quantitative proof exists on thermal ionization equilibria; only a few experiments carried out in a special furnace show qualitative agreement with a relation which will be found later for excitation equilibria. There is, however, ample evidence obtained indirectly from investigations on arcs which supports the view that the thermodynamic treatment is justified. This applies even to cases where, instead of a pure thermal equilibrium a steady state is maintained: for example, where a small energy flow passes through the hot gas which is now no longer isolated.

The equation for the dissociation equilibrium for 1 mole of gas heated to a sufficiently high temperature is

1 mole neutral gas $\rightleftarrows$ 1 mole electrons $+$ 1 mole ions,

which expresses the conservation of mass. The law of mass action for this reaction states that

$$\frac{p_e p_i}{p_a} = K = Ce^{-W/RT}T^{\frac{5}{2}},\tag{3.56}$$

where the p's are the partial pressures. The right-hand side of (3.56) is the thermodynamic expression for the dependence of the equilibrium constant K on T, and W the 'heat of reaction' of the substance (here $W = eV_i$). Also with p_a, the partial pressure of the neutrals

$$p = p_a + p_i + p_e \quad \text{and} \quad x = p_e/p_0 = p_i/p_0, \quad 1-x = p_a/p_0,$$

where p_0 is the pressure at T when ionization is absent. Substituting x in (3.56) and determining the chemical constant C from Nernst's heat theorem (see advanced textbooks on heat and thermodynamics) we obtain finally with the statistical weight ratio $g_{\text{ion}}/g_{\text{gas}}$:

$$\frac{x^2}{1-x^2}p = 4\cdot 9g_{\text{ion}}/g_{\text{gas}}\cdot 10^{-4}T^{\frac{5}{2}}e^{-eV_i/kT}.\tag{3.57}$$

Here $x = p_e/p_0 = N_e/N_0$ is the degree of ionization, that is the ratio of the concentration of electrons to molecules initially present, p the actual gas pressure (not the density) in mm Hg, T the absolute temperature, and V_i the ionization potential ($eV_i/kT = 11\ 600V_i/T$ in volts and °K). When $x^2 \ll 1$, (3.57) shows that x is approximately proportional to

$\exp(-eV_i/2kT)$ and to $1/\sqrt{p}$ or $N_e \propto N_0^{\frac{1}{2}}$. For large x, (3.57) cannot be used because of the simplifying assumptions made. The numerical factor in (3.57), and hence x, becomes smaller when excited states are allowed for. This seems rather curious, since slow electrons can now ionize excited atoms. The explanation is that a large fraction of atoms become excited instead of ionized; however, the reduction of the number of atoms which originally had the chance to become ionized exceeds the increase due to ionization of excited atoms.

The function $x = f(T)$ has an s-shape (Fig. 41 a) which shows also $x\sqrt{p}$ for gases with V_i between 5 and 25 V. For example, in N_2 ($eV_i = 15\,\mathrm{eV}$) at 1 atmosphere about 10 per cent of the gas molecules are ionized at 13 000° K, whereas Cu vapour with $eV_i = 7\cdot5$ eV at the same pressure is ionized to the same degree at 7000° K (Fig. 41 b).

It is possible to derive equations for mixtures of gases in a similar way, but this will not be discussed here. However, it is important to remember that at high temperatures molecular gases may be partially dissociated into atoms and thus the mean ionization potential to be used in the relation (3.57) will lie between that for the atom and the molecule. Also chemical reactions between the constituents have to be allowed for; this is the case, for example, with air at $T > 4000°$ where NO is produced with an ionization potential of $9\cdot5$ eV as compared with 13 to 15 volts for N_2 and O_2.

The relative number of atoms in an excited state at a given gas temperature in thermal equilibrium follows from the chemical equation

1 mole unexcited molecules + excitation energy for 1 mole

$\rightleftarrows$ excited molecules.

Again, the degree of thermal excitation, that is the ratio N_n/N_0 of the concentration of molecules in the nth excited state to the molecules initially present, is found from the law of mass action to depend on the ratio of the potential energy of the nth state to the average kinetic energy of a gas molecule:

$$N_n/N_0 = (g_n/g_0)e^{-eV_n/kT}. \tag{3.58}$$

g_n and g_0 are the statistical weights of the nth and the ground state, that is the probability of finding a state under identical conditions. A term (describing a state) with angular momentum J has a statistical weight $g = (2J+1)$; (g_n/g_0) is an integer of the order 1. If $eV_n \ll kT$, $N_n/N_0 = g_n/g_0$ and thus there are roughly the same number of molecules in the nth state as in the ground state, but this does not hold if several states exist. For $eV_n \gg kT$, the ratio of statistical weights in (3.58) can be approximated to 1 and N_n/N_0 follows from the Boltzmann distribution. From (3.58) it follows that in Hg vapour at 5000° K about 1 out

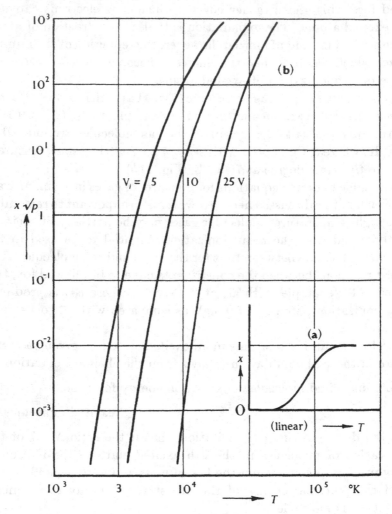

FIG. 41. Degree of ionization x in a gas of ionization potential V_i as a function of its temperature T and pressure p. (a) linear plot of $x = f(T)$; (b) logarithmic plot of $x\sqrt{p} = f(T)$.

of every 10^5 atoms originally present are in the resonance state (V_n about 5 eV).

For accurate calculations of thermodynamic equilibria Landolt–Börnstein's tables (volume supplement 2/2, p. 1252) should be consulted.

It is interesting to inquire into the contribution to ionization by electron, molecule, and quantum collisions. From (3.47), (11.9), and (12.5) of Appendixes 3, 4, the three ionization rates for Cs have been calculated; they are shown in Fig. 42 as a function of T. It follows that

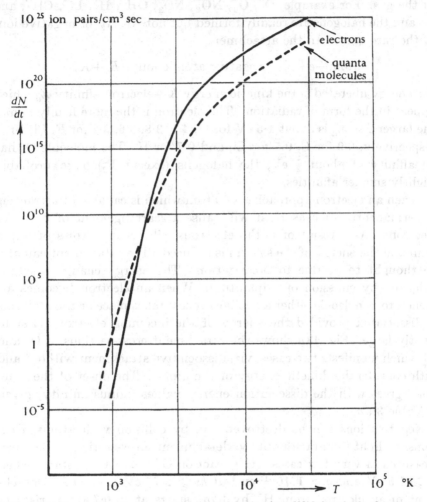

FIG. 42. Comparison between the rate of ionization dN/dt by single electron collisions, intermolecular collisions, and quanta as a function of the temperature T for Cs at 1 mm Hg.

at low T quanta and molecules are most effective whereas at high T electrons are most effective. The information enables us also to obtain an estimate of the least time necessary to set up ionization equilibrium: this time is of the order final concentration of ions divided by the ionization rate. It is herein assumed that the gas temperature is suddenly increased to the value T.

(f) *Formation of negative ions.* When an electron collides with a neutral gas atom or molecule it may become attached and form a negative ion. This process depends on the energy of the electron and the nature of the gas. For example, O^-, O_2^-, NO_2^-, NO_3^-, OH^-, H^-, Li^-, CH^-, and C^- and the halogens are readily formed but not N^-, N_2^- or negative ions of the rare gases. In the attachment

$$atom + e + K \rightarrow \text{negative atomic ion} + (E_a + K)$$

the energy liberated is the kinetic energy K+electron affinity E_a which appear in the form of radiation. The electron is the more firmly bound the larger E_a. E_a is about 1·5 eV for O, 4·1, 3·8, 3·6, 3·2 for F, Cl, Br, I respectively, 0·9 for C, 0·5 for Li, and 0·7 for H. The molecule O_2 has an affinity of about $\frac{1}{2}$ eV, the halogens (except F_2) have probably slightly smaller affinities.

When an electron approaches an atom which is capable of attracting it permanently such as H, it will cause a rearrangement of its outer electrons. As a result of it the electrons will occupy orbits of larger radii and the energy of the system is reduced. The attachment can also be thought to be due to polarization. The energy change must be balanced by emission of a quantum. When an electron becomes attached to a molecule either a similar process takes place or the molecule is dissociated provided the energy of the incoming electrons is sufficiently large. Fig. 43a shows the attachment cross-sections in O_2 and H_2 which manifests two cases, viz. dissociative attachment without and with considerable kinetic energy of the atoms. The onset of the reaction agrees with the dissociation energy values (minus affinity) given in Table 3.5a.

Negative ions can be destroyed e.g. by collision with atoms, electrons, or light quanta: the photo-detachment cross-sections have been measured in very few cases. The reaction $O^- + h\nu \rightarrow O + e$ starts when $h\nu \geqslant 1·5$ eV and $q \approx 1.10^{-17}$ cm² at $h\nu \geqslant 2·75$ eV ≈ 4500 A. Detachment of an electron from H^- by light starts at $\approx 0·7$ eV; it rises to $4·5.10^{-17}$ cm² at $h\nu = 1·5$ eV and decreases at higher energies (98a). The atomic detachment cross-section is probably of order 5.10^{-18} cm² at room temperature (100d).

The mechanism of attachment can also be visualized in the following way (Fig. 43b): an electron which approaches a molecule temporarily polarizes it, as a result of which the minimum in the P.E. curve moves up and to the right, and becomes shallower (case 2). Whether this leads to a complete disappearance of the minimum (cases 1 and 3) or not

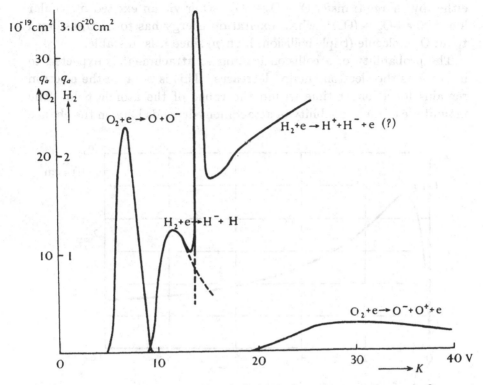

FIG. 43 a. Measured electron (dissociative) attachment cross-sections q_a in O_2 and H_2 (100 e). The origin of the high second peak in H_2 is not yet known.

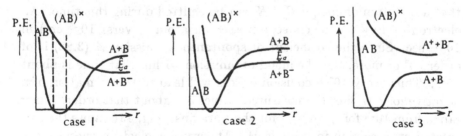

FIG. 43 b. Various types of attachment of electrons to diatomic molecules (100). P.E., potential energy; r, interatomic distance; AB, AB*, molecule in ground state and electronically excited state respectively; E_a, attachment energy; dashed lines, Frank–Condon range. Use these models with caution.

depends on the detailed structure of the molecule and the final products. Only in the latter case is a stable negative molecular ion formed. It is evident that the affinity energy E_a equals the energy difference between the dissociated pairs of particles.

Besides O^-, O_2^- has been observed. This is thought to be formed either by charge transfer $O^-+O_2 \rightarrow O_2^-+O$ or via an excited molecular ion, viz. $e+O_2 \rightarrow (O_2^-)^*$, whose excitation energy has to be transferred to an O_2 molecule (triple collision, high p) since it is unstable.

The probability of a collision leading to attachment is expected to increase as the electron energy decreases. This is because the electron remains for a longer time within the range of the atomic field. The magnitude of the probability of attachment depends here on the chance

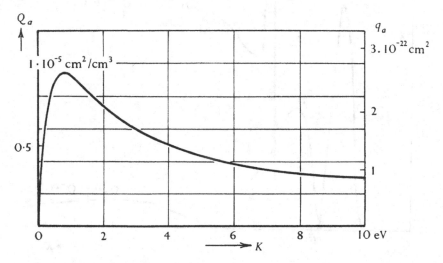

FIG. 43c. Theoretical cross-section Q_a (at 1 mm Hg) and q_a for attachment of electrons of energy K to H atoms (100).

that a quantum of energy E_a+K can be emitted during this time. An electron of 1 eV (6.10^7 cm/sec) crosses an atom (several 10^{-8} cm) in 10^{-15} sec. Since the coefficient of spontaneous emission A (3.22) is of order 10^8 or more, the likelihood of emission within that time is about 10^{-7}. Thus every 10^7th collision or less will lead to the formation of a negative ion. In fact the measured values are about that order. They range from 10^{-4} for Cl to 10^{-8} for the paraffins. Fig. 43c shows the calculated cross-section for atomic H. At very low electron energies the repulsive forces seem to make a capture more difficult. H^- is not only important in discharges (99) where the affinity spectrum from E_a to ∞ has been discovered, it also plays a major role in the continuous emission spectrum of the sun. The strong absorption of light in the photosphere for $\lambda > 5000$ Å is due to detachment of electrons from H^- (100). When an electron becomes attached to a molecule the energy liberated in this process may be converted to potential energy: for example, the molecule is first excited to a higher state and then dissociates into an excited atom and a negative atomic ion. For measurements of E_a and the probability of attachment see (100).

(g) *Ion attachment.* When a slow positive or negative ion passes a molecule of zero electric dipole moment but finite polarizability it is attracted and can become attached to it. There are different processes by means of which energy and linear momentum can be conserved which are analogous to those in electron recombination and electron capture, viz. radiative, dissociative attachment, etc. We shall treat here only the case in which capture of the ion occurs when weak forces are effective.

Equation (4.8) describes the force between an ion and a molecule of a gas of dielectric constant D and of polarizability α_0, where

$$(D-1)/4\pi = N\alpha_0,$$

N being the number of molecules per cm³. The attractive potential P per molecule is therefore by integration

$$P = \frac{e^2\alpha_0}{2r^4}. \tag{3.58 a}$$

Assuming that attachment occurs when this potential is equal to the kinetic energy of the ion (the latter can be stored as potential energy of the final product) then with a critical radius r_c the capture cross-section is

$$q_{att} = r_c^2\,\pi \approx \frac{e\pi}{v}\frac{\alpha_0}{m}. \tag{3.58 b}$$

From a more rigorous treatment q_{att} is found to be twice as large (100 c). Thus the cross-section is proportional to the square root of the ion energy. For example for $m = 10^{-23}$, $\alpha_0 \approx 7.10^{-23} \sum_n f_n \lambda_n^2 \approx 10^{-23}$ cm³, where f is the oscillator strength and λ the wavelength in μ, we have, for $v = 10^5$ cm/sec, $q_{att} \approx 2.10^{-14}$ cm².

B. EMISSION OF CHARGES FROM SOLIDS

5. THERMIONIC EMISSION OF ELECTRONS

Many substances with a high melting-point emit electrons at temperatures at which their rate of evaporation is only feeble. The emission of electrons from conductors, originally thought of as a classical evaporation process, has proved to be understandable only in terms of quantum mechanics. The essence is that electrons are not free in the same sense as are the molecules in a liquid but their potential energy in the metal is high; in fact they are so strongly bound that only an insignificant number can contribute to the specific heat of the metal (22).

When the temperature of the conductor is sufficiently raised, those electrons which are in the highest states of energy can overcome the attractive forces from the other charges and escape from the metal. The minimum energy required to escape from matter in bulk is the work function φ, which corresponds to the ionization potential V_i of single atoms. It can be shown that $\varphi < V_i$. Assume first a charged sphere of radius r. The work done to remove unit charge from r to ∞ is e/r. Now take unit charge r cm from a plane solid conductor. Because of the image force the energy to remove it to ∞ is $e/4r$. Hence $\varphi \approx V_i/4$. This is only approximately true. For C we have $V_i = 11\cdot2$, $\varphi = 4\cdot4$, but for K, $V_i = 4\cdot3$, $\varphi = 2\cdot2$, all values in volts (see also p. 59).

A calculation which is based on the Fermi–Dirac energy distribution of electrons in metals (103) gives for the emission current density j as a function of the absolute temperature T and the work function φ

$$j = AT^2 e^{-e\varphi/kT}, \tag{3.59}$$

where A and φ are in A/cm^2 and V respectively, and kT/e in V by the conversion $11\,600^\circ$ K = 1 eV. Values for A and φ are given below. The values of φ used here are the same as those obtained from photo-electric emission experiments. Actually φ is slightly dependent on T.

6. THERMIONIC EMISSION OF POSITIVE IONS

When a metal is heated to a sufficiently high temperature it can be shown by means of a mass spectrograph that besides neutral atoms

TABLE 3.12

Work function φ, emission constant A, and current density j for various emitters (102), (102 a); see also Table 3.14

Cathode	φ (V)	A	\multicolumn{2}{c}{j (A/cm^2) at T° K}
W	4·5	70	0·27	2500
W—O—Ba . .	1·3	~ 3	10	1000
Ba oxide . .	1·7	~ 40	0·5	1000
Th oxide . .	3·1	~ 3	2	1900
Th carbide . .	3·5	550	4	2000
C	4·4	48	0·15	2400

positive ions are 'evaporated' (101). Singly charged ions of W, Mo, Ta but also of metals with lower melting-points like Cu, Ag, Fe, Ni were found at temperatures near the melting-point. The ion current obtained follows an equation of the type (3.59), viz.

$$j^+ = C_1 T^2 e^{-11600\varphi^+/T}, \tag{3.60}$$

where φ^+ is the work function for positive ions which is larger than φ for electrons. $\varphi^+ = 6\cdot5$ and $6\cdot1$ V for W and Mo respectively and C_1 is of order $0\cdot1$ in A/cm² (°K)². At 2800° K about every 4000th evaporated W atom is an ion.

Larger currents of positive ions are emitted from heated salts of K and Na and halide negative ions from oxide cathodes and from lithium aluminium silicate emitting several mA/cm² (103 b). A positive ion emitter is the Kunsman anode, a metal strip coated with ferric oxide + 1 per cent Al_2O_3 + 1 per cent Na or K oxide or nitrate bound by paraffin wax; the coating is fused in air at 900° C and reduced in H_2. Such an electrode will give a current of 10 mA/cm² at dull read heat in vacuum.

7. SECONDARY ELECTRON EMISSION BY ELECTRONS

A primary electron which falls on the surface of a solid either returns to the gas (or vacuum) or penetrates into the solid and releases secondary electrons, a process often accompanied by X-ray emission. This holds whether the solid is a metal or an insulator, or has a composite surface. Little is known about the interaction between the primary electrons and the electrons in the levels of the energy bands of solids and a satisfactory theory of the secondary electron yield is still lacking.

For a secondary electron to be released the energy of a primary has to be larger than the work function φ of the solid. There seems to be no simple relation between the yield and φ. The spectrum of the emitted electrons is the sum of the secondaries and the reflected primaries and with slow electrons it is difficult to discriminate between them. In general a large fraction of them is reflected from most surfaces. However, the angle of incidence is not equal to the angle of reflection as in the case of light.

Fig. 44 shows the electron distribution, for electrons of 160 eV falling on Au. Clearly the average energy of the emitted electrons is of order of a few eV, but there are also 160 eV reflected (primary) electrons present. The former are the secondaries ejected from the metal, but Fig. 44 shows superimposed on these two main groups a small number of inelastically reflected primaries.

The emission process is still uncertain. At first it was thought that primaries with energies well above ionization potential are absorbed by the atoms, thereby emitting soft X-rays which in turn eject secondary electrons. In this case the yield as a function of the speed of primaries ought to show a fine structure corresponding to discontinuities in absorption which have only been occasionally found where adsorbed gas atoms were probably present. Other theories ascribe the emission to the transfer of energy from the primaries to the valency electrons; or to interaction between the primaries and the free electrons in a Fermi distribution whereby multiple elastic collisions then enable the secondaries to pass through the boundary into the vacuum.

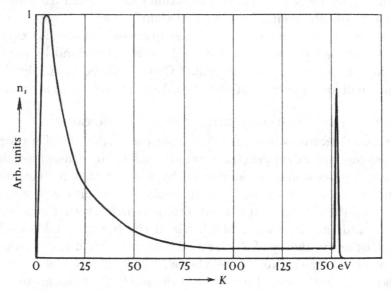

FIG. 44. Energy distribution n_s of the secondary electrons when 160 eV electrons fall on gold (105); maximum value of n_s arbitrarily taken to be unity.

The yield or coefficient of secondary emission δ, that is the number of secondaries per incident primary for normal incidence at various surfaces of solids bounded by high vacuum, is shown in Fig. 45 and in Table 3.13. All these yield curves show an increase of δ with the energy of the primaries, then a maximum and a decrease of δ at still larger energies of the primaries. The explanation is that at relatively low energies the depth of penetration of the primaries is small and thus the chance large for secondaries to escape. With increasing energy the depth increases and the chance of escape decreases, with the result that at large energies large numbers will be accelerated internally but only few escape from the solid. This view is supported qualitatively by the fact that for a given energy the yield δ increases with increasing angle of incidence. It seems therefore as if an electron beam which strikes a surface obliquely penetrates less deeply than in normal incidence and thus more secondaries are emitted. The depth of penetration of the primary electrons (range) or the depth of origin of the secondaries, for example in Pt, is about 30 and 100 layers of atoms for 1 and 2 keV electrons respectively.

Advances have been made in recent years to increase the yield by using composite surfaces (106); it is, however, wrong to believe that there is a simple and close relationship between secondary emission and

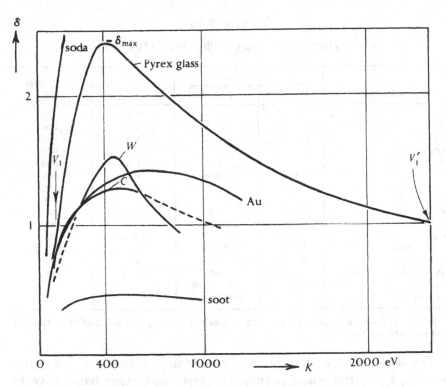

FIG. 45. Secondary emission coefficient δ as a function of the energy K of the primary electrons for different substances (104, 107).

photo-electric emission. Surfaces with a high maximum yield emit at room temperature up to 10 to 20 secondaries per primary electron. They consist of Ag–O–Cs, Sb–Cs, oxidized surfaces of Ag plus a few per cent Mg, Cu, or Be. The energies of the primaries to attain δ_{max} are between 500 and 1000 eV and more. Fig. 45 shows that normal metal surfaces have a maximum δ of order 1 whereas glasses and probably other insulators have $\delta = 1$ at energies of primaries below 100 eV. δ for Aquadag is low; it can be further reduced by using rough deposits. It is thought that the electrons have little chance to escape from these miniature crevasses because of the several reflections they suffer. Similar results have been obtained by artificial roughening of metal surfaces.

There exists also an angular distribution which for metals has a maximum lying in the normal to the surface; more complex patterns have been observed with secondaries emitted from insulators. The time interval between the striking of the primary and leaving of the secondary is found experimentally to be of order 10^{-12} sec or smaller.

TABLE 3.13

Secondary electron emission coefficients δ (104, 105, 107b)

Substance	$\delta = 1$ at		δ_{max}	V volts for δ_{max}
	V_1 volts	V_1' volts		
Li	..	..	0·5	85
K	..	..	0·7	200
Cu	> 100	..	1·3	600
Ag, Au . . .	..	..	1·5	800
W	..	..	1·5	500
C	160	∼ 1000	1·3	600
Soot	..	..	0·4–0·8	500
Pt	..	..	1·6	800
Mo	140	1200	1·3	350
NaCl	∼ 20	1400	6–7	600
MgO	..	..	2·4–4	400
Pyrex glass . .	30–50	2400	2·3	300–400
Soda glass . .	30–50	900	∼ 3	300
Oxide cathode BaOSrO	40–60	3500	5–12	1400
ZnS	..	6000–9000	..	..
Ca tungstate . .	..	3000–5000	..	..

For Na and Zn the four corresponding values are: —, —, 0·8, 300 and 100, 400, 1·1, 200 respectively (107 a).

The large yield from insulators has been explained as follows: a secondary electron when passing the first conduction band may be trapped in one of its levels. The potential of this level, however, may be at or even above the potential of the surface barrier and hence the chance of the electron to escape is large (105).

8. SECONDARY EMISSION OF ELECTRONS BY POSITIVE IONS AND METASTABLE ATOMS

When positive ions impinge on the surface of a solid or liquid conductor, charged and uncharged particles are emitted. The uncharged particles are often found to be molecules of the gas in contact with the surface; they are either positive ions which have become neutralized by extracting electrons from the surface of the conductor or they are single or groups of atoms which were ejected from the surface with relatively low speed (sputtered). Sometimes these neutral particles are metastable atoms; they were originally positive ions which have been neutralized and then reflected from the surface. The charged particles are either slow secondary electrons (few eV) from the metal or in rare cases negative or reflected positive ions; the efficiency of the latter processes seems to be small. Hardly anything is known about the secondary emission of electrons from insulators, but it can be assumed that it is not essentially different from conductors.

A secondary electron can only be ejected from the surface by a positive ion if the sum of its kinetic (K_i) and potential energy (eV_i) exceeds $2e$ times the work function φ of the metal; this follows from the conservation of energy and the fact that for each ejected electron another electron has to escape to neutralize the positive ion. Thus if the energy of the ejected electron is $\geqslant 0$

$$K_i + eV_i \geqslant 2e\varphi. \tag{3.61}$$

This condition is nearly always fulfilled even with $K_i \approx 0$. It is, however, not possible to draw general conclusions from (3.61) about the efficiency of this process.

The number of secondary electrons emitted per incident positive ion depends on the nature and energy of the ion and the metal and particularly on the state of its surface. For convenience slow and fast ions will be discussed separately; in both cases it can be stated that the mechanism of transfer of energy from the ion to the metal is still a matter of speculation (109 b).

A slow rare gas ion can be thought of as taking out an electron from the metal into one of its higher levels when it has approached to within several atomic radii of the surface. The atom then remains for a short time in a metastable state and as it moves closer to the surface transfers its excitation energy to the metal, which results in the emission of a photo-electron. Against this view it is held that the velocities of the secondary electrons should conform with Einstein's equation (3.49) whereas the observed velocities appear to be smaller.

Fast positive ions, like α-particles with energies $> 10^6$ eV, liberate electrons from a solid at a rate of perhaps 10 to 30 per incident particle. The energy of the secondary electrons can reach several thousand eV if metal foils are bombarded. There is also definite evidence that the yield of secondary electrons increases with increasing ionization energy of the gas atom forming the ion.

Fig. 46 shows the coefficient of secondary emission γ_i, that is the number of secondary electrons released per incident positive ion in vacuum as a function of the kinetic energy K of the ion for various metals and ions of uniform energy. As expected γ_i increases with K and seems to approach a maximum at somewhere about 100 keV, which for Li$^+$–Ni is ≈ 3 electrons/ion. The cause of the maximum is probably that at large speeds the incident ions penetrate the metal to such depths that only a fraction of the secondary electrons can escape. Results for slow rare gas ions are also given in Fig. 46, showing that γ_i is larger the

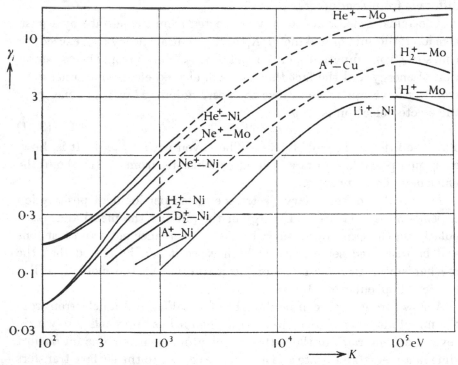

FIG. 46. Secondary emission coefficient γ_i for ions of energy K falling on the surface of various substances (110, 111, 112). For Hg ions see (114 a).

higher the ionization potential of the parent gas and the lower the work function of the metal. The numerical results depend strongly on the cleanliness of the surface and the amount of adsorbed gas. Gas-free surfaces show a lower value of γ_i (5, 203 b). γ_i appears to be large if the incident ion is excited.

Some of the incident positive ions are reflected the more often the larger the angle of incidence. For example, 10–80 per cent. of rare gas ions of energy 500–1500 eV are reflected from Ni into vacuum but only 1/100 or less of alkali ions. The reflection coefficients in the presence of a gas are probably smaller. Accommodation coefficients are treated in (18).

In gas discharges the positive ions arrive at the cathode with different speeds accompanied by photons and possibly by metastable atoms from the gas. In interpreting these measurements one has to remember that the number of secondary electrons is counted as if due to one incident ion and the associated processes. Fig. 47 shows γ as a function of the numerical value of the field X/p for various molecular gases and metals

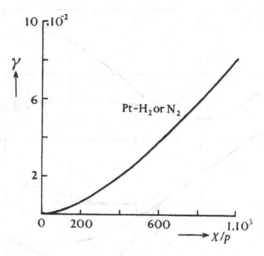

FIG. 47. Approximate values of the secondary emission coefficient γ as a function of the reduced field X/p (114). A peak in γ near $X/p \approx 100$ found earlier is erroneous.

derived from breakdown measurements (Chapter 7). At low X/p the ratio of excitations to ionizations by electron collisions is large, the absolute number of excitations (per electron per cm in field direction) small but increasing with X/p, and so γ_i rises. At higher X/p the contribution due to the faster positive ions becomes more prominent, and γ rises, since an increasing number of electrons in the tail of the energy distribution ionizes rather than excites gas molecules. At large values of X/p only a few quanta are produced in the gas, but the kinetic energy of the ions becomes appreciable and the number of electrons released rises in accordance with Fig. 46.

In rare gases (Fig. 48) γ, again obtained from breakdown measurements, first decreases and then increases with X/p. This can be interpreted as being due to quanta and metastables which are copiously produced in the gas but decrease in absolute numbers as well as relative to the ions as X/p rises. Also at larger X/p when the ions are the controlling factor γ decreases with the mass of the ion; this is because—apart from their lower potential energy—their kinetic energy is smaller because of the smaller mean free path. The physical reasons why γ depends on X/p are probably that the ion energy distribution depends on it as well as back-scattering of the secondary electrons. Also note that ions can return from the cathode as ions, neutrals, or excited species.

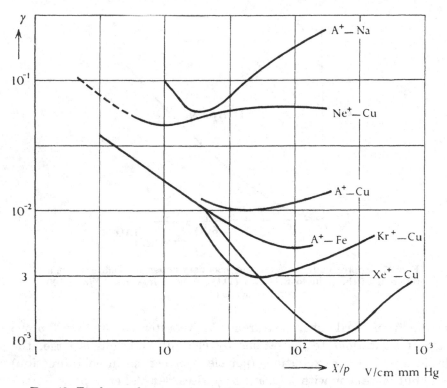

FIG. 48. Total secondary emission coefficient γ for rare gas ions and various cathode substances as a function of the reduced field X/p (2, 113). The decrease of γ with X/p at low fields may be due to the lack of metastable atoms relative to ions.

From the foregoing it can be seen that it is convenient to write γ as the sum of four terms—due to ions, photons, metastables, and neutrals —namely

$$\gamma = \gamma_i + \gamma_p + \gamma_m + \gamma_n. \qquad (3.62)$$

However, this is only true if the four processes are independent. For given X/p, γ_i could be derived from Fig. 46 if the energy distribution of ions were known; further $\gamma_p = f_p \, PN_p$, where f_p is the fraction of photons (a geometrical factor) falling on the cathode, P the photoelectric yield in electrons per quantum (section 5), and N_p the number of quanta produced per ion; finally $\gamma_m = f_m m N_m$, whence f_m depends on the geometry as above, m is the number of secondary electrons per metastable atom, and N_m the number of metastables per ion. The values of P for Ne and Kr are about 3.10^{-3} and 5.10^{-4} respectively, but the true values are likely to be larger; N_p, for example, in A at $X/p = 10$, 30, and 100 V/cm mm Hg is about 150, 10, and 2 respectively. This decrease is to be expected since at small values of X/p very few electrons have sufficient energy to ionize atoms but a large number can excite them.

A few remarks should be made concerning observations of very low and very large values of γ. From measurements with Geiger–Müller counters γ has been found (110) to have values of order 10^{-10} for ethyl-alcohol positive ions impinging on metal or glass. γ is also low for water. It has been observed too that γ is smaller for larger molecular ions because the total ion energy is the more easily converted into vibrational or dissociation energy of the molecule (rather than transferred to the electrons in the solid) the more numerous the modes of vibration. Large values of γ (0·5 to 3) have been found in glow discharges and particularly in the hollow cathode type, where the number of ions produced in the dark space and arriving at the cathode is small compared with the number of quanta. Thus here $\gamma \approx \gamma_p$. Some values of γ in Hg vapour have been derived from measurements of the sparking potential at low p: for cathodes of Pt, Fe, Al, and liquid Hg γ has been found to be 3.10^{-4}, 5.10^{-3}, 2.10^{-2}, and 0·1 respectively, with X/p between 10^2 and 10^3 V/cm mm Hg (cf. 114 a).

Secondary electrons are readily emitted from metal surfaces when slow metastable atoms impinge upon them. With metastable Hg atoms more than 0·1 electrons per metastable incident on Hg have been estimated. Larger yields have been found for rare gas metastables. 2 ^3S metastables of He ejected 0·24 electrons per particle from Pt, and 2 ^1S atoms about 0·5. From Cs 0·4 electrons per metastable A atom are ejected (108). The yield seems to depend more on the surface layer than on the nature of the substance. Usually the yield increases with gas contamination. The energy in excess of the work function of the surface is in general transferred to the ejected electron (109). Excited H_2 is likely to cause secondary emission as well as fast neutral unexcited particles (89).

When slow positive ions fall upon a surface the secondary electron yield is expected to increase as the potential energy of the ions rises (see above). This is clearly demonstrated by comparing γ_i for He^+, He^{++}, and He_+ on Mo (109 a). γ_i is largest for He^{++}, smaller for He^+, and still lower for He_2^+, namely 72, 25, and 13 per cent respectively and is independent of the ion kinetic energy. For singly charged ions of He, Ne, A, Kr, and Xe impinging on W the values of γ_i are about 29, 23, 10, 5, and 2 per cent.

9. PHOTO-ELECTRON EMISSION

An electron can only be emitted from the surface of a solid if the monochromatic radiation falling on it exceeds a critical frequency or the associated quantum a minimum energy, namely the work function φ (section 5). Light of lower frequency is (to a first approximation) ineffective, however large its intensity. The critical wavelength λ_c in Ångström units is related to φ in V by (3.48), viz.

$$\lambda_c \varphi = 12,400.$$

This means that for a metal with $\varphi = 4$ eV we must have $\lambda < 3100$ Å to eject photo-electrons. The following table of φ for various substances shows that φ lies usually between 1 and 10 V (115, 102 a). It can be found by thermionic, photo-electric, or contact potential measurements.

The values of φ for alkalis follow a sequence similar to that of the ionization potentials: as the atomic number increases φ decreases.

TABLE 3.14

Work function φ of various substances in V (6, 17, 118 a)

Li	.	.	.	.	2·46	Pt	.	.	.	.	5·36
Na	.	.	.	.	2·28	Ca	.	.	.	.	3·2
K .	.	.	.	.	2·25	Mg, Si	.	.	.	3·6
Rb	.	.	.	.	2·13	U, Th, Hf	.	.	3·5
Cs (monolayer) on Ag—O	0·7	Ba	.	.	.	.	2·5
Cs	.	.	.	.	1·94	WO_3	.	.	.	9·2
Au	.	.	.	.	4·7	CuO	.	.	.	5·3
Hg, Fe, W, Cu	.	.	4·5	Cu_2O	.	.	.	5·2
C .	.	.	.	.	4·4	BaO	.	.	.	1
Mo	.	.	.	.	4·2	S .	.	.	.	4·4
Ni	.	.	.	.	4·9	Ebonite	.	.	< 4
Pd, Re, Nb	.	.	5	H_2O and aqueous solu-	
Zn	.	.	.	.	3·4	tions .	.	.	.	6·1

For $\lambda < \lambda_c$ because the emission comes from different depths an energy distribution of photo-electrons is obtained which is nearly independent of λ within certain limits. Fig. 49 shows that the maximum of the distribution of the component normal to the surface lies at about 40 per cent of the maximum speed given by (3.49). This maximum is shifted towards the right for thin foils because now more of those high-speed electrons that have velocity vectors oblique to the irradiated surface can escape.

So far we have dealt with light whose λ is not too far below λ_c. When λ becomes shorter, the probability of ejecting more firmly bound electrons increases. (This is similar to what happens when an X-ray quantum $h\nu$ acts on a gas atom: instead of a valency electron one of

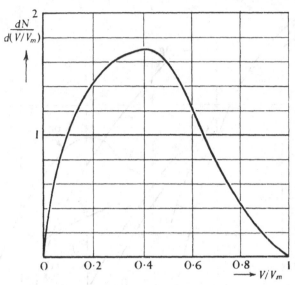

FIG. 49. Energy distribution $dN/d(V/V_m)$ of photo-electrons released by quanta ($\lambda = 2000$ to 3000 Å) from a metal surface as a function of their relative energy V/V_m. V_m is the energy corresponding to the maximum speed (17).

the inner levels is ejected.) Einstein's equation (3.49) has now to be written

$$K = h\nu - eV_i^n, \qquad (3.63)$$

where K is the kinetic energy of the emitted electron and eV_i^n the ionization energy for an 'inner' electron of quantum number n. For example, to eject a K-electron ($n = 1$) eV_i^1 is of order $10Z^2$ (eV) for an element of atomic number Z. The number of electrons emitted per quantum is larger than for ordinary light because the electrons ejected by the quanta are so fast that they produce secondary electrons in the substance; also because part of the quantum energy is 'internally converted' (section 4d (ii)).

Each quantum of sufficient energy can eject an electron but only a fraction of them are emitted. The number of electrons emitted per incident quantum is called the photo-electric yield γ_p. It is a function of the wavelength of the radiation, the substance, and the polarization of the radiation. Fig. 50 shows $\gamma_p = f(\lambda)$ for various metals. It is seen that γ_p increases with decreasing λ. In the region of the far ultraviolet γ_p reaches values of the order $0.1-1$, but it falls again in the X-ray region, apart from the well-known discontinuities (Fig. 51).

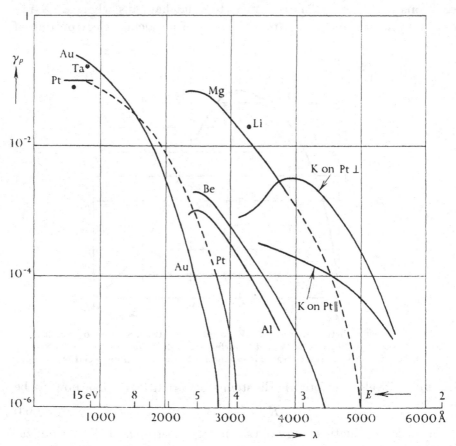

FIG. 50. Photo-electric yield γ_p as a function of the wavelength λ of the incident light (energy E of quantum) for various substances (17, 117, 114 b, 118). Photo-electric limit for Au 4·5 eV, for clean Pt 5·4 eV, but here Pt apparently contaminated. $\perp$ and $\parallel$: Linearly polarized light falls on potassium with the electric vector normal and parallel to its surface respectively. Borosilicate glass and 2537 Å light gives $\gamma_p \sim 10^{-4}$, soda glass 6 . 10^{-4}. Soda glass and $\ll 1250$ Å light 10^{-3} and probably 10^{-2} respectively (118 b).

The incident quanta are partly reflected and a large fraction of them heats up the substance. This is why γ_p is small for ordinary light. γ_p depends strongly on the condition of the surface, which affects φ or γ_p or both. γ_p is large for electro-positive surface layers, that is if the layer has a lower φ than that of the substrate: for a given radiation the layer loses more electrons than the base substance and charges up positive with respect to the substratum. The resulting electric field helps the electrons to escape (see section 6). These conditions are obtained when, for example, oxidized metals are covered with thin layers of alkali (see Table 3.14). Such photo-electric emitters show maxima in $\gamma_p = f(\lambda)$ (selective photo-effect). With polarized light this effect shows up only with an electric component normal to the emitting surface. The effect

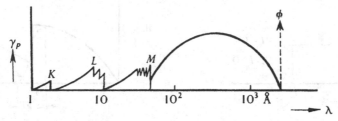

FIG. 51. Photo-electric yield, in electrons per incident quantum, for a solid as a function of the wavelength λ of radiation (schematic). $\gamma_{p\,max}$ lies between 0·1 and 1. φ is the photo-electric limit.

is of course absent with ordinary light at normal incidence because of the absence of a component of the electric vector normal to the surface.

Methods of obtaining photo-electric surfaces with a large γ_p and small φ are given in (116). For recent advances see (121 c).

10. FIELD EMISSION

When an electric field of order 10^6 V/cm or more acts on the surface of a metal *in vacuo* making it a cathode, then an electron current flows from the metal to a positive electrode. The electron theory of metals explains this 'field emission' in the following way. The electrons have potential energies which correspond to the Fermi–Dirac distribution. Fig. 52 shows the probability of finding an electron in the energy range E to $E+dE$ and Fig. 53 the relative number of electrons in that energy range. (Thus the area under the curve is the total number of electrons present.) The relative number $dN(E) = f(E).g(E)$, the latter being the density of states. The function $g(E)$ has the same form as the curve in Fig. 53. At zero temperature the probability of an electron being in a certain energy range is the same for all energies up to μ— the Fermi limit—which for Li, K, Na, Rb, Cs, Cu, Ag, Au is 4·7, 3·1, 2·1, 1·8, 1·5, 7, 5·5, 5·5 V respectively; this in contrast to the classical theory according to which the electrons have a Maxwellian distribution of energy. Also the nature of the energy of electrons is different: the modern theory assigns to electrons a finite potential energy at any temperature (including zero) whereas the classical theory assumes that electrons in a metal have kinetic or thermal energies like the molecules of a gas which become zero at zero temperature.

It has been shown in section 6 that a certain amount of energy is necessary to transfer an electron from the bulk of the metal into the vacuum. The 'heat of evaporation' of an electron is called the work function φ. In order to explain field emission in terms of energy we shall

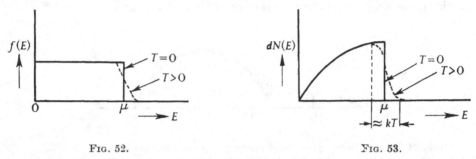

FIG. 52. FIG. 53.

FIG. 52. Fermi-probability function $f(E)$ for different temperatures T. μ is the Fermi limit.

FIG. 53. Energy distribution $N(E)$ according to Fermi for different temperatures T:

$$dN(E) = \{E^{\frac{1}{2}}/(1+\exp[-(E-\mu)/kT])\}\, dE.$$

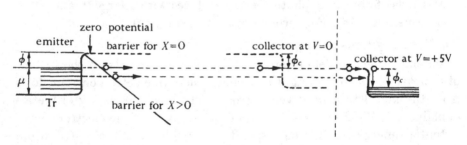

FIG. 54. Field emission of electrons and their collection (see Fig. 55).

use a model as shown in Fig. 54 where the electrons are thought to be in an energy trough Tr. The horizontal lines represent energy levels. These are filled with electrons up to a height corresponding to the energy μ; zero energy is here assumed at the top of the trough. At low temperature no electrons are found in levels above μ and thus an energy $e\varphi$ is needed to eject an electron from the level μ. Wave-mechanics shows that the wall of the trough cannot be penetrated by an electron if its thickness is infinite but it transmits electrons (de Broglie waves) when an external electric field X is applied; this can be thought to change the thickness of the wall (Fig. 54), the wall becoming thinner the larger the applied field. The probability Π of an electron 'tunnelling' through the wall or barrier follows from the solution of the wave equations (119):

$$\Pi \approx Ce^{-\frac{1}{2}K\varphi^{3/2}/X},\tag{3.64}$$

whence $K^2 = 8\pi^2 m/h^2$ and C is a function of μ and φ. An increase in X which reduces the thickness of the wall improves the chance for an electron to escape. At temperatures at which thermionic emission is negligible, the number of electrons per cm² sec of energy E which corresponds to the component of velocity normal to the surface is

$$N(E) = \frac{4\pi m}{h^3}(\mu - E).\tag{3.65}$$

The emission current is proportional to this number times the chance of escape through the potential barrier. Expressing the current density in A/cm², X in V/cm, and μ and φ in volts, we have from (3.64) and (3.65)

$$j = \int_0^\infty N(E)\Pi \, dE = 6.10^{-6}\frac{(\mu/\varphi)^{\frac{1}{2}}}{\mu + \varphi}X^2 e^{-6\cdot8\cdot10^7\varphi^{3/2}/X}.\tag{3.66}$$

For example, with $\varphi = 4$ eV, $X = 3.10^7$ V/cm, $\mu = 10$ eV, we obtain

$$j = 8 \text{ A/cm}^2.$$

Because field emission requires large values of X at the cathode, fine points and thin wires are used as emitters. Uncertainties in X and φ are introduced through lack of definition of the surface caused by small projections and cavities, crystalline structure, etc.

The existence of a field emission seems now to be established beyond doubt. It has two distinctive features: the energy distribution of the emitted electrons is different from that of thermionic origin, and no cooling of the cathode is observed when 'field electrons' flow from a cathode, in contrast to thermionic currents.

Take first a thermionic cathode of 1 cm² area at temperature T with heat losses by radiation only. With an electron current at the cathode of density j and a work function φ the power expended is $j\varphi$. If the electric power p balancing the heat losses of the cathode for $j = 0$ is kept constant, a new equilibrium is reached at a lower temperature $T - \Delta T$ which follows from

$$p = \text{const} = a\sigma T^4 = a\sigma(T - \Delta T)^4 + j\varphi.\tag{3.67}$$

Expanding to the second term the relative change in temperature is found to be

$$\Delta T/T = j\varphi/4p,\tag{3.68}$$

taking $\varphi = 4$ eV, $T = 1000°$ K, $p = 2\cdot5$ W, $j = 0\cdot1$ A/cm² we obtain $dT/T = 4$ per cent. Therefore a current of a few mA of that density will produce a decrease in cathode temperature of 40°.

In the case of field emission no change in cathode temperature can be expected for the following reason: the electrons of highest energy μ escape through the wall under the influence of the field, but the energy distribution of the electrons in the metal remains the same. Field electrons do not acquire an energy φ when emitted since they leave the trough by moving horizontally in our model, that is along lines of constant energy. Though (3.64) contains φ, this does not imply that the energy $e\varphi$ has to be expended; φ enters only in the probability of escape (121). Since the energy of the electrons in the metal is not affected (to a first approximation) by those escaping, no change in temperature of the metal surface should accompany the flow of a field current in agreement with observations. Unlike thermionic electrons, the field electrons draw their energy from the electric field only.

The other point of interest is the energy distribution of the field electrons. It is obtained by measuring the collector current i_c while a retarding potential is applied to the collector (Fig. 55). The emitting cathode is surrounded by a mesh anode followed by a collector. The anode is at a high positive potential which produces a reasonable field current i_c. When the collector is at cathode potential electrons cannot enter it because they have gained on their way to the anode and lost in their flight to the collector equal amounts of energy and arrive at the collector with zero energy. Since the collector has a large radius of curvature, the electric field at its surface is negligible. Thus the electrons have to go over a barrier, equal to φ_c of the collector substance, in order to set up a current in the collector circuit. As Fig. 55 shows, a collector potential of about 5 V positive with respect to the cathode is needed to bring the electrons over the top of the trough. Since the collector current is the integral of all the electrons of the sought distribution between the limits V_c and infinity, where V_c is the retarding potential, the differential coefficient of the collector current with respect to V_c gives the energy distribution which is shown in Fig. 55 for a constant value of X. Most of the electrons seem to have zero energy and thus come from the top level of the Fermi distribution, but there are others of less energy originating probably from lower levels. With increasing field strength the distribution maximum shifts towards faster electrons. Secondary emission from the collector is a disturbing factor and attempts have been made to allow for it by using an auxiliary electrode.

Excellent vacuum and absence of surface films are essential here. It is thus not surprising that field emission measurements have not yet been carried out with insulators.

The strong electric field needed to extract electrons can be set up either by an external electrode at high potential or by intense space charges in the gas or at the surface of the emitting body. It has to be remembered that in order to make field emission possible a strong field has to act over a finite distance. It is for this reason that the applied or

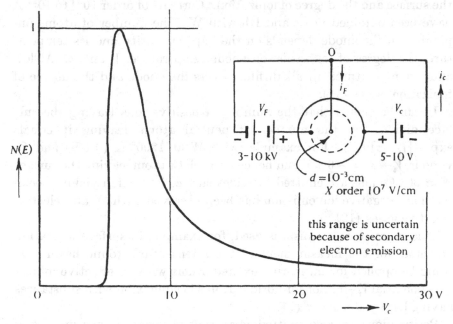

FIG. 55. Energy distribution $N(E)$ of field electrons. V_c is the retardation potential. The distribution maximum is taken as unity (120).

available potential difference must be sufficiently high. Field emission from thin films (Malter effect) has been observed with $Al-Al_2O_3-Cs_2O$ cathodes. It is thought that the large emission currents which were obtained when a primary beam of electrons of a few hundred eV was falling on the cathode are due to positive charges residing at the cathode surface because of its high resistivity as well as to electric polarization of the oxide (104).

In fields applied to anodes positive ions are emitted when $X > 10^8$ V/cm (121 b). However, the ion emitted is a gas and not the anode metal ion.

11. SECONDARY EMISSION OF POSITIVE IONS

When neutral unexcited atoms of a gas of ionization potential V_i impinge upon a metal surface of work function φ and if $\varphi > V_i$, electrons are liberated from the atoms and enter the lattice of the metal while

the positive atomic ions return to the gas. This has been observed for
Cs vapour in contact with W electrodes where $V_i = 3.9$ and $\varphi = 4.5$ V.
If a positive potential is applied to the W electrode and a second metal
electrode of any material is provided, then a positive ion current will
flow in the gas. Its magnitude is given by the number of atoms hitting
the surface and the degree of ionization. Currents of order 10^{-4} to 10^{-6} A
have been obtained in Cs and Rb with W. The number of atoms impinging on the anode depends on the vapour density and its temperature, the degree of ionization on anode temperature, V_i and φ. At low
anode temperatures an alkali film covers the anode and the degree of
ionization is low (17).

It can be shown that the number of positive ions leaving the surface divided by the number of neutral atoms leaving it equals
$\exp(-[V_i-\varphi]/kT)$. For example, with W at $1200°$ K $(0.1$ V) and Cs
vapour, $V_i-\varphi = -0.6$ V and hence for each Cs atom leaving the surface
$e^6 \approx 500$ Cs ions are emitted. Besides surface ionization yielding positive ions, negative ion emission has been observed with W and electronegative gases ($121\,a$).

This emission mechanism is used, for example, in surface ionization
detectors for the measurement of the intensity of atomic beams. It
could be applied for measuring excited atoms with an effective ionization potential V_i-V_{exc} or for normal atoms in connexion with substances
having large values of φ (74).

Positive ions are also emitted as a result of negative ions impinging
on a neutral surface, but this seems to be a rare process ($103\,a$, $107\,a$).

QUESTIONS

(1) A sphere of 1 cm² area is heated to 3000° K and enclosed in a quartz
envelope which transmits radiation down to $\lambda = 1800$ Å. This light source is
placed 100 cm from a photoelectric cathode of surface area 10 cm², enclosed
in an evacuated quartz bulb. Find the saturation current to a neighbouring anode
in the bulb. The photoelectric yield varies according to $\gamma_p = c_1 - c_2\lambda^2$, where
$c_1 = 1.8 \cdot 10^{-2}$ and $c_2 = 2 \cdot 10^{-9}$, with λ in Å. The light source is assumed to be
an ideal black body. Also, find the work function φ of the surface used.

(2) Write an essay on the experimental methods of verifying the Maxwellian
velocity distribution of neutral particles of a gas emphasizing the underlying
physical principles.

(3) From what measurements are the mean free path values obtained and why
do they differ in value according to the experimental method used? Give an
account of Born's direct experimental proof of the correctness of the theoretical
distribution of free paths.

(4) A thin cylindrical beam of electrons of 20 eV energy and 1 mA intensity moves through helium at $10\,\mu$ pressure and 20° C. When the cross-section to excite atoms to the metastable state $2\,^3S$ is $q_e = 2\cdot5 \times 10^{-17}$ cm^2 for electrons of that energy and if the metastable atoms are assumed to be destroyed on hitting the wall, which is 3 cm from the beam, and if collisions between metastables and atoms in the ground state have no effect, find the equilibrium concentration of metastable atoms in the beam.

(5) Calculate the (first) ionization potential V_i of helium, remembering that the electron to be removed to infinity is in a force-field which is 'screened' by the other electron; this reduces the nuclear charge Z to the 'effective' charge $Z - S$, S being the screening constant. Start with the equation of the force acting on one electron in the shell and assume the electrons to be on opposite sides of the nucleus and equidistant from it. Then compare the result with

$$V_i = (Z - S)V_{iH},$$

where V_{iH} is the ionization potential for hydrogen. Also, find the second ionization potential of helium. What effect has been neglected in determining V_i and why does this calculation give a value slightly smaller than V_i observed?

(6) Outline the physical principles applied in the measurements of the critical potentials for excitation and ionization. Explain the methods used to discriminate between excitation and ionization. Why has the onset of excitation not been observed by the onset of light emission?

(7) By what means is it possible to obtain a beam of slow mono-energetic electrons of variable energy? How is the beam's 'spread' measured? Indicate the difficulties encountered (e.g. contact potentials) and the limits so far obtained experimentally. What are the theoretical limits?

(8) Express the Maxwellian velocity distribution in terms of $dN/d\epsilon = f(\epsilon/\epsilon_m)$, where ϵ_m is the most probable energy, and find the fraction of particles with energies between $3\epsilon_m$ and ∞ and between 0 and ϵ_m. Calculate the average electron cross-section $\bar{Q}$ for He and A as a function of the mean energy $\bar{\epsilon}$ using Fig. 21. Plot $\bar{Q}$ against $\bar{\epsilon}$ up to $\bar{\epsilon} = 10$ eV and comment.

(9) Find the variation of the cross-section q_e with energy for argon from (3.9) for $\eta_e = a\epsilon$, with $a = 1/16$ and ϵ in eV, assuming that the total value of q_e is twice that for $l = 1$. Calculate and plot the results and compare the curve obtained with that in Fig. 21.

(10) If a beam of electrons of energy $K = 10$ keV passes through hydrogen, its total ionization is Z ion pairs. Calculate Z when the ionization process is accompanied by emission of secondary electrons of 5 eV average energy; what other types of energy loss may be effective?

(11) The ionization in the E-layer of the ionosphere is said to be due to ionization of the tenuous atmosphere by ultra-violet and mainly soft X-ray radiation from the sun balanced by electron–ion and ion–ion recombination. Assume only N_2 to be present (at a pressure of order 10^{-4} mm Hg) at a height of $h = 200$ km. The absorption coefficient in cm^{-1} is given by $\mu = a(h-x)^3$ with $a = 3.10^{-28}$,

x being measured from the ground. Take the radiation intensity at h to be $L_0 = 10^{-4}$ erg/cm^2 sec, and $\rho_e = 6 \times 10^{-12}$ cm^3/sec and $\epsilon = 150$ V per ion pair.

Calculate the variation of electron density with x, find the maximum as regards position and density. Explain why these assumptions give rise to a maximum and what the possible causes are which may produce maxima at different heights. Find L_0 if dissociative recombination of O_2 with $\rho_i = 10^{-8}$ cm^3/sec would occur only.

(12) Find the saturation current density drawn by two plane parallel electrodes 1 cm apart and 10 cm long in a vessel filled with Hg vapour at $p = 10 \mu$ and 100μ and $T = 300°$ K. The gas is uniformly irradiated by a beam of resonance light (2537) parallel to the electrodes of intensity 10 erg/sec. Calculate the number density of excited atoms. Assume the life of Hg* ($^3P_1$) to be $\tau = 10^{-7}$ sec and the absorption cross-section for resonance light to be $1 \cdot 4 . 10^{-13}$ cm^2 and neglect absorption of resonance radiation. Why are the observed current densities higher than this calculation suggests?

(13) Discuss under what circumstances collisions between excited particles can lead to ionization. Give examples for such collisions of the second kind and explain the role of the dissociation energy of molecules.

(14) Write an essay on the space charge detector. Illustrate by means of a numerical example the least number of positive ions which are easily measurable. Remark on the instability of the device and refer to the results obtained when light is absorbed by alkali vapours. Suggest methods to trace such oscillations.
[References: F. L. Arnot (19) and Ditchburn (94).]

(15) Derive from first principles the equation of motion of two equal particles of mass m and charge e which, when at great distance from one another, move each with velocity v along parallel straight lines b cm apart. Show that for scattering collisions through an angle $\frac{1}{2}\pi$ the maximum potential energy equals their kinetic energy. Derive the scattering cross-sections for electron–electron and ion–ion collisions assuming that it is given by the least distance of approach and comment on its dependence on the relative velocity.

(16) Give a critical account of experimental methods of measuring the work function φ of simple substances and compare the results obtained from the various methods. What is the reason for $\varphi < V_i$ and how is φ affected by a polarized layer?

(17) Explain qualitatively how an isolated electrode acquires a potential when a beam of fast electrons hits its surface. What is the sign of the potential and its magnitude for different electron energies? To what electrode will the secondary electrons move? Discuss the sign of the potential of a glass wall near the cathode of a (glow) discharge at such low p that fast positive ion beams are present. A beam of electrons of energy 10 kV is emitted from the cathode of an evacuated tube and impinges on an isolated metal screen. If the secondary emission coefficient δ of the screen varies with the electron energy V according to

$$\delta/\delta_0 = \sqrt{(V/A)}e^{-V/A}$$

with $\delta_0 = 4$ and $A = 1000$ V, find the potential of the screen in the steady state.

(18) Determine the maximum positive ion current which can be drawn to an electrode when a beam of Cs atoms hits a W electrode of 1 cm² area. The beam intensity is 10^{12} atoms/cm² sec and thermal equilibrium exists between the positive ions and the surface kept at $T = 1200°$ K (≈ 0.1 V) preventing condensation of Cs.

(19) From the equations of motion of an ion of mass m and relative velocity v which approaches a molecule of polarizability α_0 derive the critical radius, i.e. the impact parameter which describes the limit between scattering and capture.

(20) From information given in the book and other references plot a graph of the absorption and scattering cross-section of quanta in Cs as a function of wavelength λ from $\lambda = 0.1$ to 10^4 Å. Point out in which respect and why this differs from Fig. 38. What is the fate of the quanta which pass through the gas?

(21) From (3.24) and (3.25) find a relation between r and n. Calculate the cross-section πr^2 for H in the first resonance state. Compare the value obtained with the cross-section for ionization by electron collision from the excited state and the cross-section for momentum exchange and comment.

4

MOBILITY AND CHARGE TRANSFER

1. MOBILITY OF IONS

If ions or electrons form a swarm so that the velocities of individual particles are equally distributed in all directions about an average velocity, then an electric field acting on the swarm will move it bodily. The average speed with which the centre of the swarm moves in the direction of the field is called the drift velocity. This picture can also be applied to a continuous stream of particles with an average random velocity well above the drift velocity.

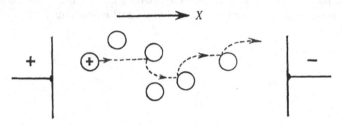

FIG. 56. Motion of a positive ion through a gas in an electric field.

Let an ion move in its own gas in a uniform and relatively weak electric field, sufficiently weak that only elastic collisions between ions and gas molecules occur, viz. no excitation and ionization. Treating the motion of a single ion implies that interaction between this ion and other ions can be neglected. Fig. 56 shows the imaginary path of an ion which starts its journey with a certain random velocity in a general direction. After an impact the initial velocity of an ion is supposed to be the same as if it were a neutral gas molecule. The electric field, however, tends to drive the particle downwards, so that the ion moves along a parabola, exactly as a ball thrown in a gravitational field. When the ion has moved along a path whose average length equals a mean free path, it collides with another gas atom and rebounds at random. Though it may happen that the ion starts moving against the electric field, it proceeds on the average in the field direction.

This is the picture which conforms to Langevin's first mobility theory (11, 13). The ionic mobility is defined as the drift velocity, i.e. the velocity component in the field direction, in unit field (X). It is easily found by assuming that the random velocities of ions and gas molecules are the same (which means that the ions are in thermal equilibrium with the gas) and that the drift velocity after a collision is zero. This is simply the same as saying that the drift velocity is small compared with the random velocity. Since τ, the time interval between two collisions, does not depend on X, we have at constant temperature:

$$\tau = \lambda_i/\bar{c}, \tag{4.1}$$

where λ_i is the ionic mean free path and $\bar{c}$ the mean velocity of the ion. During this period the ion is accelerated by the electric field X, and since it moves as if it were in high vacuum, its acceleration $a = eX/M$, where e and M are the charge and mass of the ion respectively. If we neglect, as in the kinetic theory of gases, the time of impact, the ion travels in τ seconds a distance

$$s = (eX/2M)\tau^2, \tag{4.2}$$

and hence the drift velocity becomes

$$v_d = (eX/2M)\tau = (e\lambda_i/2M\bar{c})X = kX. \tag{4.3}$$

Because of the statistical distribution of the mean free path, k, the mobility, is approximately twice as large, namely

$$k \approx e\lambda_i/M\bar{c}. \tag{4.4}$$

From (4.4) for an H_2^+ ion at $p = 1$ mm ($\lambda_i \approx 1\cdot2.10^{-2}$ cm, $\bar{c} = 1\cdot9.10^5$ cm/sec at $T = 300°$ K) k^+ is $3\cdot1.10^4$ cm/sec whereas the measured value is $1\cdot1.10^4$ cm/sec. $\bar{c}$, the mean velocity, is defined by

$$\tfrac{1}{2}M\bar{c}^2 = kT(4/\pi). \tag{4.5}$$

From (4.4) it follows that because $\lambda_i \propto 1/p$, at constant gas temperature

$$kp = \text{const.} \tag{4.6}$$

If the initial velocity immediately after a collision is not taken as zero (persistence of velocities), as previously assumed, k is larger by a factor $1\cdot5$. The mobility $k \propto \lambda_i/\bar{c} \propto \tau$, where τ is the collision interval. The reason is that each time a collision occurs the acceleration in field direction has to start anew; hence the longer the interval between successive collisions the greater the progress made by the ions.

From (4.4) we deduce that there should be no difference between the mobility of positive and negative ions. The mobility should vary inversely with the pressure. Applying this equation to the mobility of electrons, we should expect k to increase by a factor 1840 times the

molecular weight of the gas in question. Observations show that there
is no difference between the mobility of positive and negative ions of the
same gas unless the negative ion exists for a certain fraction of its life
as a free electron (see below). The fact that reliable measurements of
the mobilities of positive ions yield results which are 3 to 5 times smaller
than the figures obtained from (4.4) is, however, most important. In
Table 4.1 are given the results of measurements on various ions in their
own gas, in Table 4.2 (see p. 121) the mobility of alkali ions in rare gases.

TABLE 4.1

Mobility k^+ of ions in their own gas (17, 70a, 122, 123, 128, 128a, 128b)
at 1 mm Hg and 0° C

Ion/gas	k^+ in units of 10^3 cm/sec per V/cm	k^+ = const. for
He$^+$—He	8	
He$_2^+$—He	15·4 (13)	$X/p \leqslant 10$
Ne$^+$—Ne	3·3	
Ne$_2^+$—Ne	5	$X/p \leqslant 8$
A$^+$—A	1·2	$X/p \leqslant 40$
A$_2^+$—A	2	
Kr$^+$—Kr	0·69	$X/p \leqslant 30$
Kr$_2^+$—Kr	0·92	
Xe$^+$—Xe	0.44·	$X/p \leqslant 40$
Xe$_2^+$—Xe	0·6	
H$^+$—H$_2$	11·2 (?)	
H$_2^+$—H$_2$ (often H$_3^+$)	10	
H$_3^+$—H$_2$	> 10	
D$_2^+$—D$_2$	5	

		k^- in 10^3 cm^2/V sec
N$_2^+$—N$_2$ (often N$_3^+$ or N$_4^+$)	2	
Air (N$_2$, O$_2$)	1·4	1·9
O$_2^+$ in O$_2$	1·0	1·4
CO$^+$—CO^2	0·84	0·87
CO$_2^+$ in CO$_2$	0·73	0·73
H$_2$O at 100° C	0·47	0·43
Cl$_2^+$ in Cl	0·56	0·56
C$_2$H$_5$OH	0·27	0·27
Hg$_2^+$ in Hg at 500° K	0·3	
Hg$^+$ in Hg „ „	0·23	

Taking air as an example, the negative ion mobility is slightly larger
than the positive one. The explanation is that the (20 per cent) oxygen
molecules attach electrons very strongly whereas the formation of negative ions in nitrogen is a rare process. The latter is shown by mobility
measurements in pure nitrogen: at 1 atmosphere and in weak fields of
the order of 1 to 10 V/cm, the negatively charged particles have a mobility of up to 20 000 cm/sec/V/cm and are therefore free electrons. Thus
in air it seems as if the majority of electrons become attached to oxygen

molecules but that a sufficient fraction remains free to give rise to a slightly increased average mobility.

The nature of the ions is not always clear. For example H_2^+ in H_2 seems to change quickly into H_3^+ at larger p since the reaction

$$H_2^+ + H_2 \rightarrow H_3^+ + H$$

has a large cross-section (see page 53).

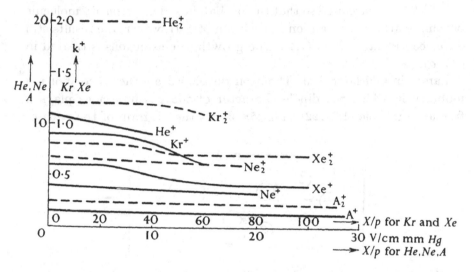

FIG. 56 a. Mobility k^+ of rare gas ions as a function of the reduced field X/p at 1 atm and 0° C (124 b).

The mobility of rare gas ions as a function of the field is shown in Fig. 56 a.

Thus Langevin's first theory seems to be faulty in the following respects:

(a) the theoretical value of the ionic mobility is up to 5 times larger than the observed value;

(b) the observed mobility is not proportional to $\sqrt{T}$ as it would appear to be from (4.4), but $k \propto 1/\text{density} \propto T/p$;

(c) the observed mobility is not $\propto 1/\sqrt{M}$ but $\propto 1/M$;

(d) the mobility of 'negative ions' in rare gases, N_2 and others, is larger than the positive ionic mobility of the respective ions; also the mobility of ions depends on their age;

(e) the ionic mean free path depends not only on the type of ion but also on the gas in which the ions move.

The fact that the measured mobility is smaller than the calculated one has been explained as due to the formation of cluster ions: M is not

Fig. 57.
Clustering.

the mass of the single ion but that of the ion plus several 'impurity' molecules surrounding it (Fig. 57). Also since the effective diameter of the ion is now increased, the mean free path of the ion has become smaller. The formation of clusters in rare gases of high purity can be avoided, particularly if ions of small age are used so that the probability of an impurity molecule becoming attached to an ion is very remote. However, the results still differ considerably from (4.4). The growth of cluster ions is treated in (124 a).

Langevin and later J. J. Thomson developed another theory of ion mobility in which the dipole character of the gas molecules plays a fundamental role (13, 122). Fig. 58 shows the diagram of the path of

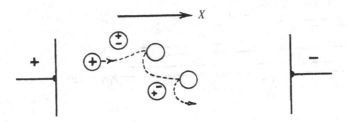

Fig. 58. Motion of a positive ion in an electric field in the presence of induced electric dipoles showing deflections and 'direct' collisions.

a positive ion in a weak electric field. While the ion moves through the gas, it induces temporarily an electric dipole moment in neighbouring molecules. According to the old idea (Fig. 56) a positive ion would travel in the field direction colliding with a molecule 'in its path', but because of a distant molecule the ion approaching it displaces its electrons with respect to the positive nucleus inducing an electric dipole in the molecule. Hence the ion is attracted by the negative end of the dipole and deflected towards this molecule. So—without colliding in the classical sense—an exchange of momentum between the positive ion and the gas molecule takes place. It appears as if this would simply mean that the mean free paths of the positive ions were shortened. In fact, however, the theory is fundamentally different from the old one, as will be shown.

Let μ be the induced electric dipole moment per molecule which is connected with the dielectric susceptibility κ and the dielectric constant

$D = 1 + 4\pi\kappa$ by the equation

$$\mu = \frac{\kappa}{N} X_i = \frac{D-1}{4\pi} \frac{X_i}{N} = \frac{D-1}{4\pi N} \frac{e}{r^2}, \qquad (4.7)$$

where X_i is the field of the ion and N the number of molecules per cm³. Then, in analogy to the field produced by a magnetic dipole at sufficiently large distances r from its centre, the relation $X = 2\mu/r^3$ holds, and the field acting on the ion is

$$X = \frac{(D-1)e}{2\pi N} \frac{1}{r^5}, \qquad (4.8)$$

which shows that the force law is an inverse fifth-power law. If the ions gain only a small amount of energy between two impacts compared with the average energy of a gas molecule—an assumption which is the more correct the smaller the energy acquired along a mean free path ($\propto X/p$)—then an approximate relation for the ionic mobility for ions moving in their own gas can be easily derived.

From (4.1, 4.4) it follows that the mobility is proportional to the time interval between two collisions:

$$k = (e/M)\tau. \qquad (4.9)$$

τ can be obtained by assuming it to be equal to the time required to cover the average distance R between two molecules, viz.

$$R = N^{-\frac{1}{3}}, \qquad (4.10)$$

and hence

$$\tau \approx \int_{r=0}^{r=R} \frac{dr}{v_d} = \int_0^R \frac{dr}{kX} = \int_0^R \frac{2\pi N r^5 \, dr}{k(D-1)e} \approx \frac{1}{eNk(D-1)}. \qquad (4.11)$$

By substituting (4.9) in (4.11), we have with $NM = \delta$, the gas density,

$$k \approx \left[\frac{1}{\delta(D-1)} \right]^{\frac{1}{2}}. \qquad (4.12)$$

Since $(D-1) \propto \delta$, it follows that in agreement with observations

$$k\delta = \text{constant}. \qquad (4.13)$$

The exact treatment leads to an equation similar to (4.12), viz.

$$k = A \left[\frac{1 + M/M_i}{\delta(D-1)} \right]^{\frac{1}{2}}, \qquad (4.14)$$

where M and M_i are the mass of the molecule and ion respectively and A is a constant dependent on D, on the sum of the radii of the ion and the molecule and on the gas temperature. For large D, small T, and small ions and molecules, A approaches 0·52. A decreases to about 0·1 when the radii and T are large and D small. The numerical agreement between theory and experiment is reasonably good but not perfect. However, (4.14) indicates that k^2 against M/M_i should give a straight line. Indeed measurements with ions of the alkalis moving in rare gases, H_2 and N_2, confirm the dependence on M/M_i.

One might ask whether there is sufficient time available for the electron cloud of the molecule to be displaced to the same extent by the field of an ion passing it as it would be in a constant electric field of equal magnitude. The answer is that any dispersion should appear at much larger 'frequencies': the ion has a speed of order 10^4 to 10^5 cm/sec and passes the molecule (10^{-8} cm) within a time of 10^{-12} to 10^{-13} sec which is long enough to shift the electron-cloud into an equilibrium position.

Another important point is the reduction of the mobility due to charge transfer, which will be discussed more fully later on. Ions moving in their own gas often become neutralized when colliding with neutral atoms while the atoms lose electrons. This results in an increase in the speed of the neutrals and a decrease in the speed of the ions.

It has been shown that ions moving in moderate fields exhibit constant mobility. This, however, is not so in stronger fields. Here an ion may pick up perhaps 0·1 to 1 eV along a mean free path and though $v_d < \bar{c}$ our original assumption $v_d \ll \bar{c}$ becomes invalid. It is more correct to assume that an ion loses in one collision on the average a certain constant fraction of its mean energy. Calling this constant fraction κ, we obtain by equating the energy gain per second to the loss suffered in elastic collisions, $\bar{c}/\lambda$ being the collision rate:

$$\tfrac{1}{2}\kappa M(\bar{c})^2 \bar{c}/\lambda \approx eXv_d. \tag{4.15}$$

The drift velocity v_d from (4.3), where k is now a variable, is

$$v_d \approx (\kappa/2)^{\frac{1}{4}}(e\lambda_1/M)^{\frac{1}{2}}(X/p)^{\frac{1}{2}} = c_M(X/p)^{\frac{1}{2}}, \tag{4.16}$$

where $\lambda_1 = \lambda p$ and κ is about 0·5 for ions in their gas. (Classical collision of equal masses.) c_M is the mobility coefficient.

Fig. 59 shows the drift velocity of ions in Kr and Xe measured in large fields as a function of the square root of the energy acquired along a mean free path; the linear graph confirms the validity of (4.16) and c_m is of the right order. At still higher fields, v_d should increase faster than $(X/p)^{\frac{1}{2}}$ because of inelastic collisions (see v_d of electrons). In argon and the lighter rare gases the mobility at lower X/p is constant and after passing a weak maximum decreases with X/p; at larger values of X/p the mobility should follow (4.16), but in strong fields the energy distribution is likely to be affected. The dependence of v_d on X/p over its whole range is shown schematically in Fig. 59. For example, the mean kinetic energy of argon ions rises to twice that of atoms at $X/p = 45$ V/cm mm Hg and to 5 times at $X/p = 150$. Thus the 'ion temperature' at these fields is considerably above the gas temperature.

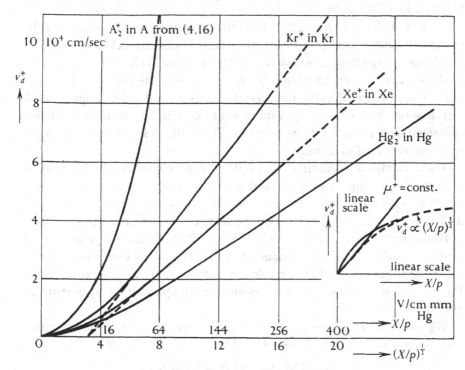

FIG. 59. Drift velocity v_d^+ of ions in their own gas as a function of the square root of the reduced field $\bar{X}/p$ (128). *Inset*: $v_d^+ = f(X/p)$ plotted linearly showing the transition from the constant mobility to the root law.

Another problem is the motion of ions in non-uniform fields, when $(dX/dx)\lambda$ is not negligibly small compared with X. It has to be remembered that the mobility of an ion is only clearly defined if equilibrium conditions are satisfied. Such conditions are, for example, that the ions must have the same random velocity as the gas molecules, or they must lose in a collision on the average a certain fraction of their energy and replace it from the field. Ions starting with zero velocity have to move a certain distance and collide a few times before they are in equilibrium; practically speaking, it means that the mobility observed in uniform fields near the starting-points appears to be higher since the ions are not scattered but accelerated in the field direction. Similarly the ions moving in non-uniform fields are never in equilibrium with the field. In decreasing fields, for example, the drift speed will in general be much higher than the value corresponding to the field at that point (see electrons in non-uniform fields). In strongly non-uniform fields the ion drift velocity is probably determined by the potential the ions have fallen through in a free path.

So far ions have been regarded as entities which move independently of one another. This picture requires amplification. At large current and large charge densities (say above 10^{12} ions/cm^3) the ions are usually in the company of an equal number of electrons. The result is that each ion attracts a considerable number of electrons and drags them with it (for a short time) until regrouping occurs. The ion is thus virtually scattered and its mobility is reduced. This effect is greater for larger concentrations of charges.

There is also a reduction in mobility to be expected in the case of ions moving in highly excited (and ionized) gases. If the degree of excitation is high, there is a good chance of a collision between an ion and an excited atom, and because of the larger cross-section of the latter, a transfer of energy on to the ion is possible, particularly if a third body is nearby; thus the random energy of the ion is increased and the mobility diminished. Similarly one has to expect a reduced mobility for excited ions whose presence is shown by spark lines appearing in the spectrum.

N_3^+ and N_4^+ ions are produced by a stage process. For example, $N_2^+ + e \rightarrow (N_2^+)^*$ when $K > 22 \cdot 1$ eV. Then

$$N_2 + (N_2^+)^* \rightarrow (N_4^+)^* \rightarrow N_3^+ + N.$$

In Table 4.1 above measured values of k^+ for ions in their own gas are given. Where such values do not exist estimated ones have been included. In a number of cases it was not certain what type of ion (molecular, atomic, or mixture of both) was used. At $p >$ order 1 mm Hg rare gas atomic ions collide with excited atoms forming molecular ions. Also it is not always known over what range of X/p, k^+ can be assumed to remain sensibly constant. Atomic (or molecular) ions moving in their own gas show a mobility which is reduced through charge transfer (section 3); no charge transfer occurs for atomic ions in a molecular gas or vice versa, as can be seen from recent measurements (Fig. 60). For similar reasons the mobility of alkali ions in rare gases (Table 4.2) is seen to be always larger than the mobility of rare gas ions in their own gas.

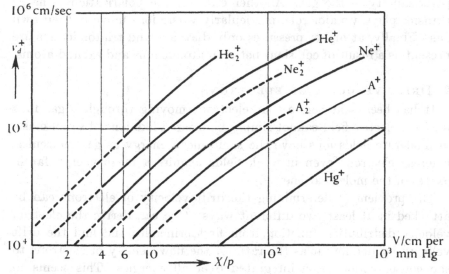

FIG. 60. Drift velocity v_d^+ of atomic and molecular ions in their own gas as a function of the reduced field X/p (124, 70 a). For Kr, Xe, and N_4^+ ions see (128 b); Hg: large X/p see (128 b, c); above $X/p \approx 100$, Hg⁺ only. Drift velocities for N_2^+ at $X/p = 20$ and 800 are 2·5 and 30.10^4, for N_3^+ at $X/p = 6$ and 300, 1·5 and 50.10^4 ((128 b); see also (128 d)).

TABLE 4.2

Mobility of alkali ions in rare gases (122); *see also* (6 b)

Ion/gas	k^+ at 1 *mm* Hg and 0° C in units of 10^3 cm/sec per V/cm
Li⁺ in He	25
Na⁺ in Ne	8·9
K⁺ in A	2·8
Rb⁺ in Kr	1·6
Cs⁺ in Xe	1·0

So far the steady motion of positive ions in an electric field has been discussed. If, however, the ions start from rest, they have to move a certain distance before their drift and random velocities have reached equilibrium values. The time interval (time constant) corresponding to that distance can be found in the same way as in the case of electrons starting from rest. In (4.19) the mass of the ion has to be substituted for m and the value of the collision coefficient $\kappa = \frac{1}{2}$ since equal masses collide.

One may summarize by saying that the mobility is determined by four effects; the size of the ion, that is an effective radius which defines when a classical collision takes place; on this picture is based the old Langevin theory and it will be the more accurate the smaller the polarizability of the gas. Another effect is the polarization. Charge transfer plays a major role, particularly where ions move in their own gas. Finally, at higher pressures only diatomic and triatomic ions are present as a result of collisions between atomic ions and excited atoms.

2. DRIFT VELOCITY OF ELECTRONS

It has been said before that electrons moving through a gas in a uniform electric field cannot simply be treated like ions of small mass: in moderate fields ions have the same mean energy as gas molecules, whereas electrons even in weak fields acquire a mean energy far in excess of the molecular energy.

The problem of determining the drift velocity of electrons can be attacked in at least two different ways. One can derive the electron velocity distribution function from first principles and find the drift velocity by defining it as the ratio of the flow density of electrons to the concentration, both integrated over all energies. This seems to be the most rigorous treatment, but it can only be applied when either the energy acquired along a mean free path is small enough to exclude inelastic collisions or when numerous simplifying assumptions are made about the dependence of the inelastic cross-sections on the electron velocity.

A less rigorous but mathematically easier approach is to use average values for all variables right from the start. This means, for example, that the velocity distribution is taken as invariant. Though it is known that the deviations from a standard distribution can be considerable when the field is varied, we shall nevertheless adhere to this view since it gives results which in many cases have proved to be reasonably correct and simple.

Let us assume that electrons are moving independently in a weak uniform electric field X. They collide elastically with molecules and lose on the average a certain fraction κ of their mean kinetic energy $mc^2/2$ (c is used instead of $\bar{V}$); this amount is equal to the energy which they have acquired at the end of a (constant) mean free path, taken parallel to the field.

If the field X, which acts along the x-axis, is suddenly applied, an electron which starts from rest takes a certain time to reach a velocity equilibrium with the field. If its drift velocity at time t is v, then it gains a certain amount of kinetic energy which is the difference between that which it picks up in moving $dx = v\,dt$ cm in the field direction and that which it loses, that is $\kappa mc^2/2$ times the collision frequency.

Thus, provided $c \gg v$, we have, with

$$v = (e\lambda_e/mc)X, \qquad (4.16\,a)$$

$$d(mc^2/2) = eXv\,dt - \kappa(mc^2/2)(c/\lambda_e)\,dt. \qquad (4.17)$$

By substituting v from above, this equation can be integrated by parts. With $y = c/c_\infty$, where c_∞ is the value of c when $t \to \infty$, we have the solution

$$t/\tau = \tfrac{1}{2}\ln[(1+y)/(1-y)] - \tan^{-1}y, \qquad (4.18)$$

where τ is a time constant;

$$\tau = (\tfrac{1}{2})^{\frac{1}{2}}(m/e)^{\frac{1}{2}}(1/\kappa)^{\frac{3}{4}}[\lambda_e/X]^{\frac{1}{2}} \propto (Xp)^{-\frac{1}{2}}. \qquad (4.19)$$

Its meaning is that for $t = \tau$ the random velocity c of the electrons has reached about 94 per cent of its final value; the value c_∞ is obtained by putting the left-hand side of (4.17) equal to zero:

$$c_\infty = (2/\kappa)^{\frac{1}{4}}(e/m)^{\frac{1}{2}}[X\lambda_e]^{\frac{1}{2}} \propto (X/p)^{\frac{1}{2}}. \qquad (4.20)$$

Further substituting c_∞ from (4.20) in (4.16 a), we find the final drift velocity

$$v_\infty = (e/m)^{\frac{1}{2}}(\tfrac{1}{2}\kappa)^{\frac{1}{4}}[X\lambda]^{\frac{1}{2}} = c_1(X/p)^{\frac{1}{2}} \qquad (4.21)$$

and the ratio (v/c) in the steady state is

$$v_d/c = (v/c)_\infty = (\tfrac{1}{2}\kappa)^{\frac{1}{2}}. \qquad (4.22)$$

Assuming a Maxwellian distribution the numerical factors in (4.19), (4.20), and (4.21) are changed to $(\pi/4)^{\frac{1}{2}}$, $(4/\pi)^{\frac{1}{4}}$, $(\pi/4)^{\frac{1}{4}}$ respectively.

This treatment, however, is applicable not only to electrons in weak fields but also to positive ions in moderate fields. In this case the time constant (4.19) is smaller because, though $m^{\frac{1}{2}}$ is more than 50 times larger, $\kappa^{\frac{3}{4}}$ is about 10^3–10^4 times larger. The time constant varies with p (at constant field) according to

$$\tau p^{\frac{1}{2}} = \text{const.} \qquad (4.23)$$

At constant X/p we have

$$\tau X = \text{const.} \qquad (4.24)$$

The larger p, X, and κ, the quicker is the equilibrium value c_∞ reached. The number of collisions which occur during τ seconds follows from (4.19) and (4.20)

$$Z_\tau \approx \tau c_\infty/\lambda_e = \frac{1}{\kappa}. \qquad (4.25)$$

Thus the total number of collisions necessary to approach the final value c_∞ is independent of the gas density.

A numerical example will illustrate the order of magnitude of the various parameters. For electrons in He at $p = 1$ mm Hg and $X = 3$ V/cm $= 1/100$ e.s.u., $\kappa = 2m/M = 2\cdot8.10^{-4}$ and $\lambda_e = 5.10^{-2}$ cm; from

(4.19) it follows that the equilibrium random velocity is reached after approximately $\tau \approx 1.10^{-6}$ sec which is equivalent to a distance $\int_0^\tau v\,dt$ measured in the field direction of ≈ 1 cm with v from (4.16 a). The final random velocity follows from (4.20) and is $c_\infty = 1 \cdot 5 . 10^8$ cm/sec or ≈ 4 eV. An electron moving in helium thus has to collide $1/\kappa = 3600$ times before it acquires 94 per cent terminal velocity. The final drift velocity is obtained from (4.22): $v_d \approx 2 . 10^6$ cm/sec which is less than 2 per cent of the final random velocity c_∞.

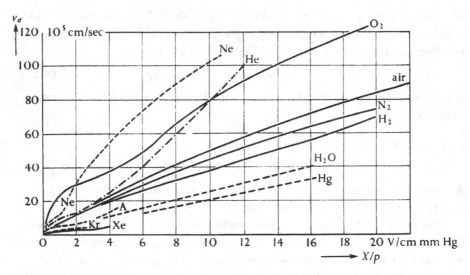

FIG. 61. Drift velocity v_d of electrons as a function of the reduced field X/p (129, 130); for higher fields see (132 a and b).

Equation (4.21) shows that because of the large mean kinetic energy of the electrons their drift velocity is proportional to the square root of the 'mean free path energy' or to the square root of the field over the gas density. This is only true if κ, the fraction of energy lost in one collision, is constant. The 'electron mobility' then varies inversely as $(X/p)^{\frac{1}{2}}$, reaching a maximum when $X \to 0$ (p. 138).

Fig. 61 shows the electron drift velocity as a function of the reduced field. It can be seen that in the range of weak fields the drift velocity is proportional to $(X/p)^{\frac{1}{2}}$ in full accordance with the theory outlined above. If the variation of the mean free path of the electron with its energy is taken into consideration, the drift speed in He, Ne, and A agrees quantitatively with (4.21), with $\kappa = (8/3)m/M$ and λ_e taken from

Fig. 21 (8/3 is taken instead of 2 to allow for the energy distribution). The value of the mean energy (the electron temperature) is taken from Figs. 129, 130 for the required value of X/p.

However, this agreement holds only up to a certain value of the field, for example $X/p \leqslant 2$ in He. Above this value the observed drift speed increases faster than expected. The reason is that now inelastic collisions become effective. This can be understood by the following simple

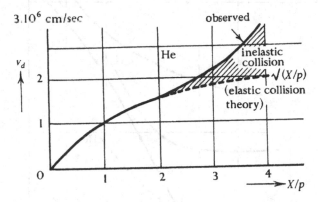

FIG. 62. Drift velocity v_d of electrons in He as a function of the reduced field X/p showing the effect of inelastic collisions and change of the energy distribution (13).

argument: assume an electron suffers an inelastic collision by which it is brought to rest. It acquires speed again by moving in the field direction only. Thus until such electrons have reached equilibrium they have temporarily a higher drift speed (Fig. 62).

If this picture were correct one would expect a departure from the $(X/p)^{\frac{1}{2}}$ relation at those values of X/p for which a sufficient number of inelastic collisions take place. Since the random energy for He at $X/p = 2$ is only 4 eV, it would seem that very few electrons are available with energies above 19·8 eV, the lowest excitation potential of He. The electron energy distribution must therefore have varied with X/p in the sense that an increasingly large proportion of fast electrons appear when X/p rises. This, in fact, is confirmed by exact calculations of the energy distribution in He which include inelastic collisions and the variation of the mean free path with energy (131, 132c).

The drift velocity of electrons in molecular gases cannot be calculated from (4.21) with $\kappa = \kappa_{\text{elastic}}$ because even at the lowest values of X/p the number of inelastic collisions is quite appreciable. This can be seen, for example, from measurements of the collision coefficient κ as a function of X/p which is shown in Fig. 63. The average energy loss in one collision in H_2 at low X/p was found to be of order 10^{-3}, whereas κ for elastic collisions $(= 2m/M)$ is about 4 to 6 times smaller. Thus even a slow

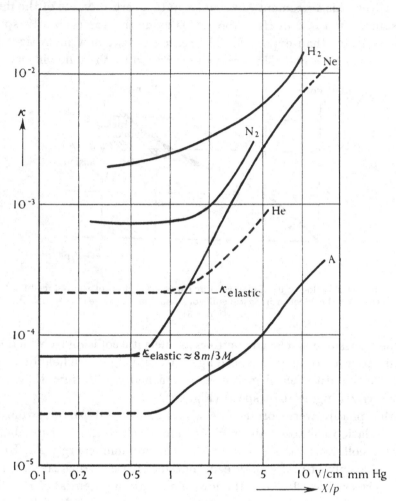

FIG. 63. Average fraction κ of electron energy lost in a collision with a gas molecule as a function of the reduced field X/p for various gases (9, 10). κ can be calculated from data on inelastic cross-sections (see 8, 43).

electron colliding with a H_2 molecule can excite it to a vibrational level, for which energies of order 0·5 eV are required. There exists a larger number of vibrational levels of the molecule which are quite evenly distributed over the whole energy scale up to the ionization energy and the probability of an exciting collision for electrons of order 1 eV is relatively high (of order 10^{-2}). Thus κ will rise with X/p, and v will increase faster than $(X/p)^{\frac{1}{2}}$. This seems to explain the character of the curves shown in Fig. 61 for various molecular gases. The drift velocity is very roughly a linear function of X/p, and the 'mobility constant' c_2 in $v_d = c_2 X/p$ for H_2 and N_2 is of the order 5.10^5 cm/sec per V/cm at 1 mm Hg over the range investigated.

When X/p is very large, v_d is high, but of course its value can never exceed that corresponding to the motion in high vacuum for a given potential difference. This means that at very low p, v_d does not depend any longer on X/p. Also, at higher values of X/p, ionization occurs and the newly-formed electrons start at the bottom of the energy distribution. At the same time the energy distribution becomes more and more anisotropic.

The considerations at the beginning of this section hold for a constant mean free path. The equations can also be solved for certain dependences of λ_e and κ on the mean energy. There is, however, a point which is of more fundamental interest as far as inclusion of the dependence $\lambda_e(\bar{c})$ is concerned. The drift velocity in general has been shown (132) to be

$$v_d = \frac{eX}{3m}\left\{\frac{2\lambda_e}{c} + \frac{d\lambda_e}{dc}\right\}. \tag{4.26}$$

It has to be remembered that c and λ_e are functions of the electron energy and hence of the field X.

The drift velocity in electronegative gases is not much different from other molecular gases. Cl_2, Br_2, and I_2 show at $X/p = 40$ and 80 drift velocities of 8 and 13.10^6 cm/sec respectively.

3. Charge transfer

When a positive ion collides with a gas molecule or atom two processes can occur. Firstly, the ion and the molecule exchange momentum and energy; at the same time their direction of motion is altered. Secondly, apart from this redistribution of energy, an exchange of charge can take place which is accompanied by preferential scattering. If, for example, fast ions are moving through a gas, a collision can result in an ion extracting an electron from a gas atom, as a result of which the 'fast' ion becomes a fast neutral atom while the 'slow' atom becomes a slow positive ion. When positive ions move in an electric field, charge transfer results in a reduction of the effective drift speed of the ions and hence their mobility appears to be diminished.

The charge transfer has played a prominent part in the first studies on 'canal rays' by W. Wien (91). Through charge transfer singly charged positive ions were found to become neutralized, doubly charged ions to become singly charged, etc., and simultaneously the reverse processes corresponding to ionization took place. However, the total (net) charge in the gas remains here unchanged. Charge transfer has not often been found to occur with negative ions, probably because of the small energy of electron attachment. It should be mentioned that collisions between very fast ions and gas molecules lead not only to charge transfer but also to excitation of neutral molecules and dissociation of polyatomic molecules (Chapter 3 A). Charge transfer is particularly pronounced in atomic gases where atomic ions are present and in molecular gases with molecular ions, as can be seen from Table 4.1 and Fig. 60.

If, for example, ions of a type A of large energy collide often enough with molecules or atoms of a type B (Fig. 64 b), we have in equilibrium

$$A_{\text{fast}}^{+} + B_{\text{slow}} \rightleftarrows A_{\text{fast}} + B_{\text{slow}}^{+}. \tag{4.27}$$

A beam of fast ions (canal rays) can be produced conveniently by a highly anomalous glow discharge at a gas pressure of 10^{-2} to 10^{-3} mm Hg, having a cathode with a small hole. Through this hole the ions emerge in a concentrated beam with energies slightly less than the potential difference between cathode and anode of the discharge tube. The ions are passed into a field-free space where, after having suffered a sufficiently large number of charge exchanging collisions (proportional to the pressure times the distance the beam has travelled), an equilibrium is established between the number of positive and neutral particles in the beam. If n^{+} is the number of ions per second per cm² and λ_{+0} the mean distance after which neutralization by collision occurs, then the number of such collisions per cm³ per second is n^{+}/λ_{+0}; in equilibrium the reverse process, namely ionization, occurs at the same rate. Denoting the cross-sections by Q and the flux density of neutral atoms by n_0, we have

$$n^{+}/n_0 = \lambda_{+0}/\lambda_{0+} = Q_{0+}/Q_{+0}. \tag{4.28}$$

This relation shows that the ratio of positive particles to the number of neutrals is equal to the inverse ratio of the corresponding mean free paths for charge transfer. It will be shown later how the mean free paths for charge transfer or the corresponding cross-sections Q vary with the speed of the ions. n^{+}/n_0 increases with increasing beam energy because Q_{0+} rises with the latter.

The phenomenon of ionization by collision between neutral molecules was shown by deflecting magnetically the positive ions in the field-free section of a canal ray tube. The ion beam passes two strong transverse magnetic fields (Fig. 64 a). With the field H_b on alone, ions are driven to the wall of the tube producing a fluorescent spot. If now H_a is also applied—thus removing all ions from the beam before they reach H_b—the fluorescent spot on the glass wall remains stationary, independent of the intensity of the field H_a; only its brilliance varies a little with the field. Since all the ions have been removed from the beam by the field H_a, it follows that new ions must have been produced in the beam by fast neutral particles colliding with stationary neutrals. (The effect of

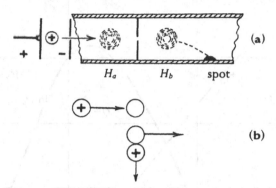

FIG. 64. Illustrating the charge transfer effect in canal rays.

secondary electrons was eliminated.) Thus it is necessary to assume a collision process of the type (see Chapter 3 A)

$$\text{molecule}_{\text{fast}} + \text{molecule}_{\text{slow}} \rightarrow \text{ion} + \text{electron} + \text{molecule}_{\text{slow}}.$$

Returning to the question of charge transfer, illustrated in Fig. 64 b, from (4.28) it follows that since the path associated with neutralization is shorter than the one corresponding to ionization, in equilibrium the number of ions in a canal ray is smaller than the number of their neutral satellites. The dependence of the charge transfer cross-sections on the energy of the ions for fast protons moving in H_2 and for He^+ in He is shown in Fig. 65. With increasing energy the neutralization cross-sections decrease except for H^+ in H_2 where a maximum occurs around 7 keV. H^+ in Kr and Xe show maxima for 100 to 150 eV ions. Their origin is not known; it may be associated with inelastic collisions between ions and atoms.

Another important fact, found early, was that protons moving in H_2 have charge transfer cross-section orders of magnitude larger than for protons in any other gas. This problem has been studied and explained more recently by means of wave mechanics. Roughly speaking, charge transfer processes are more likely to occur the nearer they are to 'energy resonance', that is to say, the smaller the energy which is exchanged in such a collision. This energy is a minimum for ions moving in their own gas and where charge transfer is strongly pronounced. Table 4.3 is self-explanatory. The amount of transferred energy ΔE is equal to the sum of the kinetic and ionization energy (eV_i) plus the excitation energy of the hit molecule (including rotation and vibration) minus the neutralization energy which for ions in their gas equals eV_i.

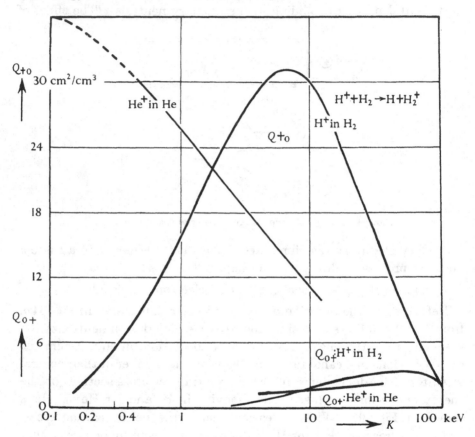

Fig. 65. Neutralization cross-section Q_{+0} and ionization cross-section Q_{0+} for He and H ions as a function of the kinetic energy K of the particles at 1 mm Hg and 0° C (18). For further results see (134. d)

Fig. 66 shows the charge transfer cross-sections for slow positive ions in various gases as a function of their energy. The shape of this curve is qualitatively in agreement with wave-mechanical calculations except for the range of very small energy. The classical kinetic theory cross-sections are also given in the figure, showing that they are of the same order as the transfer cross-sections; for He they are approximately 15 times the area of the first Bohr orbit. The results at very low speeds are

TABLE 4.3

Magnitude of charge transfer (24)

V_i	gas	Hg^+	O_2^+	CO_2^+	A^+	N_2^+	Ne^+	He^+
10·4	Hg	st	..	..	..	..	..	..
12·2	O_2	..	st	..	..	w	..	..
14	CO_2	..	..	st	..	..	..	..
15·8	A	..	..	..	st	st	w	w
16·3	N_2	..	..	..	m	st	..	..
21·6	Ne	..	..	..	..	w	st	..
24·6	He	..	..	..	..	w	..	st

st = strong; m = moderate; w = weak

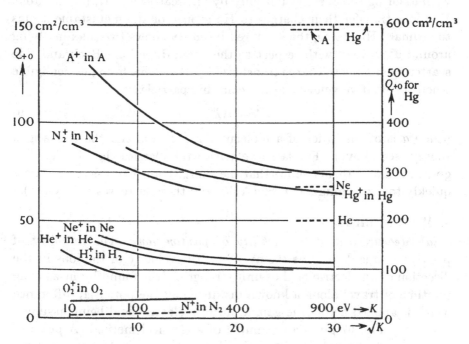

FIG. 66. Charge transfer neutralization cross-section Q_{+0} as a function of the energy K in various gases at 1 mm Hg and 0° C (76, 89, 133, 134, 134 a). Dotted lines on right: classical cross-sections.

not very reliable. Comparison of these curves with the curves of Fig. 36 show that the cross-section for neutralization is much larger than that for ionization by ion bombardment in agreement with older measurements, and also that only ions with an energy above a minimum energy are capable of ionizing gas molecules. However, the accuracy of these cross-section measurements is not sufficient to determine the critical potentials for ionization of molecules by ions.

Recent work has been concerned with an analysis of the various forms of charge transfer reactions and their final products. It transpires that by distinguishing only between ionization and neutralization a too crude physical picture is obtained. In fact a charge transfer collision may be accompanied by dissociation, excitation of the ion or the neutral particle (and subsequent light emission), ion formation, etc. (134 b). For example, the emission of light by canal rays and its association with various types of ions has been ascribed to a charge exchange reaction (134 c).

Whether or not an ion has exchanged its charge cannot be decided by measuring the energies but only by the scattering. If, for example, He$^+$ ions transfer their charges to He atoms, or the metastable atoms to ordinary He atoms, the scattered intensity shows two maxima lying around 0° and 90° with respect to the beam direction. Back and side scattering are due to charge transfer (Fig. 64 b). This becomes negligible when the relative velocity of the colliding particles

$$v \gg a \, \Delta E / h, \tag{4.29}$$

where a is of the order of a molecular radius and ΔE the exchanged energy (see above). For large velocities the classical law holds. For given ΔE and a, theory shows that the charge transfer cross-section rises quickly to a maximum at $v \approx a \, \Delta E / h$ and then decreases slowly (18).

4. Measurements

(a) *Measurement of the mobility of positive ions.* The mobility of positive ions is defined as the average drift velocity of the ions in the direction of an electric field of unit strength. It is found by measuring the time of travel along a known distance and the potential difference which is applied across the gap, a uniform electric field being assumed (13, 122). Fig. 67 gives an example of a classical method: A pulse of X-rays or of α-particles ionizes a thin layer of gas near the left (positive) electrode; simultaneously an electric field of intensity V/d is applied by closing the switch S for a certain time. If the time interval between the ionizing flash and the removal of the electric field (by opening S) is long enough, then ions can reach the negative electrode and a significant current will flow through the electrometer M. The ion transit time is found by reducing the interval to the smallest value Δt for which the electrometer M indicates the onset of a current. The mobility is thus

$$k^+ = \frac{v_d}{X} = \frac{d^2}{V \Delta t}, \tag{4.30}$$

where Δt is the time of flight of the ions across d. Since the onset is not very sharp, the determination of k^+ is not accurate. In addition this method is fundamentally objectionable, cf. page 23 (25 a).

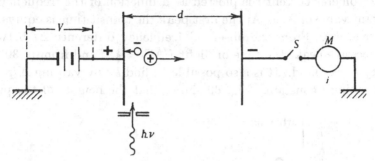

FIG. 67. Classical measurement of the ion mobility.

The electric shutter method is not only more accurate but has also a large resolving power which allows discrimination between ions of slightly different mobility. The working principle is shown in Fig. 68.

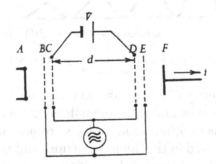

FIG. 68. Electric shutter method for measuring ion mobilities.

A is a positive ion emitter, for example a Kunsman electrode, consisting of Fe oxides and K or Na salts, but a glow discharge can be used instead. Now BC and DE are two pairs of grid electrodes acting as electric shutters. A small high-frequency potential is applied between each pair which alternately stops the ions and allows them to pass these gates. Thus bunches of ions enter the space CD and move in a uniform field which is maintained by the constant potential difference V applied between C and D; they can reach F only if they find the second gate open.

An ion covers that distance in a time d/v_d. Hence only those ions whose transit time is an odd multiple of d/v_d and thus corresponds to $\frac{1}{2}, \frac{3}{2}, \frac{5}{2}, \dots$, cycles of the applied high frequency will reach the collector F.

Since the periodic time varies as $1:\frac{1}{3}:\frac{1}{5},...$, the frequencies at which ions reach F vary as $1:3:5,...$, and hence the frequency difference between successive current peaks is constant. This is shown in Fig. 69, where the electrometer current i is plotted as a function of the frequency for a constant value of X/p. At the first peak the transit time is equivalent to $\frac{1}{2}$ cycle, etc. From the observed frequency difference Δf between two successive peaks the time of flight $i/\Delta T = \frac{1}{2}\Delta f$ and from (4.30) the mobility k^+ is found. It is also possible to find k^+ by varying X/p or d while keeping f constant. Fig. 69 shows that the heights of the peaks

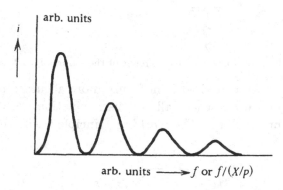

FIG. 69. Collector current i in Fig. 68 as a function of the shutter frequency f or the frequency per unit field (X/p).

decrease with increasing f because as f is increased the 'entrance gate' is open for a shorter time and thus a smaller number of ions can reach the collector. If ions of different mobility were present, several sets of peaks would be observed in the 'ionic spectrum' and they would appear in the zero as well as in higher orders (134 e).

In the mobility tubes used (Fig. 68) d was a few cm, X of order 10 to 100 V/cm, p between 5 and 30 mm Hg, and f between 3 and 30 kc/s. Results of these measurements have been shown in section 1.

(b) *Measurement of the drift velocity of electrons.* Electrons are emitted from an electrode C (Fig. 70) which is either a hot filament or an illuminated plate. The electrons move through a uniform electric field which is obtained by using a series of equidistant annular plates which are connected to appropriate potentials. Finally, the electrons fall on to a positive electrode which consists of three parts. In the absence of disturbances causing asymmetry, the partial currents to 1 and 3 measured with the electrometer M are equal: $i_1 = i_3$. When a uniform magnetic field perpendicular to the plane of the paper is applied, the beam of electrons is deflected. Let a be the width of the central part of the

anode and d the distance CA. If for a given value of X the intensity
of the magnetic field H is such that the beam is deflected by $\frac{1}{2}a$ cm,

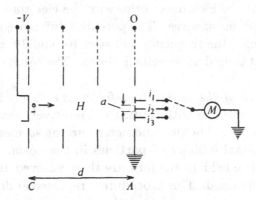

FIG. 70. Measurement of the electron drift velocity.

then the ratio of the (transverse) magnetic force to the (longitudinal)
electric force X ($= V/d$) is a measure of the drift velocity v:

$$\tfrac{1}{2}a/d = \tan \alpha = \frac{(e/c)Hv}{eX} = Hk^-/c, \qquad (4.31)$$

where k^- is the 'electron mobility' and c the velocity of light. k^-, H,
and X are expressed in c.g.s. units. For the lateral deflexion ($\frac{1}{2}a$), half
of the total electron current flows to one external section, the other
half to the two other sections, provided the beam current and its distribution is not altered by the magnetic field. Thus to find k^- that value
of H is to be found for which $i_3 = i_1 + i_2$.

The experiments have shown that the mobility k^- is a function of
X/p; in certain cases (section 2) the relation $v = c(X/p)^{\frac{1}{2}}$ is obeyed. Some
of the results obtained by this method have been included in Fig. 61.

The same method can be used to determine the drift velocity of positive ions in which case C is a Kunsman electrode or a glow discharge.
However, in order to produce sufficiently large deflexions relatively
strong magnetic fields have to be used.

The drift velocity of electrons can be measured accurately by the
electric shutter method as described in Fig. 68. In order to achieve very
high resolving power, the electrodes BC and DE are each arranged
in a plane. By connecting alternate wires to the terminals of a high-
frequency source, a lateral electric field is produced between them.
Thus electrons can, for example, only pass either shutter when the

shutter field is about zero. The 'gate' is therefore only open when the instantaneous value of the high-frequency potential difference between the wires is of the same order as the volt equivalent of the thermal energy of the electrons; otherwise the electrons are driven to one of the two systems of wires. The potential difference between them is of order 100 volts, the frequency between 10^4 and 10^7 cycles. A uniform electric field is used as described above. The results are shown in Fig. 61.

(c) *Measurement of the charge transfer cross-section.* In a classical method a beam containing both neutral and positively charged particles falls on a thermopile. The galvanometer current so measured is proportional to the total number of particles in the beam. By means of a transverse electric field all the ions are then removed at $x = 0$ and a lower reading is obtained. This procedure is repeated at different values of x. From the results Q_{+0} and Q_{0+}, the neutralization and ionization cross-sections are calculated. Instead of the thermopile a photographic plate has been used to measure the beam intensity.

Modern measuring methods for charge transfer cross-sections can only be indicated (17): a beam of ions of uniform speed, free from neutral atoms or molecules, enters a chamber through an aperture which can be blocked by a movable electrode to measure the total primary beam current i_0. In the 'collision chamber' containing gas at low pressure cylindrical electrodes of length l are arranged and the electron saturation current i_2 to a positive collector, caused by ionization by ion collisions, is measured. When the polarity of the collector electrode is made negative, a current i_1, which is the ion current i^+ carried by slow ions (produced by charge transfer) plus the electron current i_e, again due to ionization of atoms by fast primary ions, is collected. Thus both the ionization Q_{0+} and charge transfer cross-section Q_{+0} as a function of the ion (beam) energy can be measured simultaneously from

$$i_1 = (i^+ + i_e) = i_0\,pl(Q_{+0} + Q_{0+}) \qquad (4.32)$$

and
$$i_2 = i_e = i_0\,pl\,Q_{0+}. \qquad (4.33)$$

Recent work deals with discrimination of the final products of the transfer reactions and the detailed mechanism of the processes involved (134 b).

QUESTIONS

(1) Discuss the essential difference between Langevin's ion mobility theory based on a constant collision time with that based on collisions at a distance due to the electric dipoles of the gas molecules. Substitute the equation for the electric polarizability $\alpha_0 = (D-1)/4\pi N$ into (4.14). Compare the corresponding factors in the two mobility equations. How does the result compare with the capture cross-section (page 89)? In particular explain the experimental results when ions move through a 'foreign' gas (100 c).

(2) Give the physical reasons for assuming a constant collision time in low fields (per unit gas density) and constant fraction of energy lost per collision in moderate fields. At what values of the relative field do we expect the latter to fail and what variation of v_d with X/p is likely to occur at larger values of X/p?

(3) Outline a modern method of measuring positive ion mobilities. How would you find what types of ions are present? What conclusions would you derive from results in which the values of the mobility are found to vary with p at constant X/p?

(4) Write an essay on the appearance of molecular and atomic ions in various ion sources and list the important collision processes which account for the production of molecular ions of inert and common gases.

(5) Given an electron swarm which moves through a gas. Calculate its random energy E as a function of the position when the electron starts at $x = 0$ with zero energy in the field direction. Find the 'final' energy when $p = 10\mu$, $X = 10$ V/cm, and $\lambda_1 = 5 \cdot 10^{-2}$ cm. Assume the fractional energy loss $\kappa = 10^{-3}$. At what distance would it obtain 95 per cent of its final drift velocity?

(6) If electrons drift in He in an electric field show that $v_d \propto \sqrt{(X/p)}$ up to $X/p \approx 2$. For larger values of X/p, v_d rises faster. This is thought to be due to the onset of inelastic collisions. Show that this is the case using the electron energy distribution for He (fig. 41, p. 549, Hdb. d. Phys. 21, 1956) and an approximate total excitation cross-section $q_{exc} = 1 \cdot 10^{-18}(K-20) + 4 \cdot 10^{-18}$ for $K \geqslant 20$ V, assuming that for lower energies only elastic losses occur. The condition for departure is that the losses by inelastic collision be 50 per cent of the elastic ones (Fig. 21).

(7) How are electron drift velocities in various gases measured and what difficulties are encountered at very low and very high values of X/p? Give an account of earlier work too and comment on the results obtained in non-uniform electric fields. To what extent is the idea correct that in this case α/p loses its meaning? Estimate the non-uniformity at which this difficulty would arise.

(8) Write an essay on the production of fast neutral particles by charge transfer and explain how a beam of neutral particles in the ground state can be obtained without metastable 'fellow travellers'. How would you show that metastables are present and what methods of separation exist?

(9) A discharge tube is filled with H_2 at 0·1 mm Hg and has a cathode with a hole of 1 mm diameter. If the cathode current density is 10^{-5} A/cm² and 95

per cent of the current is carried by positive atomic ions, find the number of ions/cm^2 sec as a function of the distance from the cathode if charge transfer occurs from the hole onwards. Assume $Q_{+0} = 18$ and $Q_{0+} = 1$ cm^{-1} at $p = 1$ mm Hg. Also calculate the equilibrium between neutrals and ions at 0·1 mm Hg and the distance at which 90 per cent of the equilibrium value of neutrals is obtained.

(10) Give an account of the experiments which suggest charge transfer collisions. Also show how it is possible to distinguish between large and small charge transfer cross-sections. Include work which indicates other processes than ionization in this type of collision. Discuss under what conditions an atomic ion of small kinetic energy can make a charge transfer collision in a foreign atomic gas remembering that the transfer of the electron is an inelastic process.

(11) Calculate from first principles the mobility k_e of electrons in a very weak electric field, remembering that the random electron energy depends here on the temperature difference between electrons and gas, which in turn is related to the energy acquired along a mean free path in the field direction, and assuming losses by elastic collisions (κ). Start with Langevin's equation and neglect factors of the order 1. Evaluate the result for electrons in N$_2$ and compare it with observations, viz. for $p = 1$ atm, 0° C, $X/p = 10, 6, 2, 1 \times 10^{-2}$ V/cm mm Hg, $k_e = 5, 7, 13,$ 16×10^3 cm^2/V sec respectively. Calculate k_e in zero field for N$_2$ at 1 atm and He at 1 mm Hg, both at 0° C, and for Hg at 10^{-2} mm Hg, 47° C, and compare with experimental results. Give reasons why in stronger fields the calculated mobility differs from observation.

5

DIFFUSION AND MUTUAL REPULSION

A. DIFFUSION

1. INTRODUCTION

Charged or uncharged particles are said to diffuse in a gas if they move from points of high concentration to those of low concentration. The origin of this motion is a purely thermal one; thus the actual path of an individual particle is a zigzag line (135). Diffusion of electrons or ions moving in their own or in a foreign gas was originally regarded as completely analogous to self- and mutual diffusion of gas molecules. If N is the concentration of particles, i.e. their number per unit volume, and if N is a function of position only, the velocity of diffusion is simply

$$\mathbf{v} = -(D/N)\mathrm{grad}\,\mathbf{N}, \tag{5.1}$$

where D is the diffusion coefficient in cm^2/sec, the negative sign indicating that the motion occurs in the direction of decreasing concentration. Diffusion, in one direction only, follows from (5.1)

$$v = -\frac{D}{N}\frac{dN}{dx}. \tag{5.2}$$

vN is the density of flow, for example, the number of particles per sec per cm^2 of cross-section which move in the x-direction. Taking, for example, ions of one sign moving in their own gas, we have from kinetic theory

$$D_i = \frac{\bar{c}\lambda_i}{3}, \tag{5.3}$$

where λ_i is the mean free ionic path and $\bar{c}$ the mean velocity. For example, in N_2 at room temperature $\bar{c}$ is of order 4.10^4 cm/sec and λ_i at 1 mm Hg about 5.10^{-3} cm, hence $D_i \sim 50$ cm^2/sec. Since $\lambda_i \propto T/p$, we have $D_i p = $ const, and because $\bar{c} \propto T^{\frac{1}{2}}$ we have $D_i \propto T^{\frac{3}{2}}$. It is tacitly assumed that the average kinetic energy of an ion is equal to that of a gas molecule and the velocity distribution of all particles is Maxwellian (7). $\bar{c}$ is defined by $m(\bar{c})^2/2 = (4/\pi)kT$ (Chapter 3). At lower temperatures quantum-mechanical calculations lead to lower values of λ_i.

When diffusion occurs in a gas-filled vessel, particles travel in the direction of the negative concentration gradient and the time required to move between two points in a general direction can also be found from kinetic theory (71, 72). The average (r.m.s.) displacement of an individual particle in one coordinate is given by

$$\bar{x} = \sqrt{(2Dt)}. \tag{5.4}$$

An ion in N_2 at 1 mm Hg which is at zero time at the axis of a tube is found 1 cm from the axis after 10^{-2} sec ($D_i = 50$); the total path travelled is of order $4.10^4.10^{-2} = 400$ cm. Again from (5.4), τ, the average life of an ion, can be found:

$$\tau \sim \frac{x^2}{D_i}, \tag{5.5}$$

where x is, for example, the tube radius where this limits the life of an ion. The correct expressions for an infinite cylinder of radius R, an infinite strip of width L, and an infinite tube of rectangular cross-section (a, b) respectively are given by

$$\tau D_i = (R/2 \cdot 405)^2, \quad (L/\pi)^2, \quad a^2/\pi^2[1 + (a/b)^2]. \tag{5.6}$$

The numerical results are of course of the same order as those obtained from (5.5). For larger X/p ($\bar{v}_i > \bar{v}_{gas}$), D_i is larger (139 b).

Here are a few data obtained from diffusion experiments (which are only moderately accurate) in comparison with data obtained indirectly from mobility measurements (Chapter 4).

TABLE 5.1

Diffusion coefficients D_i for ions in their own gas in cm^2/sec at $p = 1$ mm Hg (6, 11)

Gas	D^+ (meas.)	D^- (meas.)	D^+ calc.
H_2 . . .	98	110	110 (180, pure)
O_2 . . .	21	32	24
N_2 . . .	23	31	24 (47, pure)
Air . . .	22	33	24 (33, pure)
CO_2 . . .	18	20	18
Hg . . .	..	..	~ 12
He . . .	..	..	380 (very pure)
Ne . . .	..	..	~ 120 ,,
A	..	..	47 ,,
Kr . . .	..	..	17 ,,
Xe . . .	..	..	12 ,,

The table shows that for gases of the same purity reasonable agreement between theory and observation exists; however, highly purified gases would give larger values for D_i as can be seen from the data for a few molecular gases.

Comparison of D_i of Table 5.1 with the values of D for molecules diffusing in their own or a foreign gas show that $D_i < D$ by a factor of 4 to 5. This cannot be explained if (5.3) is applicable: with $\lambda_i = \lambda\sqrt{2}$, $\bar{c}_i = \bar{c}$, we should expect $D_i > D$. This discrepancy can be resolved by assuming that the ions exchange momentum with neutral molecules over distances which are larger than their diameter by inducing dipoles in the gas molecules (see Chapter 4). These distant scatterers reduce the 'effective' mean free path according to their degree of polarizability and account for the low values of D_i. In certain cases charge transfer (Chapter 4) reduces the diffusion of ions.

If electrons diffuse through a gas the coefficient of diffusion is

$$D_e = \frac{\bar{c}\lambda_e}{3}, \tag{5.7}$$

where $\bar{c}$ is their mean velocity (which is usually many orders of magnitude higher than that of neutral atoms) and λ_e the mean free electronic path (which is a function of the electron velocity). Since $m\overline{c^2}/2 = \frac{3}{2}kT_e$ and 1 eV corresponds to 7740° K, we obtain from (5.7) and (3.3)

$$D_e \approx 2.10^5\lambda_e \sqrt{T_e} \approx 2.10^7\lambda_e \sqrt{V}. \tag{5.8}$$

In Ne at $p = 1$ mm Hg, $\lambda_e = 0.15$ cm and assuming $V = 4$ eV ($T_e = 30\,000°$ K) we have $D_e = 6.10^6$ cm²/sec. The dependence of D_e on gas temperature and pressure is the same as for D_i. The actual values of D_e are likely to be much smaller because of collisions through polarization (see above).

There is an important relation between ion mobility and diffusion coefficient. Unfortunately it is often based on the first of Langevin's mobility theories which leads to mobilities exceeding those observed by a factor 3 to 5. From (5.3) and (4.4) the ratio of mobility and diffusion for positive ions

$$\frac{k^+}{D^+} = \frac{e\lambda}{m\bar{c}} \frac{3}{\bar{c}\lambda} = \frac{3e}{m\bar{c}^2} = \frac{e}{kT} = \frac{eN}{p}, \tag{5.9}$$

where e is in e.s.u. and $p = NkT$ in dynes/cm²; at N.T.P. eN is equal to the Loschmidt number ($2.7.10^{19}$) times the electronic charge ($4.8.10^{-10}$ e.s.u.) and thus is related to the Faraday constant which refers to 1 mole. At $T = 300°$ K and with $k = 1.37.10^{-16}$ ergs per °K and per particle we have

$$k^+/D^+ \doteq 43, \tag{5.10}$$

where k^+ is expressed in cm/sec per V/cm (k^+ in e.s.u. is 300 times larger).

A derivation free from the above objection is obtained by taking ions diffusing with a drift velocity which is balanced by their motion in an electric field. We have

$$v = -\frac{D^+ dN^+}{N^+ dx} = -k^+ X. \qquad (5.11)$$

In thermal equilibrium the partial ion pressure $p^+ = N^+ kT$, the gas pressure $p = NkT$ and $dN^+/dx = (1/kT)dp^+/dx$. The force acting on the ions contained in 1 cm³ is $dp^+/dx = eN^+ X$. Hence from (5.11) we have

$$\frac{k^+}{D^+} = \frac{1}{N^+ X}\frac{dN^+}{dx} = \frac{1}{N^+ X}\frac{1}{kT}eN^+ X = \frac{e}{kT} = \frac{eN}{p}. \qquad (5.12)$$

We conclude therefore that this relation holds, whatever the field of force of the atoms acting on the ions. It is, however, not applicable to electrons for at least two reasons: electrons drift with velocities which in general are not proportional to the applied field and their mean free path depends on their speed.

Equation (5.12) only holds when the electric field in which the ions move and diffuse is moderately strong: in stronger fields v is proportional to the square root of the field. From the few existing observations we can assume that (5.12) is fulfilled in weak fields.

2. DIFFUSION OF IONS IN AN ELECTRIC FIELD

Let ions of one sign be present which diffuse through their own gas in the x-direction and let there be an electric field whose direction is parallel to x. The velocity of the ions is

$$v = v_{\text{diff}} + v_{\text{field}}. \qquad (5.13)$$

A case of special interest is that of zero resultant velocity, which has been treated before. Thus starting from (5.11), we find by integration taking $V_x = 0$ at $x = 0$ and with $X = -dV/dx$

$$N_x/N_0 = e^{-(k^+/D^+)V_x}. \qquad (5.14)$$

This means that the numerical ion concentration decreases with rising potential or decreasing field and approaches zero when V_x approaches infinity. The conditions are easily visualized. Positive ions are provided coming from the region of negative x. If the concentration at $x = 0$ were kept at the constant value N_0 and no other field were acting, a concentration distribution would ensue which would be controlled by the diffusion coefficient only. This would result in a migration of ions in the x-direction. By applying an electric field (not necessarily a uniform one) of such direction that the potential increases with x, the ions are driven back towards the origin and an equilibrium distribution is established as described by (5.14).

In a field of moderate strength the ions are in thermal equilibrium with the surrounding gas. The ratio k^+/D^+ can thus be substituted from (5.9) in (5.14) and we have

$$N_x/N_0 = e^{-eV_x/kT}. \qquad (5.15)$$

Hence the concentration at a point which has the potential V_x is uniquely defined by the ratio of electrostatic (potential) energy to kinetic (thermal) energy (5.15), known as the Boltzmann distribution. The result is formally the same as that for the density distribution of molecules in an isothermal atmosphere: the gravitational field attracts the molecules towards the earth's surface, increasing their concentration at low levels, and diffusion tends to reduce differences in concentration and drives them up into the atmosphere.

3. Diffusion of Ions and Electrons in an Electric Field (Ambipolar Diffusion)

When electrons and positive ions of equal concentration are present in a gas on which an electric field acts, it is found that their motion can often be treated as if the assemblage of charges diffused with a common velocity.

Let the net space charge at any point be zero. Then from Poisson's equation

$$N^+ = N^- = N. \qquad (5.16)$$

We assume that everywhere

$$dN^+/dx = dN^-/dx = dN/dx. \qquad (5.17)$$

This applies particularly to the positive column in glow discharges in a certain range of pressure, etc., and is equivalent to saying that N is to be sufficiently large; for differences in N^+ and N^- would produce net space charges and thus such changes in fields which would tend to annul the originally assumed difference of concentration.

Only the steady state will be treated here. Further it is assumed that the gas density is large enough to provide sufficiently numerous collisions between the charged particles and molecules along the path considered so as to establish statistical equilibrium. Since constant mobility is attributed to the motion of electrons (which is contrary to the facts), this treatment applies only to cases where either the variation in the electric field is small enough or where the curvature of the function: drift velocity against X/p is small, as in molecular gases.

In the common electric field which is due to the different 'mobilities' of electrons and ions, the electrons will tend to move ahead of the ions, the latter pulling the electrons back. In the steady state the number of electrons crossing unit area in unit time at any point will be equal to the number of ions and hence the drift velocities will be equal:

$$v_d^+ = v_d^- = v_a. \tag{5.18}$$

If we assume that both electrons and ions move in the direction of the field, we obtain from (5.18), (5.11), and (5.13) the 'ambipolar speed'

$$v_a = -\frac{D^+}{N}\frac{dN}{dx} + k^+X = -\frac{D^-}{N}\frac{dN}{dx} - k^-X. \tag{5.19}$$

By eliminating X from these two equations we have

$$v_a = -\left[\frac{D^+k^- + D^-k^+}{k^+ + k^-}\right]\frac{1}{N}\frac{dN}{dx} = -D_a\frac{1}{N}\frac{dN}{dx}. \tag{5.20}$$

The coefficient of ambipolar diffusion is a mean diffusion coefficient averaged in the ratio of the mobilities. D_a can often be approximated when $k^- \gg k^+$ and $T_e \gg T_i$; from (5.9) it follows that $D^-/k^- \gg D^+/k^+$ and thus

$$D_a = \frac{D^+k^- + D^-k^+}{k^+ + k^-} \approx \frac{D^-k^+}{k^-} = \frac{kT_e}{e}k^+. \tag{5.21}$$

When $T_e = T_i$ we find that $D_a = 2kT_ek^+/e$, i.e. twice the value of (5.21). If in (5.19) $c\sqrt{X}$ instead of k^-X is used, the resulting expression for v_a becomes cumbersome and not convenient for physical interpretation. One weakness of this treatment is the use of the 'electron mobility' (see Chapter 4); the other is associated with the transition from ambipolar to electron diffusion at low N, see (135 a).

The ambipolar diffusion coefficient has been measured in H_2 (135 b), He (135 c), and N_2 (135 b) and found to be $D_a.p \approx 124, 550$ and 700 ± 50 cm²/sec respectively. Assuming e.g. in He $k^+ \approx 15.10^3$ at $p = 1$ mm Hg and $T_e \approx T \approx 300°$ K $\approx 3.10^{-2}$ eV, the observations agree with theory. Care has to be exercised in using (5.21). For example as p is raised, atomic ions are gradually replaced by molecular ions. It is also known that T_e can vary with position. At low p diffusion ceases to occur (wall collisions > volume collisions) and ambipolar diffusion is to be replaced by free fall in a space-charge field. The distribution of charge concentration is such that again equal numbers of positive and negative charges arrive at the insulating wall.

Fig. 70 a shows D_a as a function of p in various gases, obtained from measurements in a positive column. D_a decreases as p is increased and as the molecular weight is raised (30).

Another point of interest is the range of applicability of the concept of ambipolar diffusion. It has been said before that it applies when the concentration of charges $N^+ = N^-$ is large. On the other hand it cannot apply when the degree of ionization is very high, i.e. the number of neutral particles so small that collisions between ions and electrons

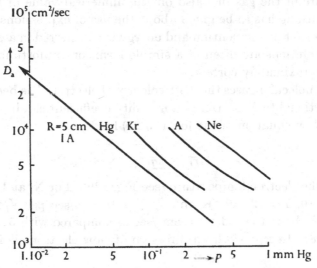

FIG. 70 a. Ambipolar diffusion coefficient D_a as a function of the gas pressure p. $(D_a = 10^5$ cm²/sec calculated at $p = 2\mu$ in Hg) (30).

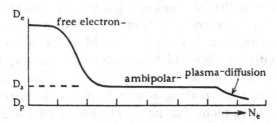

Fig. 70 b. Free electron (D_e), ambipolar (D_a), and fully ionized plasma (D_p) diffusion coefficients as a function of the electron concentration N_e at constant electron temperature T_e. $D_p . p \propto (kT)^{7/2}$, from the electric conductivity $\sigma \propto T^{3/2}$ and Wiedemann-Franz's law or (5.9) and p. 257.

exceed those with neutral particles. There are thus two transitions (Fig. 70 b) separating the three regions.

4. DIFFUSION OF ELECTRONS IN AN ELECTRIC FIELD

This problem cannot be discussed here quantitatively since its analysis is much too involved. Qualitatively it can be stated that the average drift velocity of electrons due to a concentration gradient together with an electric field is not simply given by the sum of (4.4) and (5.1) because the electrons have a mean energy (or temperature) which is far above that of the gas. In addition there is in many cases some doubt whether the use of the Maxwellian velocity distribution is permissible and represents a 'good approximation'. Further, one has to remember—as shown in Chapter 4 on mobility—that the energy distribution depends not only on the nature of the gas but also on the numerical value of the field (X/p). A warning has to be given about the use of expressions derived by assuming that no momentum and energy is transferred in a collision; the formal solutions are often of a simple form, but unfortunately are not even approximately correct.

For some molecular gases the drift velocity of electrons has been found to be proportional to X/p. Hence a mobility coefficient can be assigned to them and an equation of the form (5.21) holds, viz.

$$\frac{k_e}{D_e} = \frac{e}{kT_e},\tag{5.22}$$

where T_e is the electron temperature (see Fig. 129). For N_2 at 1 mm Hg and $X/p = 10$, $T_e \doteq 2.10^4$ °K and $k_e \doteq 5.10^5$ cm/sec per V/cm (Fig. 61). D_e would be of the order 10^6 cm²/sec as compared with 5.10^5 cm²/sec from (5.8). Recent work on diffusion of slow electrons is found in (136, 137).

5. DIFFUSION OF ELECTRONS IN A MAGNETIC FIELD

Let us assume that the electrons in a gas have a constant component of velocity c in the xy-plane and that a uniform magnetic field H is applied in the z-direction. The electrons will be forced to move along arcs of circles with a radius which is found by balancing the Lorentz force eHc and the centripetal force mc^2/r, giving $r = mc/eH$.

Since the electrons move along a curved path their mean free path λ (projected onto the xy-plane) has now to be measured along an arc, and so the distance s between two successive points of collision and the arc length λ are given by

$$s = 2r\sin\tfrac{1}{2}\alpha, \qquad \lambda = r\alpha,\tag{5.23}$$

where α is the angle through which the electron has turned. The number of collisions for a given distance is increased by the magnetic field; this is equivalent to an increase in gas density. At constant gas temperature with $\lambda - s \ll \lambda$ we have

$$\Delta p/p = \Delta\lambda/\lambda.\tag{5.24}$$

But from (5.23) we have, by expanding in a power series,

$$\Delta\lambda/\lambda = (\lambda - s)/\lambda \propto \lambda^2/r^2,\tag{5.25}$$

and also

$$r \propto \frac{mc}{H} \propto \frac{T_e^{\frac{1}{2}}}{H}.\tag{5.26}$$

From (5.24), (5.25), and (5.26) we have

$$\frac{\Delta p}{p} \propto \frac{1}{mT_e}\left(\frac{H}{p}\right)^2. \qquad (5.27)$$

The effect is strong for electrons at $p < 0.1$ mm Hg and $H \approx 100$ Oe; it is negligible in the case of ions. Though the above treatment only applies when the arc along which the electrons travel is small, a more rigorous treatment leads to the same result for any value of α (10). This apparent increase in gas density affects all transport phenomena like drift velocity or mobility, diffusion, etc. Also an ionized gas can be regarded as an isotropic medium when magnetic, electric, or other fields are absent, but the application of a field gives the ionized gas crystal-like properties.

When a magnetic field is acting in the z-direction, all velocity components along z will remain unchanged and so will the coefficient of diffusion D_z which defines diffusion in this direction. However, a charge moving along x or y (or with velocity components in x or y) is accelerated along y or x respectively and describes the arc of a circle until deflected by a collision. Thus the charge proceeds along x or y at a slower rate than in zero magnetic field, that is the diffusion coefficients D_x and D_y are reduced. However, by including the constant in (5.27) a result wrong by a factor ≈ 100 is found. The rigorous treatment (138 a) shows the cause. An abbreviated derivation follows.

Consider electrons in a cylinder in an axial field. Their drift velocity can be expressed by a force $x\tau/m$, τ being the collision interval. For the Lorentz force $eHc_\perp$, with the angular velocity ω and $D_0 = D_{H=0}$, the radial and rotational components of drift are

$$v_r = -(D_0/N)dN/dr - eHv_\theta\,\tau/m, \qquad (5.28)$$

$$v_\theta = eHv_r\,\tau/m. \qquad (5.29)$$

Substituting (5.29) in (5.28), with e in e.m.u. and

$$\omega = eH/m, \qquad (5.30)$$

the diffusion coefficient D_H in a direction normal to H is

$$D_H = D_0/[1+(\omega\tau)^2]. \qquad (5.31)$$

Since $\omega \propto H$ and $\tau \propto 1/p\sqrt{T_e}$, $(\omega\tau)^2 \propto H^2/(p^2T_e)$ in conformity with (5.27). This relation has been tested experimentally by measuring the currents flowing to disk and ring electrodes on which an electron swarm falls. The ratio of the currents determines the radial spreading of the beam.

Fig. 71 shows that D/D_z varies linearly with $(H/p)^2$ as (5.31) predicts. In large magnetic fields and in plasmas the phenomena observed become fairly complex (30 b, 138 b, 139 c).

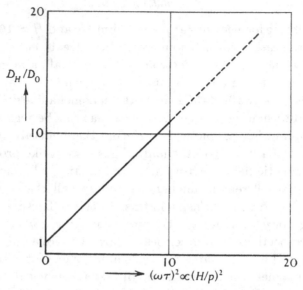

FIG. 71. Ratio of the electron diffusion coefficient D_H normal to the applied magnetic field H to the diffusion coefficient D_0 as a function of the angle $\omega\tau$ or the reduced field H/p (138). Note that $\tau = $ constant.

B. MUTUAL REPULSION

When free electric charges are uniformly distributed throughout space, each charge is in an electric field which is given by the vector sum of the fields of all the other charges; thus the force exerted on a charge is zero. Although the average distance between the particles and their concentration will not change with time, the individual particle because of its thermal energy will change its position perpetually.

If, however, charges of one sign are confined to a restricted volume and then suddenly released, the electric field of the charges will exert a force on a single one which in general is not zero. The ensuing motion leads to a redistribution of charges which comes to a halt when the concentration has become uniform. This process is called mutual or self-repulsion. It can also be visualized as an attraction between the opposite charges at infinity and the single charges in the volume.

6. MUTUAL REPULSION OF CHARGES IN VACUUM

In the absence of neutral particles the force acting on a charge will accelerate it in a direction of decreasing field and hence the charge

density decreases with increasing time and distance. Assuming that
sinks and sources are not present, i.e. that no volume recombination or
ionization occurs, the problem (in one coordinate) is described by the
partial differential equation

$$-\frac{\partial \rho}{\partial t} = \text{div}(\rho v) = \frac{\partial(\rho v)}{\partial x} = \rho\frac{\partial v}{\partial x}+v\frac{\partial \rho}{\partial x}, \tag{5.32}$$

where ρ is the charge density ($\rho = eN$, N being the concentration) and
v the drift velocity. In the steady state $\partial\rho/\partial t = 0$, and replacing the
partial differentials by the total ones we have

$$\frac{1}{\rho}\frac{d\rho}{dx}+\frac{1}{v}\frac{dv}{dx}= 0, \qquad \text{or} \qquad -\frac{d\rho}{\rho}=\frac{dv}{v}, \tag{5.33}$$

and the solution is $v\rho = \text{const.}$ (5.34)

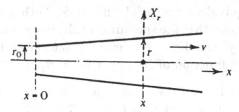

FIG. 72. Spreading of an electron beam in vacuum by mutual repulsion.

This result is useful in that it allows us to predict the variation of ρ and v
with the distance: a large speed is associated with a small charge density
and vice versa.

Next we discuss fully the divergence of an electron beam of circular
cross-section. Let the electrons have uniform speed v in the x-direction
(Fig. 72) and the beam radius be r_0 at $x = 0$. The electrostatic repulsion
accelerates the electrons radially and so the radius of the beam increases
with x. We neglect the variation of the electronic mass with speed as
well as the electromagnetic forces acting between current elements which
produce forces radially inwards.

By Gauss's theorem the radial electric field (X_r) of a cylinder of unit
length, containing the charge $q = i/v$, where i is the beam current, is

$$X_r\, 2\pi r = 4\pi i/v \quad \text{or} \quad X_r = 2i/vr. \tag{5.35}$$

The radial acceleration of an electron or the force per unit mass is, from
(5.35), $$\ddot{r} = (e/m)X_r = 2ei/mvr = c_1/r. \tag{5.36}$$

Multiplying (5.36) by $\dot{r}$ and introducing the numerical radius $R = r/r_0$
we have $$(r_0\dot{R})^2 = 2c_1 \ln R \tag{5.37}$$

subject to the condition $x = 0$ when $R = 1$ and $\dot{R} = 0$. The solution of (5.37) is

$$t = \frac{r_0}{2}\left(\frac{mv}{ei}\right)^{\frac{1}{2}} \int\limits_{R=1}^{R=R_1} \frac{dR}{(\ln R)^{\frac{1}{2}}}, \tag{5.38}$$

and with the initial current density $j = i/\pi r_0^2$

$$x = vt = \frac{1}{2}\left(\frac{v^3}{j\pi e/m}\right)^{\frac{1}{2}} \int\limits_{1}^{R_1} \frac{dR}{(\ln R)^{\frac{1}{2}}}. \tag{5.39}$$

By substituting $z = \ln R$ and integrating, the value of the integral is found to be approximately equal to R_1 for $R > 2$. Thus the spreading of the beam per unit length is

$$R/x \approx (j^{\frac{1}{2}}/v^{\frac{3}{2}})(4\pi e/m)^{\frac{1}{2}} \approx 8.10^{-5}j^{\frac{1}{2}}/V^{\frac{3}{4}}, \tag{5.40}$$

where V is the equivalent potential difference, both j and V being in e.s.u. The result shows that the spreading of electrons moving in vacuum is proportional to $j^{\frac{1}{2}}$ and $V^{-\frac{3}{4}}$. For more exact calculations the value of the integral as a function of its upper limit R_1 is (139):

$R_1 =$	1	2	3	4	6	8	10	15
	0	2·1	3·2	4·1	5·7	7·1	8·5	11·6

For example, the initial radius of a beam of current density

$$j = 10^{-2}\,\text{A/cm}^2 = 3.10^7\,\text{e.s.u.}$$

with an accelerating voltage $V = 10^4$ volts $= 33$ e.s.u. is increased by a factor 3 at a distance of $x = 100$ cm (139 a).

7. Mutual repulsion of ions in a gas

When ions of one sign are present in a gas and their mobility is constant, the variation of charge density with time can be easily found. First let us assume a sheath which may expand in the x-direction only. We have Poisson's equation

$$dX/dx = 4\pi\rho. \tag{5.41}$$

The drift velocity of ions is given by

$$v = kX, \tag{5.42}$$

and the equation of continuity is

$$-\partial\rho/\partial t = \partial(\rho v)/\partial x.$$

Taking ρ as independent of x we obtain by substitution

$$-\partial\rho/\partial t = k\rho(\partial X/\partial x) = 4\pi k\rho^2. \tag{5.43}$$

With the condition that for $t = 0$, $\rho = \rho_0$ we find

$$\frac{1}{\rho} - \frac{1}{\rho_0} = 4\pi kt. \tag{5.44}$$

Thus the charge density at a given point x varies as $1/t$ for large t; the rate of change of charge in the early phases, when $\rho \approx \rho_0$ in (5.44), is larger for higher mobilities and for larger initial charge densities. Because of (5.42) the results hold for charged particles which are in temperature equilibrium with the gas. The charge distribution does not enter here because $\partial\rho/\partial x$ was assumed to be small. From (5.44) it also follows that after a long time ρ_0 becomes numerically unimportant and $\rho t = $ const. In a gas of $p = 1$ mm Hg with $k = 10^3$ cm/sec per V/cm and $\rho = 10^8 e$, the time required to reduce the density to $10^6 e$ is about 6.10^{-4} sec.

8. MUTUAL ELECTROSTATIC REPULSION AND DIFFUSION

It has been pointed out before that allowing for self-repulsion leads to partial differential equations which are mathematically inconvenient, and this is even more true when repulsion and diffusion occur simultaneously. In the majority of cases, however, it is only necessary to know which of the two processes is the main one. The answer can be obtained in the following way. From kinetic theory the root mean square displacement of a molecule in one coordinate as a function of time is given by

$$\bar{R} = \sqrt{(2Dt)}. \tag{5.45}$$

If we take a plane sheath of charges confined to a space of thickness $2x_0$, at any point x outside the space charge we have

$$X = 2\pi\sigma, \tag{5.46}$$

where σ is the surface density of charge. The velocity of drift is $v = kX = 2\pi\sigma k$ and the path travelled is

$$x - x_0 = 2\pi k\sigma t = \bar{R}. \tag{5.47}$$

By equating (5.45) and (5.47) we find the surface density of charge for which the two processes are comparable, viz.

$$\sigma = \frac{1}{\pi k}\sqrt{(D/2t_1)}. \tag{5.48}$$

Let $\sigma = 2x_0\bar{\rho} = 2.1 \cdot 5.10^{-10}.10^8 = 0 \cdot 1$ e.s.u., $D = 4.10^{-2}$ cm²/sec at 1 atm, and $k = 3$ cm/sec per V/cm, then $t_1 \approx 2.10^{-2}$ sec.

This means that after $0 \cdot 02$ sec an ion has travelled the same distance by diffusion as by drift in its own electric field. The larger σ or ρ, the shorter is that time; for larger t self-repulsion essentially controls the motion (Fig. 73). Since Dp and kp are invariant, (5.48) shows that

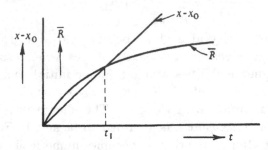

FIG. 73. Self-repulsion $(x-x_0)$ and diffusion distance $\bar{R}$ as a function of the time t.

$\sigma \propto \sqrt{(p/t)}$; therefore the time, at which repulsion produces the same displacement as diffusion, will rise with the gas pressure and, taking equal time intervals, self-repulsion becomes less important as p is increased. In the case of an expanding sphere of ions it can be shown similarly that repulsion exceeds diffusion for time intervals larger than

$$t \approx r_0^6 p(kp)^2/(Dp)^3, \tag{5.49}$$

where r_0 is the initial radius of the sphere. Thus this critical time is proportional to the pressure. This relation holds for the usual size of discharge vessel provided $N < 10^8$ ions/cm³. For large N the instantaneous radius of the sphere due to self-repulsion is given by

$$r^3 = 3kQt, \tag{5.50}$$

where Q is the constant charge present.

9. MEASUREMENTS

(a) *The coefficient of diffusion of ions.* A gas enters a chamber (Fig. 74) in which it is feebly ionized, for example by X-rays. The ions formed are separated by means of an electric field F normal to the flow of gas. The gas carries the ions of one sign through two narrow tubes; some of them will diffuse to the walls of the tube while the rest enter the right chamber where they are attracted by the collector C. The number of ions reaching C is measured by an electrometer. Theory (7, 11) shows that the number increases as the speed of flow of the gas V_1 and the tube radius a increase and the tube length l decreases. By using two sets of tubes of length l_1, l_2, D_i is found from the measured ratio of the number of ions collected in unit time

$$n_1/n_2 \approx e^{-c_1 D_i (l_2 - l_1)/a^2 V_1}, \tag{5.51}$$

where D_i is the diffusion coefficient and c_1 a known numerical constant ($\approx 3\cdot6$). This method is not very accurate. The most reliable results of D_i are obtained indirectly by measuring the ion mobility (Chapter 4).

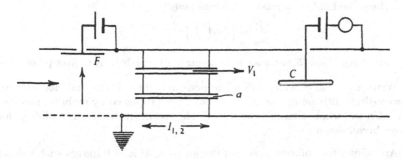

FIG. 74. Measurement of the diffusion coefficient D_i of ions in gases.

(b) *The coefficient of diffusion of electrons.* From a photo-electric cathode P electrons are released by suitable illumination (Fig. 75). The beam of electrons entering the right chamber through the slit S is driven

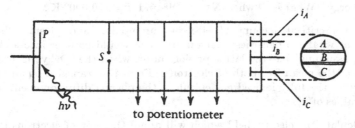

FIG. 75. Measurement of the electron diffusion coefficient D_e.

through the gas by a uniform axial electric field and falls on to the split anode. Because of diffusion the beam becomes wider by an amount which can be measured from the distribution of currents which reach the three parts of the anode. The spreading of the beam increases with increasing energy (random velocity) of the electrons and decreasing energy (temperature, random velocity) of the gas molecules. Thus, by observing the ratio (9, 10)

$$R = i_B/(i_A+i_B+i_C)$$

and comparing it with a theoretically derived value, the ratio of the mean electron energy to the molecular energy for a given value of X/p is found. (5.8) gives the coefficient of electron diffusion D_e. This method can also be used to determine D_i.

QUESTIONS

(1) Show analytically that in vacuum mutual repulsion of charged particles of constant mass forming a beam of cylindrical cross-section is balanced by electromagnetic attraction when the beam velocity $v = c$, the velocity of light.

Does the same hold for a relativistic mass $m(v)$

$$\frac{m(v)}{m_0} = \left\{ 1 - \left(\frac{v}{c}\right)^2 \right\}^{-\frac{1}{2}} ?$$

(2) Derive equation (5.6) for an infinite strip of width L from first principles.

(3) Write a critical essay on diffusion measurements of ions and electrons and show why direct diffusion measurements have been replaced by mobility measurements. Under what circumstances will the relation between mobility and diffusion break down?

(4) An infinite slab of ions (e, m^+) of thickness d_0 at $x = 0$ moves with velocity v in the x-direction through vacuum. Find the variation of d with x by self-repulsion.

(5) Derive from first principles the (Debye) distance D over which a finite field exists in a plasma with charge densities N^+ and N^- when a small number of electrons is displaced to one side of it, using one-coordinate geometry; equate the work done to move an electron through the space charge region to its mean kinetic energy. What is D when $N^- = 10^{10}$ and $T = 10\,000°$ K?

(6) Give a critical account of electron, ambipolar, and plasma diffusion. Estimate at what ion densities electron diffusion would go over into ambipolar diffusion, assuming that the latter predominates when the 'Debye-length' is of the order of the mean free path of electrons. From the transport equations and the theory of the fully ionized gas calculate the plasma diffusion coefficient for various values of N_e.

(7) Calculate the electric field which will annul the flow of electrons in a gas in which a constant electron concentration gradient is maintained. Find its value in H_2 at $p = 1$ mm Hg when $N_e = 10^{10}$ cm^{-3} at $x = 0$ decreases linearly to zero at $x = 0.1$ cm and $T_e = 10^4$ °K. What conclusion can be drawn from this example?

(8) Explain in physical terms the cause of the apparent increase in gas pressure when a magnetic field is applied to electrons diffusing through a gas. Complete (5.27) and find the change in p when $p = 10^{-2}$ mm Hg, $\lambda_1 = 10^{-2}$ cm, $H = 10$ Oe, $kT_e = 3$ eV? Also find the corresponding change in the diffusion coefficient.

(9) Metastable atoms in a gas can be destroyed by collisions in the gas and by wall collisions. The total decay rate can be described by a time constant τ, τ_v and τ_w being the corresponding values for volume or wall collisions only. Derive the equation for τ in terms of d, the distance between parallel walls, the gas pressure p, etc., assuming an infinite slab of gas. Find the condition for a possible stationary value in τ, the mean life of the metastables.

(10) Show why the arguments leading to (5.27) give a qualitatively right but quantitatively wrong result.

6

RECOMBINATION

A. RECOMBINATION BETWEEN ION AND ION

1. GENERAL REMARKS

When ions of unlike sign move at random in a gas there is a certain chance that a positive ion will collide with a negative one. In this case the negative ion with its loosely bound electron sheds its charge which is received by the positive ion. The result is that two neutral molecules or atoms are formed (Fig. 76).

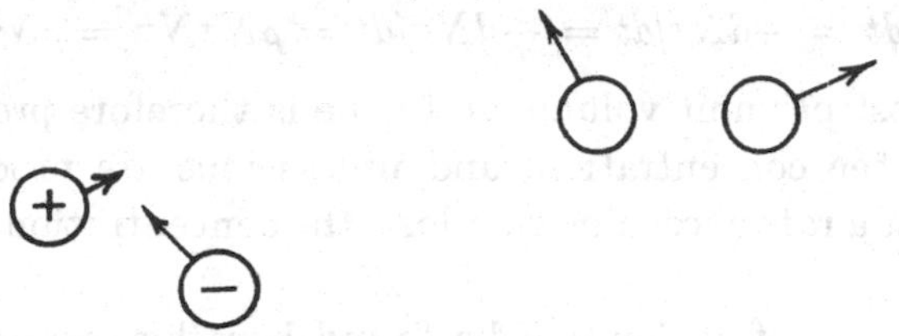

FIG. 76. Recombination of ions.

The energy which is evolved by a recombination of slow ions is approximately equal to ionization energy minus electron affinity, which is of course the sum of energies required to produce the positive and negative ion respectively from the neutral molecule. The recombination energy can appear as a rise in the kinetic energy of the two molecules; it can appear as excitation energy or as radiation emitted during the actual recombination period. The increase in kinetic energy is probably a rare process; the reason is that linear and angular momentum and spin must be conserved. The recombination spectrum of ions has a series limit which lies usually in the far ultraviolet. Only those recombination processes will be treated here in which singly charged ions are involved.

It can be easily shown that ion recombination particularly at higher pressure is controlled by the electric field of the charges on the ions. The number of collisions between uncharged neutral molecules equals concentration $N \times$ collision frequency v/λ per cm³ sec which at $p = 1$ atm is of the order $10^{-11}N^2$, whereas the number of collisions between ions of the same concentration is $10^{-6}N^2$.

The probability of recombination depends on the relative speed of the ions and decreases the faster the ions move relatively to one another.

This is because the time interval during which the particles are close together and which is available for neutralization decreases with increasing speed. It is unlikely that the probability will depend much on the nature of the molecule, for neither ionization potentials nor excitation levels vary in order of magnitude with the type of molecule. However, the probability of recombination increases when the effective diameter of the ions increases and should be larger for excited ions.

Two ions may approach one another for various reasons. Either their thermal or random motion brings them together or they are attracted by their electrostatic field. The number of collisions which lead to neutralization is proportional to the concentrations of positive and negative ions, that is $dz \propto N^+N^- dt$, and with the recombination coefficient ρ in cm^3/sec the rate of recombination is, when $N^+ = N^-$:

$$dz/dt = -dN^+/dt = -dN^-/dt = \rho N^+N^- = \rho N^2. \qquad (6.1)$$

The number lost per unit volume and time is therefore proportional to the square of the concentration, and unless ions are produced in the same volume at a rate exceeding that loss, the concentration will decrease with time.

The average life of an ion can be found here by assuming that the rate of recombination is just balanced by an equal rate of ion production

$$dN/dt = \rho N^2. \qquad (6.2)$$

The average life τ of an ion at the concentration N_0 follows from (6.2) by writing $dN/dt = N_0/\tau$ or

$$\tau = 1/\rho N_0. \qquad (6.3)$$

Hence the 'recombination life' of an ion at 1 atmosphere with $\rho \approx 10^{-6}$ and $N_0 = N^+ = N^- = 10^6/cm^3$ is of the order 1 sec.

2. Recombination at high gas pressure

What is termed high pressure here is about 1 atmosphere and above. We assume that two ions with unlike charges move at random through their gas before they meet by chance. If the distance between them becomes smaller than the distance between either of them and some other ion, that is if they move under the influence of the electric field of their own charges only, then they attract one another with a force corresponding to their field

$$X = e/r^2, \qquad (6.4)$$

where r is the instantaneous distance. Since pressure and hence collision frequency are large, the relative speed v_r between the ions will be small compared with the thermal speed and is given approximately by

$$v_r = (k^+ + k^-)X = (k^+ + k^-)e/r^2 = -dr/dt. \qquad (6.5)$$

By integration we obtain the time of approach t_1, the limits being r_0 and zero. The initial distance r_0 can be estimated from the average distance between two ions when the ions are assumed to be uniformly distributed throughout the volume. Taking, for example, a sphere filled with ions of concentration N per unit volume, the volume per ion pair is

$$1/N = 4\pi r_0^3/3. \qquad (6.6)$$

Integrating (6.5) with r_0 from (6.6) and equating it to the average life of the ion as given by (6.3), we obtain with $\rho = \rho_i$

$$\tau = t_1 = 1/4\pi Ne(k^+ + k^-) = 1/\rho_i N. \qquad (6.7)$$

Thus the recombination coefficient is

$$\rho_i = 4\pi e(k^+ + k^-). \qquad (6.8)$$

We conclude that in the region of high pressure the ratio of recombination coefficient and mobility is a constant. Since the mobilities are inversely proportional to the density it also follows that $\rho_i \delta = $ constant. The observed values of ρ_i are in satisfactory agreement with the theory proposed by Langevin. The right-hand branch of the curve in Fig. 77 confirms the predicted dependence of ρ_i on p. The dependence of the recombination coefficient on the nature of the gas follows from its mobility (Chapter 4). For example, O_2 ions at 1 atmosphere (Table 4.1) have a mobility $k^+ + k^- = 3 \cdot 2$ and hence $\rho_i = 5 \cdot 7 . 10^{-6}$ as compared with $\rho_i = 1 \cdot 5 . 10^{-6}$ measured. This agreement is very satisfactory taking into account the uncertainty of (6.8). If, for example, instead of a sphere a cube had been chosen, the volume per ion pair and ρ_i would have been smaller by a factor 4. Another source of uncertainty is that sometimes the negative ions lose their electron for a short time with the result that the measured negative ion mobility appears to be a little higher. There may also be a deviation from the assumed constant mobility which comes into effect when the distance between two unlike ions becomes smaller than 10^2 to 10^3 atomic diameters.

The recombination energy of the molecular oxygen ions is about $12 \cdot 2 - 1 = 11 \cdot 2$ eV, that is the difference between ionization and affinity energy. If we assume that recombination is followed by dissociation into excited species—a process which has a much higher probability than one in which the excess energy is converted into kinetic energy—we would expect to find as final products $(O + O^*) + O_2^*$. This is due to an elevation of one molecule into an excited level ($^3\Sigma_u^-$) requiring about

9 eV, whereby the molecule dissociates (7 eV) subsequently into an un-excited (^3P) and an excited atom (^1D, 2 eV), while the second molecule is excited to a lower level ($^1\Sigma_g^+$) subsequently emitting radiation of about 2 eV, or this amount is imparted to the molecules as kinetic energy. Observations show that this is a likely mechanism (84).

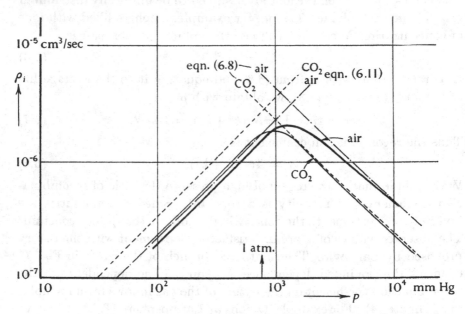

FIG. 77. Recombination coefficient ρ_i for ions in different gases as a function of the pressure p (17).

Table 6.1 shows some values of ρ_i at 1 atmosphere for different gases.

TABLE 6.1

Recombination coefficient ρ_i of ions in their own gas (1)

$p = 760$ mm Hg; $T = 0°$ C; age of ions ≈ 0.1 sec

Gas			ρ_i in units of 10^{-6} cm³/sec
O_2	.	.	1.5 (0.3 in cavity resonator)
Air	.	.	1.4
CO	.	.	0.85
CO_2	.	.	1.7
SO_2	.	.	1.4
N_2O	.	.	1.4
H_2O	.	.	~ 0.9 at 100° C

3. RECOMBINATION AT LOW GAS PRESSURE

At high pressure the motion of ions is governed by the drift velocity which they acquire in their mutual electrostatic field. Since their collision frequency is high, the velocity at any point is in equilibrium with the field. However, at low pressure the collision frequency and losses in collisions are low and therefore the relative velocity of ions is much higher. Because of these higher ionic velocities a large number of encounters between ions of opposite signs occur which do not lead to recombination. Collision without recombination will take place when the direction and magnitude of the ion velocities is such that their attracting field is not powerful enough to deflect the ions and make them move closely together over a sufficient distance. If, however, the ions strike a neutral gas molecule while they are fairly close together and are thus slowed down nearly to thermal velocity, then the electric field of the

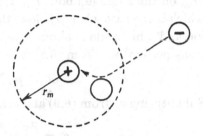

FIG. 78. Recombination of ions in the presence of a third body.

ions will keep them together for a sufficient time for recombination to occur (Fig. 78).

If we assume such a process of transfer of ionic energy to the neutral gas of temperature T, then recombination should only take place if average kinetic energy of the ion after collision with the molecule is smaller than the electrostatic energy of the ion pair. This means that

$$\tfrac{3}{2}kT < e^2/r \quad \text{or} \quad r_m \leqslant 2e^2/3kT, \tag{6.9}$$

where r_m is the largest permissible distance of approach for ions to recombine; it is of the order 10^{-6} cm or 100 atomic diameters.

Thomson's theory (11), which is given below, is therefore based on the concept that two ions are only capable of recombination if an uncharged molecule or atom is sufficiently near, a process which is often referred to as a three-body collision.

According to Fig. 78 the positive ion is assumed to be at the centre of a sphere of radius r_m, the negative ion crosses the boundary, suffers a collision with the neutral molecule within the sphere, thereby losing energy, and now neutralizes its charge with the positive ion. The number of times per second that a negative ion enters the sphere of radius r_m (neglecting the radius of the molecule) is qv_rN, where

$$q = \pi r_m^2, \qquad v_r = \left\{\left(\frac{1}{m^+}+\frac{1}{m^-}\right)3kT_i\right\}^{\frac{1}{2}}$$

is the relative velocity of the ions with the masses m^+ and m^-, and T_i the temperature of the ions which is assumed to be above gas temperature.

The expression $\left(\dfrac{1}{m^+} + \dfrac{1}{m^-}\right)$ is the inverse reduced mass. Since there are also N positive ions per unit volume, the total number of collisions between ions is N times the above value. However, not every ion entering the sphere causes recombination. Assuming a straight path of the ion, the length of path within the sphere will be between zero and $2r_m$, on the average about r_m. If the mean free path of an ion $\lambda_i \gg r_m$, which is the case when p is low, the probability of a collision between an ion and a molecule is about r_m/λ_i, and hence the number of recombinations per cm^3/sec from (6.2) is

$$dN/dt = qv_r N^2 r_m/\lambda_i = -\rho_i N^2. \qquad (6.10)$$

Substituting r_m from (6.9) and v_r and q from above with $m^+ = m^- = M^\pm$

$$\rho_i = \frac{8\pi}{27} \frac{e^6}{(kT)^3} \frac{1}{\lambda_i} \sqrt{(6kT_i/M)}. \qquad (6.11)$$

Since $\lambda_i = \sqrt{2}\lambda \propto 1/p$, (6.11) shows that for $T = $ constant we have $\rho_i/p = $ constant. ρ_i depends strongly on T, viz. $\rho_i/p \propto T^{-\frac{5}{2}}$. The value of T_i is not clearly defined, but it is not likely to be very much larger than T. ρ_i is smaller for ions of larger mass. (6.11) has been confirmed experimentally at pressures below atmospheric and at temperatures above room temperature. For example, the recombination coefficient of oxygen molecular ions at $p = 100$ mm Hg and room temperature ($M = 32.1\cdot66.10^{-24}$ g, $T_i \sim T \sim 300°$ K, $\lambda_i = 4\cdot9.10^{-3}.\sqrt{2}$ cm at $p = 1$, $e = 4\cdot8.10^{-10}$ e.s.u., $k = 1\cdot37.10^{-16}$ erg$/°$K) follows from (6.11) as $\rho_i = 2\cdot3.10^{-7}$.

Fig. 77 shows the results from theory and experiment of ρ_i as a function of p in air and CO_2 with $T = $ constant which shows satisfactory agreement. (6.11) also indicates that at low pressure the recombination process is closely related to diffusion. Since $D_i \propto v_i\lambda_i \propto \sqrt{(T_i/m)}\lambda_i$ we obtain from (6.11)

$$\rho_i/p \propto (D_i p)/T^3. \qquad (6.12)$$

At given temperature the value of ρ_i at unit pressure is thus increased as the diffusion coefficient of the ions at $p = 1$ is increased. As long as $\lambda_i \gg r_m$, λ_i can be taken to be independent of T.

As the energy liberated in the recombination process is shared by three bodies, the energy is not likely to reappear as kinetic energy but rather, if circumstances permit, as excitation and dissociation energy of the molecules.

It should be added that there are other recombination processes which do not require the presence of a third body and for which ρ_i is independent of p (18). If the neutralization energy is emitted as a quantum, a value of ρ_i below 10^{-14} cm³/sec is to be expected, but if the two molecules or atoms can carry away the energy as excitation energy $\rho_i \approx 10^{-8}$ cm³/sec. These 'spontaneous' recombination processes become significant only at low (and perhaps at high) pressure, as can be seen from Fig. 79.

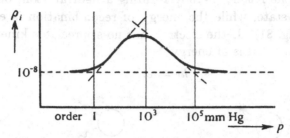

FIG. 79. Variation of the ion recombination coefficient ρ_i with gas pressure p assuming spontaneous recombination.

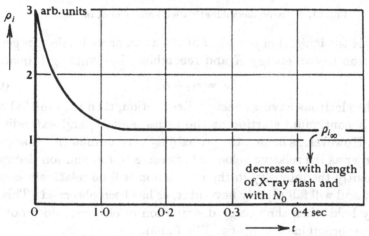

FIG. 80. Variation of the ion recombination coefficient ρ_i with time t (140).

Finally there is definite experimental evidence that ρ_i depends on the time interval between the cessation of ionization and the measurement (Fig. 80). First the ions are close together and ρ_i is large, ρ_i for $t \to \infty$ decreases with increasing length of the ionizing flash and higher intensity. This is thought to be connected with the formation of heavier ions (O_3).

B. RECOMBINATION BETWEEN IONS AND ELECTRONS

4. GENERAL REMARKS

Consider a gas which contains neutral molecules and equal numbers of ions and electrons of concentration N; the concentration of charges is small compared with that of the molecules. We define the electron recombination coefficient by

$$-dN/dt = \rho_e N^2. \tag{6.13}$$

As in the case of ion–ion recombination various neutralization processes are possible. For example, the electron can approach the ion and be caught by the latter, thereby forming a neutral atom or molecule in the ground state, while the energy of recombination is emitted as a quantum (Fig. 81). If the electron has no appreciable kinetic energy the quantum emitted is of energy

$$h\nu = eV_i, \tag{6.14}$$

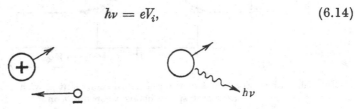

FIG. 81. Electron recombination with emission of radiation.

where V_i is the ionization potential of the atom or molecule. In general the electron has an energy K and recombines by emitting a quantum of energy

$$h\nu = eV_i + K. \tag{6.15}$$

When the electrons have an energy distribution, then the emitted spectrum is a continuum starting at the series limit eV_i and extending to infinity. However, as before, the probability of recombination is assumed to be larger as the relative velocity between electron and ion decreases; consequently the intensity in the continuum will be relatively large at the limit and will fall rapidly beyond it, as has been observed. This rule will only hold when the energy distribution of electrons does not outweigh the recombination probability function $\rho_e = f(K)$.

Recombination can also occur in stages (20): the electron is caught by the ion and forms a highly excited atom; the energy $e(V_i - V_e)$ is emitted, V_e being the excitation potential of a particular state. Then, for example, the electron falls into the ground state and a second quantum of energy eV_e is emitted. In this case light of energy below that at the series limit is produced. Fig. 82 shows an example. The recombination coefficient for the capture of an electron into an upper level of the atom will be designated ρ_e^*.

The recombination energy can be used wholly or partly to dissociate molecules, it can be converted into vibrational or kinetic energy or produce chemical reactions. Table 6.2 gives the results of observations.

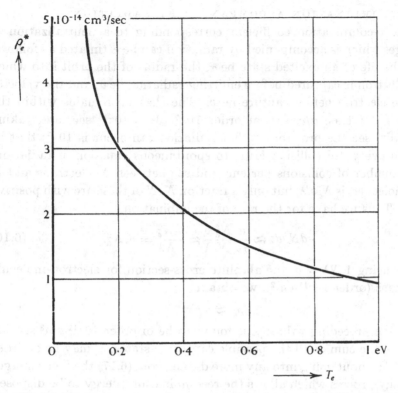

FIG. 82. Calculated spontaneous electron recombination into the 2 ^2P state of Cs. Coefficient ρ_e^* as a function of the electron temperature T_e. 1 eV = 11 600° K.

TABLE 6.2

Recombination coefficient of electrons ρ_e in various gases (13, 135b, 143)

Gas				p mm Hg	T_e eV	ρ_e cm³/sec
H	.	.	.	..	0·03	10^{-11}
He	.	.	.	order 1	0·03	$1·7.10^{-8}$
Ne	.	.	.	15–30	0·03	2.10^{-7}
A .	.	.	.	15–30	0·03	3.10^{-7}
H₂	.	.	.	3–12	0·03	$< 3.10^{-8}$
N₂·	.	.	.	2–5	0·03	$10^{-7}(?)$
O₂·	.	.	.	2–20	0·03	$2·7.10^{-7}$
Cs .	.	.	.	0·01–0·1	0·15	$3–4.10^{-10}$
Hg	.	.	.	0·3	0·15	$\sim 2.10^{-10}\,(10^{-7})$

In rare gases ρ_e increases with the gas density at $T < 300°$ K; $\rho_e \propto \sqrt{p}$.

They seem to fall into two groups: those giving ρ_e of order 10^{-10} sec and are associated with atomic ions (section 5) and those giving values of $> 10^{-8}$ sec apparently connected with molecular ions (section 6).

5. Recombination accompanied by radiation

The recombination coefficient corresponding to a neutralization of charges which is accompanied by radiation can be estimated as follows: Let the life of an excited state be τ, the radius of the orbit into which the electron is captured before emitting radiation be a, and the velocity of the electron before capture be v. The electron remains within the atom for a time a/v, say of order $10^{-7}/10^8 = 10^{-15}$ sec, and taking $\tau \sim 10^{-8}$ sec the probability that radiation can occur is 10^{-7}; that is, one in every 10^7 collision leads to spontaneous emission of radiation. The number of collisions per cm³ and sec between N_e electrons and N gas molecules is $N_e v/\lambda$, but only a fraction N_e/N of these are with positive ions. Thus we have for the rate of recombination

$$-dN_e/dt \approx \frac{a}{\tau v} \frac{N_e^2}{N} \frac{v}{\lambda} \approx \frac{aN_e^2}{\tau N \lambda} = \rho_e N_e^2. \qquad (6.16)$$

Introducing $1/N\lambda = q$, the absolute cross-section for electron-molecule collisions (order 10^{-15} cm²), we obtain

$$\rho_e \approx aq/\tau. \qquad (6.17)$$

With the preceding values ρ_e is found to be of order 10^{-14} cm³/sec. If we take the sum of all the possible capturing states, ρ_e may be of order 10^{-13}. Without going into any more detail, from (6.17) the fact emerges that any process which allows the recombination energy to be disposed of in a short time will make ρ_e large. Also it looks as if ρ_e should be independent of the speed of the electron because an increase in v reduces the time the electron is near the ion in the same proportion as it increases the number of times the electron meets an ion. This is not the case (Fig. 82); we surmise that q depends on v. Early observations have shown that ρ_e decreases rapidly as v is raised (144 a) and recent work using 'pulse-heating' of the electron gas have confirmed it (124 b). ρ_e seems to vary inversely as $T_e^{\frac{3}{2}}$. The process just described is sometimes called radiative recombination (141). See Appendix 4.

Another type of recombination makes use of double excitation (page 49). [Readers may have heard about the Auger effect, Shenstone effect, auto-excitation, excitation into states of positive energy, excitation of negative energy terms, internal conversion of energy, etc. All these are based on the same simple idea.] Consider a free electron approaching the positive ion. It is captured into one of the upper levels. The surplus energy (= recombination minus excitation energy) is here used to excite a second electron of the same atom into a higher level. The result is a doubly excited atom with two electrons in different levels. It can either lose part of its energy by radiation of a quantum, thereby forming a singly excited atom, or lose all the excitation energy and the electron and become an ion again. The recombination coefficient depends on the speed with which the first process will take place. ρ_e is estimated to be $\sim 10^{-12}$ cm^3/sec. This process is called dielectronic recombination. It is the reverse of Fig. 28a (144d).

6. RECOMBINATION OF MOLECULAR IONS AND ELECTRONS

There are recombination processes which seem to lead to much higher values of ρ_e. It was shown (144) that ρ_e is expected to be large when the time for disposal of recombination energy becomes considerably smaller

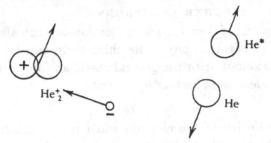

FIG. 83. Recombination between a helium molecular ion and an electron leading to dissociation into He atoms one of which is excited.

than the average life of an excited atom. However, the dissociation of a molecule occurs within a time interval of one vibration, that is $\sim 10^{-13}$ sec, while according to the Frank–Condon rule the electron changes from one to another excited state in a molecule within a time which is negligible compared to the period of vibration.

We have already shown in section 2 that $O_2^+ + O_2^-$ may give excited atoms and molecules after having exchanged the electron. Here similarly $O_2^+ + e$ may lead to the formation of an excited unstable molecule O_2^*. (The two O atoms repel one another, and separate with a speed of the order $10^{-8}/10^{-13} = 10^5$ cm/sec.) The excitation energy of O_2^* of about 12·2 eV is now shared between the two atoms O+O. For example, they could take up 2 and 4 eV, as can be seen from the level diagram of O, 5 eV being used to dissociate the unexcited molecule and ~ 1 eV going into kinetic energy.

This 'dissociative' recombination seems to explain the large values of electron recombination coefficients ($\rho_e \sim 10^{-8}$) and more which have been found in molecular as well as in rare gases. For example, it has been suggested that the He molecular ions (Fig. 83) recombine with electrons according to

$$He_2^+ + e \rightarrow He^* + He$$

forming two He atoms. He_2^+ has been found to be the principal ion in discharges at ordinary p (see Chapter 4). Since about 25 eV are available as recombination energy, one excited He atom (20 eV) is formed and one in the ground state; the excess energy is shared between dissociation energy and kinetic energy of the atoms. Rough theoretical estimates put the value of ρ_e to $\sim 10^{-7}$ cm^3/sec, though this value ought to be accepted with caution (141). Recent work indicates values of dissociative recombination coefficients in He, Ne, A, Kr, and Xe of 4.10^{-9}, 2.10^{-7}, $6\cdot7.10^{-7}$, $1\cdot2.10^{-6}$, and $1\cdot4.10^{-6}$, respectively (144c).

7. ELECTRON RECOMBINATION IN THE PRESENCE OF ATOMS, MOLECULES, OR OTHER ELECTRONS

The recombination of electrons and molecular ions can also take place in the presence of a third body. The third body can be an atom or molecule either excited or in the ground state, an ion, or an electron. Using the same ideas as in section 5, we write

$$\rho_e = \bar{c}q_0, \tag{6.18}$$

where $\bar{c}$ is the velocity of the electron when it has lost its energy in collision with the third body (assumed to be an atom or a molecule) and q_0 the cross-section for recombination. We have, as in section 3,

$$q_0 = \pi r_0^2 r_0/\lambda', \tag{6.19}$$

where λ' is the effective mean free path taken approximately equal to λ_e/κ; $\kappa = 2m/M$ for atoms but κ is usually several orders of magnitude larger for molecules (Fig. 63). Inserting (6.19) in (6.18) and expressing $\bar{c} = f(T)$, we obtain

$$\rho_e \approx \pi r_0^3(\kappa/\lambda_e)\sqrt{(3kT/m)}, \tag{6.20}$$

or by comparing with (6.9) and (6.11)

$$\rho_e \approx \rho_i \kappa^{\frac{1}{2}}. \tag{6.21}$$

For molecules, $\kappa \sim 10^{-2}$ and so $\rho_e \sim 0\cdot1\rho_i$. Here $\rho_e \propto p$. Experiments show coefficients of that order of magnitude, but their dependence on pressure, energy, and temperature have not yet been sufficiently tested.

An electron recombination process which becomes numerically significant at high electron densities is, for example,

$$H^+ + 2e \rightarrow H^* + e,$$

since the balancing of momentum and energy is greatly facilitated in the presence of a third body of small mass. H* is probably the lowest resonance state of H at ≈ 10 eV. It has been calculated that ρ_e increases with N_e when $N_c > 10^{12}$ cm^{-3}. Thus radiative recombination gives for $kT_e = 1\cdot4$ eV and $N_e < 10^{10}$ cm^{-3}, $\rho_e = 3.10^{-13}$ cm^3/sec but 2-electron recombination for $N_e = 10^{15}$ and 10^{18} cm^{-3} gives $\rho_e = 10^{-12}$ and 10^{-10} cm^3/sec respectively (100 g, 144 b). ρ_e rises as kT_e decreases.

8. THE MEASUREMENT OF THE RECOMBINATION COEFFICIENT

(a) *Ion–ion recombination coefficient.* (i) In earlier methods ρ_i is obtained in principle from the balance between the rate of ionization and of loss by recombination (25 a), viz.

$$(dN/dt)_{\text{ioniz}} = \rho_i N^2. \tag{6.22}$$

The equilibrium value N is found by measuring the saturation current and dN/dt from the slope of the current/voltage curve (Chapter 2). The gas in a parallel plate condenser is continuously irradiated with X-rays. Stray radiation falling on walls or electrodes has to be carefully avoided because all solids represent a copious source of photo-electrons. To find N the irradiated volume must be determined, for example, by photographing the cross-section of the X-ray beam.

The interpretation of the results is difficult for the following reasons: Ions of different age, and hence complexity, are present. The ionization is not uniform and thus the rate of recombination is larger where N is higher and vice versa (Fig. 80). Diffusion to the walls cannot be excluded.

An example of a modern method is now given.

(ii) ρ_i can be derived from a measurement of the rate of change of ionization. An X-ray tube (Fig. 84) with a large focal spot supplies constant radiation which passes through an ionization chamber for the measurement of the intensity of the beam, and the aperture of a rotating disk which produces X-ray flashes, and finally enters the recombination chamber which is sufficiently far away so that uniform intensity is obtained in it. The cycle of operation is as follows. The rotating disk opens and closes the aperture for the beam and after t seconds a sweep field (V_s) is applied which removes the ions. In order to avoid inductive changes in potential of the electrometer during the sweep period, a dummy chamber C_d is used. During the sweep time, C_r receives the charge of the ions and during the subsequent interval receives an equal and opposite charge from V_P through R by adjusting P, indicated by zero deflexion on the electrometer E.

ρ_i is found by integrating (6.2) with $N = N_0$ at $t = 0$, giving

$$1/N - 1/N_0 = \rho_i t. \tag{6.23}$$

ρ_i is known if N, N_0, and t are found. N and N_0, the concentrations t and zero seconds after the flash ceases, are obtained by applying the sweep field and measuring (by means of P) the current due to the ions. The latter is equal to N times the volume of C_r times the number of flashes per second. The time interval t is given by the frequency of revolution of the disk and the angle between aperture and contact S; it can be varied by changing the relative position of disk aperture and contact.

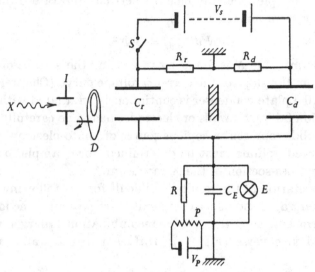

FIG. 84. Measurement of the ion–ion recombination coefficient. The intensity of the X-rays from the tube X is measured in the ionization chamber I.

(b) *Ion–electron recombination coefficient*. A spectro-photometric method has been devised to find ρ_e in the afterglow of discharges in metal vapours. (6.23) can be written

$$1/N_2 - 1/N_1 = \rho_e(t_2 - t_1). \qquad (6.24)$$

Assuming that the electric energy which is supplied to the discharge is cut off at $t = 0$ and that the following emission of light is caused by recombination only, then the intensity of light at a time t from a certain small wavelength range must be proportional both to the number of ions and of electrons present, viz.

$$I_\lambda = c_1 N^2. \qquad (6.25)$$

The constant c_1 can be found, for example, by measuring simultaneously I_λ and N_0 in the steady state; N_0 is obtained from a probe measurement (see Appendix 5). This method is confined to large concentrations of N (10^{10} to $10^{13}/cm^3$), as present in decaying discharge plasmas.

ρ_e can be obtained too from instantaneous probe measurements. Two probes are used simultaneously; one is made positive and one negative with respect to the space. The first probe gives the electron temperature T_e and the space potential, the second the electron current at space potential which combined with T_e gives $N^+ = N^- = N$. From the measurements of N at two different known instances, ρ_e is found by means of (6.24). Again ρ_e can only be determined for large values of N.

The charge concentration of decaying plasma contained in a resonant cavity can be determined by observing the change of resonance frequency which is proportional to N. The applied frequency and the geometry of the cavity determine the constant of proportionality. A probe signal of a certain frequency is transmitted most strongly at a time when the instantaneous value of the resonance frequency of the cavity is equal to that of the signal. In this way N as a function of the time is found, the time being measured by means of a calibrated time base of an oscilloscope.

The measurement of ρ_e becomes increasingly difficult at $p > 100$ mm Hg because of the attachment of electrons to impurity molecules. These form negative ions which recombine more readily and a large apparent recombination coefficient results. It has to be remembered that any electron attachment to impurity molecules will render too large apparent values of ρ_e whereas any process of ionization occurring concurrently with recombination will make ρ_e smaller than the true values.

QUESTIONS

(1) Write an essay on the collision processes associated with ion–ion recombination at low and high pressure. Show whether these conform with the principles of similarity. Find an expression for the recombination cross-section, compare the absolute values with other relevant collision cross-sections, and comment on it.

(2) Discuss the radiative electron–ion recombination process and include recombination via excited states. If $\rho_e = 10^{-11}$ cm^3/sec, calculate the partial pressure of impurity molecules, all forming negative ions O_2^- in N_2, which would give a rate of attachment dN/dt equal to that of electrons recombining with ions. Assume $p = 10$ mm Hg, $kT_e \approx 0.1$ eV, $N_e = 10^{10}$/cm^3, and an average attachment cross-section of $q_a \approx 10^{-19}$ cm^2. Comment on what the result means in terms of vacuum technique requirements for experiments of that type.

(3) If the formation of negative ions of atomic hydrogen in H_2 would require attachment of electrons of average energy 1 eV (Maxwellian) with the average cross-section $\overline{q_a}$ from Fig. 43, calculate the time which elapses to reduce the electron concentration to $1/e$ of its initial value if 1 per cent H is present. Compare the result with the corresponding time interval when attachment could take place in every collision between an electron and a hydrogen molecule at 1 mm Hg. This result indicates that the effect of impurities reveals itself only after a finite time.

(4) Write an essay on dissociative recombination and review critically the experimental work from which high values of recombination coefficients have been erroneously inferred.

(5) Give a brief account of recent experiments which support the view that radiative recombination is initiated by short pulses of an electric field which increase the mean electron energy and reduce recombination rate and light emission.

(6) Given a semi-infinite slab of gas (plasma) with ions and electrons of constant concentration $N_0 = 10^{10}/cm^3$. Its boundary is crossed by charges of both signs which enter the adjoining gas by ambipolar diffusion where they are neutralized by recombination. With the plane boundary at $x = 0$, find the spatial distribution of N using the condition $dN/dx = 0$ for $N = 0$. Find the general expression and approximation for small and large values of x and calculate for a gas density of 100 mm Hg $kT_e = 0\cdot03$ eV, $\mu_i^+ = 10^3$ cm^2/V sec, $\rho_e = 1.10^{-10}$ cm^3/sec the distance at which the charge density has reached $0\cdot1\ N_0$. Under what physical conditions could this case arise?

(7) A large vessel with plane walls $2R$ apart contains gas which is ionized at a rate $(c.N)$. Charges move by ambipolar diffusion (D_a) towards the walls where they are instantaneously neutralized but some recombine on the way (ρ). Derive an expression for $N(x)$ when ρ is small and $N(R) = 0$, $N(0) = N_0$. Discuss the result in physical terms.

(8) Write an essay on your ideas how in the steady state a long ionized cylindrical column of a glow discharge loses its charges when the gas pressure is of the order 1 atm. Consider carefully the fundamental processes which are effective and give reasons for invoking them.

7

IONIZATION AND EXCITATION IN AN ELECTRIC FIELD

1. IONIZATION IN UNIFORM FIELDS

When an electron or a positive ion moves through a gas which does not form negative ions, it is likely to excite or ionize atoms or molecules by collisions provided its energy exceeds the corresponding critical values. Such an inelastic collision can take place either in a field-free space or in an electric field. In the former case new electrons and ions or excited

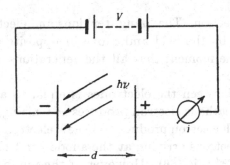

FIG. 85. Circuit for measuring the multiplication of current, the initial current being produced by a photo-cathode irradiated by light.

atoms are produced which remain for a short time in the path of the primary particle; but the collision products will not be found, and inelastic impacts do not occur, beyond the 'range' of the primary particle. In the presence of an electric field, however, excited atoms as well as new electrons and ions are found throughout the gas together with the primary particle; they may be distributed over a region which is bounded by the electrodes and the walls confining the space. The more important aspect of this process is, however, that electrons and ions formed by impact are being driven by the electric field to the respective electrodes and collide on their way with gas molecules and thereby produce new ions and electrons; the excited atoms, on the other hand, move unaffected by the field. Thus one primary electron starting at a negative electrode can be multiplied in an electric field a great many times, a process which will be discussed now in more detail.

(a) *Multiplication by collisions in the gas*. Fig. 85 shows a discharge gap with two plane electrodes of infinite size, d cm apart connected to a source of V volts producing a uniform field in the interspace of value $X = V/d$. The gas pressure p is 1 to 100 mm Hg. By irradiating with ultraviolet light or otherwise, electrons can be liberated from the cathode. Let a 'primary' electron—which after numerous collisions has acquired equilibrium energy—produce at any point between the electrodes (except very near to the cathode) α electrons and α positive ions per cm of

FIG. 86. Multiplication of charges in an electric field.

its path in field direction. The 'daughter' ions and electrons are subsequently separated by the field and move in opposite directions. We shall assume for the moment that all the generations of the electrons ionize gas molecules.

If the distance between the electrodes is equal to a few electronic mean free paths, a cathode electron produces αd ionizing collisions along d cm and since each electron produces one new electron per impact, the total number of electrons arriving at the anode per 1 electron starting at the cathode is $2^{\alpha d}$ (Fig. 86). However, if the number of free paths in the gap is sufficiently large and distributed, we must treat ionization as occurring continuously throughout the space. Thus the increase in the number of ion pairs along an element of length dx is $\alpha\, dx$ per electron, and for N_x electrons at x we obtain

$$dN = N_x \alpha\, dx. \tag{7.1}$$

By integrating between $x = 0$ and $x = d$ we obtain for the number of electrons N at d or the current i at d

$$N/N_0 = i/i_0 = e^{\alpha d}, \tag{7.2}$$

where N_0 is the number of electrons per sec and i_0 the current produced by irradiation of the cathode. The coefficient α is often called Townsend's first ionization coefficient; its value depends on the field X, the pressure p, and the gas, as we shall show later. The multiplication process is accompanied by radial diffusion of charges. For this reason the ratio i/i_0 is not equal to the current density ratio j/j_0.

We notice here that with a constant field, and constant emission from the cathode, the current increase is greater for longer distances d. This is contrary to expectations on the basis of any picture which treats the gap as an ohmic resistance in the circuit.

The form of (7.2) is not changed by writing $\alpha d = (\alpha/p)(pd)$; the factor α/p implies that we deal with ionization by single collisions and hence ionization should be proportional to the gas density; the factor (pd) is proportional to the number of molecules contained in the gap. The use of such reduced parameters like X/p, α/p, (dp), etc., is of advantage since it shows at once whether a particular relation obeys the rule of similarity or not.

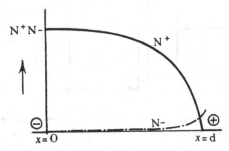

FIG. 87. Distribution in the gap of the number density of electrons N^- and ions N^+ in the steady state.

According to (7.2) the electrons liberated at the cathode are multiplied by ionizing collisions in an electric field. (To speak of an avalanche would be wrong since here we are treating steady states which exclude the time.) The electrons entering the anode are reckoned to constitute the current though their number, say, half-way between the electrodes, is very much lower. This apparent contradiction is because the picture given tells only half of the story: actually in a steady state the number of electrons reaching the anode in unit time is equal to the number of positive ions arriving at the cathode. This is because the ion concentration is larger and the velocity smaller than for electrons. Otherwise positive space charges would accumulate in the interspace and would distort and reduce the field. The spatial distribution of space charges in equilibrium is shown in Fig. 87. Another point is that for each electron starting from the cathode and for $e^{\alpha d}$ electrons reaching the anode, simultaneously $e^{\alpha d}-1$ positive ions arrive at the cathode; the difference is due to the fact that the first electron liberated at the cathode was taken to be unaccompanied by a positive ion. The spatial distribution of electrons has been recently verified experimentally (145b) by observing the far ultraviolet light emitted by electrons colliding with hydrogen molecules.

So far it has been assumed that an initial electron current is emitted by the cathode. Instead of irradiating the cathode to obtain a small number of electrons, the initial electrons can also be supplied by ionization in the gas. We assume with Fig. 88 that $dN/dt = N_0$ ion pairs per cm³ per sec are produced in the gas by irradiation (the cathode being shielded so that no photo-electrons are emitted). Counting x from the

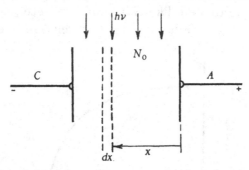

FIG. 88. Multiplication with a given rate of production of ions in the gas. Note that for convenience $x = 0$ is at the anode.

anode for convenience, each electron starting at x will be multiplied to $e^{\alpha x}$ electrons at A; radiation uniformly absorbed in the gas provides $N_0\,dx$ electrons in an element and thus the number of electrons arriving at A per cm² and sec is, neglecting lateral diffusion,

$$n = \int_0^d N_0\, e^{\alpha x}\, dx = (N_0/\alpha)(e^{\alpha d} - 1). \tag{7.3}$$

Since the saturation current in small fields ($X \to 0$) is $j_0 = eN_0 d$ (2.6) we obtain the multiplication factor

$$i/i_0 = n/(N_0 d) = (1/\alpha d)(e^{\alpha d} - 1). \tag{7.4}$$

This relation shows what would happen if an ionization chamber with plane electrodes were operated at a high potential, provided that the gas did not contain negative ions. It is clear that the factor i/i_0 will be smaller than in the case described before, because the initial electrons start this journey somewhere in the gap instead of at the cathode surface and experience on the average a smaller number of ionizing collisions.

(b) *Multiplication including secondary effects.* Up to now the positive ions have been regarded only as carriers of current; the current density j was taken to be so small that space charges could not distort the field. Furthermore, any secondary effects by the ions at the cathode have been neglected. How does this simple theory (7.2) compare with experiment? Fig. 89 shows i/i_0, the reduced value of the current, flowing through the circuit (Fig. 85) as a function of the electrode distance d in air at 4 mm Hg and a field of 700 V/cm. We find that the logarithm of the reduced current is a linear function of the distance up to about 0·5 cm and then rises faster than described by (7.2). The reason for the faster rise is

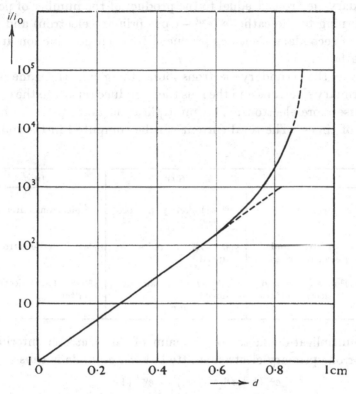

FIG. 89. Multiplication ratio i/i_0 of current densities as a function of the inter-electrode distance d for a constant reduced field $X/p = 700$ in air at $p = 4$ mm Hg (8); from (7.2): $\tan\theta = (1/d)\ln(i/i_0) = \alpha$, θ being the angle between the curve and the abscissa.

that secondary effects occur at the cathode. This has been shown by changing the material and surface properties of the cathode.

The secondary effect was originally associated with the positive ions which release secondary electrons by impinging on the cathode. We know today that the release of secondary electrons is due not only to the potential and the kinetic energy of the positive ions but also to other processes, for example to the arrival at the cathode of photons, and neutral and metastable particles (p. 79). The explanation for the deviation of the graph, Fig. 89, from a straight line is that at larger electrode distance the electrons which are released from the cathode consists of two groups; the primary ones which are liberated by irradiation from an external source forming a constant current of density j_0, and the secondary electron which for simplicity we shall regard here as being liberated by positive ions falling on the cathode. The number of secondary electrons is equal to the product of the number of positive ions returning to the cathode ($e^{\alpha d}-1$ per primary electron) and γ, the number of secondary electrons produced by each positive ion arriving at the cathode.

However, the secondary electrons ionize the gas in the same way as do the primary electrons and the ions they produce return to the cathode and release more electrons. This multiplication takes place an infinite number of times. The total current can be computed in the following manner:

Cathode	Gas	Anode
Cycle 1 1 electron starts	$(e^{\alpha d}-1)$ ion pairs produced	$e^{\alpha d}$ electrons enter
Cycle 2 $\begin{cases}(e^{\alpha d}-1)\text{ ions arrive and}\\ \gamma(e^{\alpha d}-1)\text{ electrons start}\end{cases}$	$\gamma(e^{\alpha d}-1)^2$ ion pairs produced	$\gamma(e^{\alpha d}-1)e^{\alpha d}$ electrons enter
Cycle 3 $\begin{cases}\gamma(e^{\alpha d}-1)^2\text{ ions arrive and}\\ \gamma^2(e^{\alpha d}-1)^2\text{ electrons start}\end{cases}$	$\gamma^2(e^{\alpha d}-1)^3$ ion pairs produced etc.	$\gamma^2(e^{\alpha d}-1)^2e^{\alpha d}$ electrons enter

The multiplication factor or the sum of all electrons entering the anode for one primary electron emitted at the cathode is thus

$$e^{\alpha d}(1+z+z^2+...) = e^{\alpha d}/(1-z), \tag{7.5}$$

where $z = \gamma(e^{\alpha d}-1)$.

With n_0 primary electrons/sec released at the cathode, ionization by electron collisions in the gas, and secondary electron emission by positive ions arriving at the cathode, the current at d is

$$i/i_0 = n/n_0 = \frac{e^{\alpha d}}{1-\gamma(e^{\alpha d}-1)}. \tag{7.6}$$

The experimental verification of this equation has been successful.

Returning to Fig. 89, α is found from the slope of the straight part to be about 8 ion pairs/cm and analysis of the curved portion (from 7.6) gives $\gamma \approx 8.10^{-4}$. This means that about 1000 ions release 1 secondary electron from the (Zn ?) cathode. For γ see page 94 ff. and (203 b).

It was thought for a long time that this type of multiplication applies only to relatively low pressures or large values of X/p (order 100).

Recent experiments with a Ni cathode in air and N_2 at about atmospheric pressure at $X/p = 40\text{–}45$ have revealed that (7.6) holds at large values of (pd) too. Fig. 90 gives a series of graphs for N_2 from which follow $\alpha \approx 6$ and $\gamma \approx 4.10^{-4}$ at $X/p = 45$. Comparison with Fig. 91

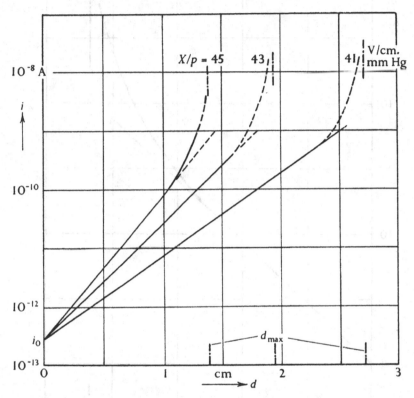

FIG. 90. Multiplication ratio i/i_0 as a function of the electrode separation d for different reduced fields X/p in nitrogen with Ni electrodes ($p = 300$ mm Hg) (147).

showing the multiplication for a Cu cathode and argon is instructive. Here the curve departs from the straight line at a lower value of d/d_{max} indicating a much larger value of γ in rare than in molecular gases. It is thought that this is due to photo-electric emission at the cathode (page 100 ff.).

The electrons which enter the anode have come to life at various points in the gap. It is permissible here to work out the current by adding all electrons entering the anode because with each electron an ion has come to life simultaneously and has migrated towards the cathode. This is equivalent to one charge moving through the whole distance d. Inspection of (7.6) shows that the current tends to infinity when the denominator approaches zero, provided this can take place

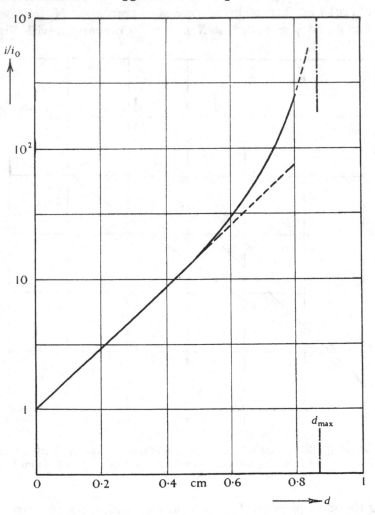

FIG. 91. Multiplication ratio i/i_0 as a function of the electrode separation d in argon with a Cu cathode at $X/p = 24$ ($p = 37$ mm Hg) (145). Note that the curve deviates from the straight line at $\leqslant \frac{1}{2}d_{max}$ as distinct from the curves in Fig. 90.

without invalidating the original assumptions. The condition for the 'breakdown' of a gap is then

$$\gamma(e^{\alpha d}-1) = 1. \tag{7.7}$$

It means that the current in a discharge becomes unstable and thus a large current may develop without a foreign agency liberating electrons at the cathode; the original discharge thus goes over into a self-sustaining discharge.

Of course this reasoning is not free from a number of objections. First of all (7.6) describes a steady state whereas the condition (7.7), which is derived from it, refers to the limit of stability. Further, in deriving (7.6) a uniform electric field was assumed and it is difficult to see how this could be true if the current were allowed to attain a very large value involving field distortion by intense space charges. The only answer to these objections is that in many cases field distortion becomes important only when the point of instability is reached. From observations it follows that the difference between the value of the applied field in which space-charge distortion becomes apparent and the maximum value—the breakdown field—is so small that the condition (7.6) represents a very good approximation for breakdown of gases even at high pressure.

Another method of determining α is from statistical variations in electron avalanches due to fluctuations in gas ionization and secondary emission. It has been shown that an electron avalanche with a total number of n electrons produces an average of γn secondary electrons at a cathode. The probability $P(\nu)$ that ν secondary electrons are released by n avalanche electrons is given by Poisson's distribution, since $\bar{\nu} = \gamma n$ is in general a small number (say 100 or less) and hence (158 a)

$$P(\nu) = (\bar{\nu})^{\nu} e^{-\bar{\nu}}/\nu!.$$

The probability that n electrons are found in an avalanche is, with $\bar{n} = e^{\alpha d}$,

$$P(n) = \frac{1}{\bar{n}} e^{-n/n}.$$

Thus from observations of $P(n)$, the frequency of occurrence of n, α can be found; the results agree with the known values.

(c) *The ionization coefficient for electrons.* In order to use (7.6) for the computation of the breakdown potential, it is necessary to know more about the dependence of α on X. The following classical argument leads to a well-known semi-empirical relation. An electron with a small energy is assumed to begin a mean free path λ. In order to be capable of ionizing at its next collision it has to gain from the field an energy $eXl \geqslant eV_i$, and hence the least distance l the electron has to move in field direction is

$$l = V_i/X. \tag{7.8}$$

The probability that the distance travelled is larger than the mean free path follows from (3.5), the statistical distribution of the free paths

$$z_l/z_0 = e^{-l/\lambda}, \tag{7.9}$$

where z_l/z_0 is the relative number of collisions having free paths $> \lambda$.

Since 1 cm of actual path contains $1/\lambda$ mean free 'ionizing' paths, the probability P of a free path longer than λ per cm of path is equal to the number of ionizing collisions per cm of path in field direction:

$$P = \frac{1}{\lambda}e^{-l/\lambda} = \alpha. \tag{7.10}$$

From (7.8) and (7.10) we obtain with $\lambda = \lambda_1/p$

$$\frac{\alpha}{p} = A e^{-B/(X/p)}, \tag{7.11}$$

where $A = 1/\lambda_1$, $B = V_i/\lambda_1$, and λ_1 is λ at $p = 1$.

Firstly, we note that the proper and convenient way to present the relation between the ionization coefficient and field strength is to write $\alpha/p = f(X/p)$ in accordance with the similarity rule (Appendix 1).

This means that the number of ionizations along one mean free path is a function of $eX\lambda$, namely the energy acquired by an electron along a mean free path in field direction. Further, (7.11) represents an S-shaped curve. At large values of X/p, α/p reaches a saturation value which, as we shall see, is not in agreement with observations. The third point is that the constants A and $B = V_i/\lambda$ of equation (7.11) which follow from our argument do not agree with the experiments. For example, for nitrogen $\lambda_1 \approx 0.03$ cm, hence we should have $A = 30$, $B = 450$, whereas in fact A and B are about 12 and 340 respectively. What are the reasons for the divergence?

One is that the electron was assumed to start each free path with an energy small compared with the ionization energy. However, the number of electrons with energies above ionization energy is quite large and this makes α too small. Secondly, the probability of ionization by an electron of energy larger than eV_i was assumed to be unity whereas, as shown in Chapter 3, only a fraction of all collisions of electrons with the required energy lead to ionization. This makes α appear too large. We have also to remember that a large number of inelastic collisions produce excited atoms and molecules and that this process occurs with a certain probability. Finally, the number of collisions an electron makes when moving unit distance in field direction is many times larger than $1/\lambda$ since many of the collisions between electrons and atoms are elastic. The modern approach is to correlate α/p with the efficiency of ionization by electrons (see p. 63 and (148 b)).

It has been indicated earlier that a value of α can be obtained from the slope of experimental curves, examples of which were given in Figs. 89, 90, and 91. From Fig. 89 for air it can be seen that (7.11) compares well with experiment up to $X/p \approx 800$. Table 7.1 gives a few values for the constants A and B in (7.11). It has to be borne in mind that (7.11) gives the correct dependence, if at all, only over a restricted range of X/p. The physical meaning of the constant A is that of a saturation value which α/p tends to at large X/p whereas B is proportional to an 'effective' ionization potential for this process which includes excitation losses, etc.

TABLE 7.1

Values of the coefficients A and B in (7.11) for various gases

Gas	$A \dfrac{1}{cm\ mm\ Hg}$	$B \dfrac{V}{cm\ mm\ Hg}$	Range of validity X/p
N_2	12	342	100–600
H_2	5·4	139	20–1000
Air	15	365	100–800
CO_2	20	466	500–1000
H_2O	13	290	150–1000
A	12	180	100–600
He	3	34 (25)	20–150 (3–10)
Hg	20	370	200–600

Fig. 92 shows α/p as a function of X/p for a number of gases plotted on a double logarithmic scale in order to cover a large range. Of particular interest is the fact that contrary to (7.11) the ionization coefficient has a maximum. This is to be expected from the curves of the ionization efficiency (page 63), which also show a maximum, lying for molecular and the majority of atomic gases somewhere between 80 and 150 eV. Now an electron acquires an energy of, say, 100 eV in a field X at the end of a mean free path λ when $X\lambda \approx 100$. Since $\lambda p = \lambda_1 p_1$, the values at 1 mm Hg (Chapter 4), this corresponds to X/p of order 10^3. This is precisely the region of α_{max} (Fig. 92).

Furthermore, the maximum number of ion pairs produced per cm should be approximately the same for an electron of energy of 100 eV or one moving in an electric field $X/p = 10^3$. Again comparing α/p and s/p (Figs. 92 and 33) the maxima prove to be nearly the same. In some cases α/p is perhaps up to twice as large as s/p because an electron moving in an electric field has a better chance of ionizing since its actual path length is increased by scattering collisions.

The ionization by collision in an electric field can also be represented by a coefficient

$$\eta = \frac{\alpha}{X} = \frac{\alpha d}{V} = \frac{\alpha/p}{X/p}. \tag{7.12}$$

η is the average number of ions produced by an electron per V of potential difference through which it has fallen. From (7.2) and (7.12) we

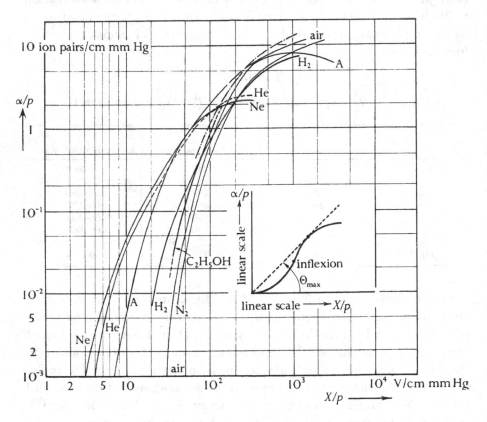

FIG. 92. Electron ionization coefficient α/p as a function of the field X/p for different gases (2, 17, 145, 147, 149 a). H_2O is slightly above Hg (155 a), for some hydrocarbons see (148 b). He at low X/p (147).

have $i/i_0 = e^{\eta V}$. Here $1/\eta$ is the average volt equivalent per ionization which, for fast electrons in most gases, has been found to be of the order 30 to 40 eV and which, because of the relatively increasing importance of excitations at smaller electron energies rises with decreasing electron energy. The maxima for all rare gases (except Rn) can be seen in the following table:

TABLE 7.2

Maximum ionizing power η_{max} of electrons in uniform electric fields

	He	Ne	A	Kr	Xe	
$\eta_{max} =$	1·2	1·5	2·2	2·4	$2 \cdot 6 \cdot 10^{-2}$	ions pairs per V
$(1/\eta)_{min} =$	83	66	45	42	38	V per ion pair
At $X/p =$	50	100	200	200	300	V/cm mm Hg

Thus for rare gases η_{max} increases for the heavier atoms and for lower V_i; its corresponding value of X/p increases with the atomic number. Conversely the energy per ion pair is a minimum (see (7.10)) when η is a maximum. $(\alpha/p)/(X/p)$ is largest when the tangent to the curve passes through the origin. In this way, or analytically, we find that, for example, from Fig. 89, for air, optimum conditions are about 66 eV per ion pair at $X/p = 340$ V/cm per mm Hg. The tables 7.2 and 7.3 give the optimum values—the Stoletow constant—for various gases.

TABLE 7.3

Stoletow constant $(1/\eta)_{min}$ for various gases

	N_2	H_2	CO_2	Air	Hg	
$1/\eta_{min} =$	75	93	62	66	80	V per ion pair
At $X/p =$	350	500	400	365	≈ 200	V/cm mm Hg

When a swarm of electrons moves in a uniform field, then those electrons with energies above a certain critical value are able to ionize the gas. Hence it should be possible in principle to evaluate α from the known data of the ionization efficiency of electrons of uniform energy, the energy distribution in the swarm, and the electron drift velocity. In addition Townsend's empirical relation (7.11), giving $\alpha/p = f(X/p)$, also follows from the rate of ionizing collisions by electrons and their drift. Let the number of ionizing collisions per electron per second be z, and v_d the distance the electron has moved in 1 sec in the field direction, then

$$\alpha = z/v_d. \tag{7.13}$$

From Appendix 3 approximately $z \propto p T_e^{\frac{1}{2}} e^{-eV_i/kT_e}$; further, we have from (8.40) $T_e \propto X/p$, and from (4.21) $v_d \propto (X/p)^{\frac{1}{2}}$; thus we find

$$\alpha/p \propto e^{-V_i/(X/p)\text{const}}, \tag{7.14}$$

which is of the same form as (7.11).

All attempts to calculate α rigorously in rare gases have been a failure, probably because the form of the distribution function of energies— assumed above to be Maxwellian—is not sufficiently well known. In molecular gases like N_2, H_2, air, etc., the agreement between experiment and theory for $X/p > 20$ is satisfactory, but at higher values the discrepancies are again considerable (146, 148 b).

It has been shown in Fig. 89 that for a given field and pressure the current rises exponentially with the electrode separation. In many experiments, however, d is constant and i_0 and V given, while p is gradually increased. How then does i vary with p?

From (7.2) and (7.11) we have with $X = V/d$

$$\ln(i/i_0) = \alpha d = A(pd)e^{-Bpd/V}. \tag{7.15}$$

It can be seen by inspection that for small as well as for large values of p (or pd) $i \to i_0$, that is multiplication ceases. Also by differentiation of (7.13) with respect to p, the maximum of i/i_0 can be found: it occurs when $pd = V/B$. Fig. 92 a shows the result. For molecular gases pd_{max} lies around 1 mm Hg cm when V is 500 V. The arguments leading to the maximum are independent of the assumptions made about the precise form of $\alpha/p = f(X/p)$. The physical explanation is the same as that given at the end of section 3 a. The variation was discovered by Stoletow (27). According to (7.15) Fig. 92 a also represents $\alpha = f(p)$.

In order to know when the use of the ionization coefficient α/p is convenient or permissible, the motion of electrons in the experiments for measuring α/p must be discussed. An irradiated cathode releases electrons which are accelerated in a uniform field forming an electron swarm. The swarm passes through a gas usually at a pressure of 1 to 100 mm Hg, thereby becoming wider through diffusion and self-repulsion. The electrons move in field direction with a drift velocity which is in general small compared with their thermal or random velocity which they acquire through frequent elastic collision or inelastic collisions with small energy loss. The random and drift velocity, $\bar{c}$ and v_d respectively, are always in a fixed ratio except near the cathode where all electrons are moving in field direction. The distance from the cathode over which the ratio varies is about equivalent to $(1-2)V_i$, that is about 30–50 V; this has to be allowed for in evaluating observations.

Fig. 61 gives the observed drift velocities v_d of electrons in various gases. Both v_d and $\bar{c}$ of course increase with X/p, but the results are obtained by assuming that the energy distribution is invariant. This, however, cannot be strictly true. A detailed discussion of that problem is beyond the scope of this book. The ratio of $v_d/\bar{c}$ as a function of X/p ($\propto X\lambda$) in several gases is shown in Fig. 93. It increases with increasing $X\lambda$, the 'mean free path voltage', because the directive effect of the

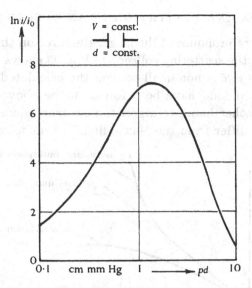

FIG. 92 a. Multiplication ratio i/i_0 as a function of the number of molecules in the gap (pd) for $X = 500$ V/cm and $d = 1$ cm in air. Actually $i/i_0 < 1$ at large p because of back-scattering of electrons in the gas near the cathode (p. 21). At $p \to 0$, ln $i/i_0 \to 0$.

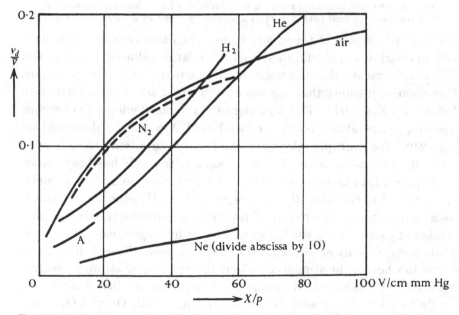

FIG. 93. Approximate ratio of drift to random velocity $v_d/\bar{v}$ for electrons in various gases as a function of the reduced field X/p (9, 158 b). ($\bar{v} = \bar{c}$.)

field becomes more pronounced the larger the force on the electron and the less frequent the scattering collisions. Fig. 93 shows that at values of X/p for which $v_d/\bar{c}$ is not small enough the calculated values of α/p would be expected (and have been found) to be above the observed values. On the other hand, in argon the electron energy distribution is most likely to differ from the Maxwellian in that fewer high energy

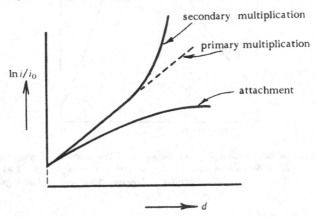

FIG. 94. The multiplication ratio i/i_0 as a function of the electrode separation d for constant reduced field X/p when a secondary effect or attachment occurs.

electrons are present in the steady state. Thus theory based on a Maxwellian distribution should (and does) give large values of α/p too, and this would explain the disagreement between theory and experiment. This discussion shows that α/p can only be used when the distribution depends on X/p solely. The dependence of the mean energy of electrons (electron temperature) on the reduced field X/p will be discussed on page 249. For multiplication in a non-steady state, see (126).

(d) *Multiplication in the presence of negative ions.* It has been shown in Chapter 3 that an encounter between a free electron and a gas molecule can lead to the formation of a negative ion. Whether or not this is possible depends on the nature of the molecule: only a relatively small number of gases like O_2, the halogens, atomic hydrogen, carbon, organic compounds, etc., are effective in attaching electrons. There is no general rule as to whether the atomic or molecular form is more strongly electronegative. For example, H atoms bind electrons readily whereas H_2 hardly forms negative ions; on the other hand, both O^- and O_2^- exist.

If a uniform field is used to measure the ionization current in electronegative gases, it is found that for a given value of X/p the ionization, instead of increasing exponentially with increasing electrode separation, rises less rapidly (Fig. 94). This is explained by assuming that electrons become attached to gas molecules and form negative ions which because of their large mass and low speed cannot ionize other molecules by collision. Another possibility is that the electron is detached from a negative ion by collision with a molecule or a free electron. However, this cannot compensate for the time during which the electron has been

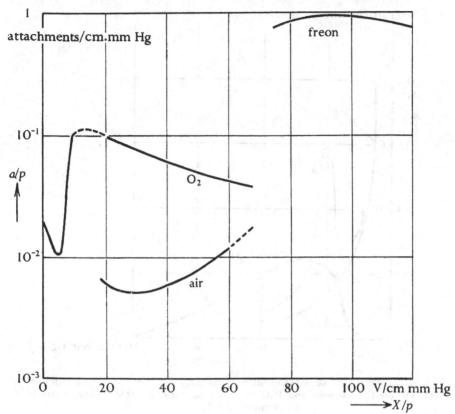

FIG. 95. Reduced electron attachment coefficient a/p in negative ions/cm mm Hg per electron as a function of the reduced field X/p for various gases (148). Freon = CCl_2F_2.

attached to a molecule, and so the net effect is still a reduction in multiplication as compared with that in ordinary non-attaching gases.

In analogy with the ionization coefficient it is convenient to define in a given field X/p a magnitude a/p which is the number of attachments of electrons to molecules per cm of path in field direction at unit pressure. Figs. 95 and 96 show a/p as a function of X/p in various gases. Though the numerical results are still somewhat uncertain, it transpires that as X/p rises a/p passes a maximum. This seems to indicate that at larger X/p an increasing number of electrons have energies which are too high to allow an attachment to take place. This maximum seems to lie at the higher X/p the larger the likelihood of attachment to the particular molecule. a/p is obtained from curves showing i as a function of d (Fig. 97). The analysis is based on 'dissociative attachment' being operative: the neutral molecule binds an electron and forms an excited

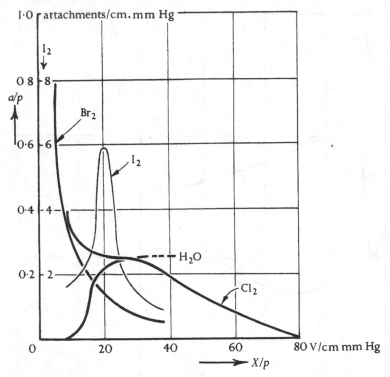

FIG. 96. Reduced electron attachment coefficient a/p as a function of the reduced field X/p for the halogens and H_2O at $p = 1$ mm Hg (26, 148). Note separate ordinate for I_2.

negative molecular ion which dissociates subsequently into a negative atomic ion and an atom. The last stage of the process can also lead to a small molecular ion or radical and a neutral molecule. Fig. 97 also shows that the departure from straight lines is more pronounced for lower X/p, which indicates that a/p becomes comparable with α/p at low X/p. It is clear that the independence of i/i_0 of the electrode separation d at low X/p arises from the fact that not only is the multiplication along the path reduced but also the number of free electrons available. This can be easily seen in the following way.

The increase in the number of electrons (n_e) produced along dx is proportional to the number of electrons n_e (all per cm² per sec) and to

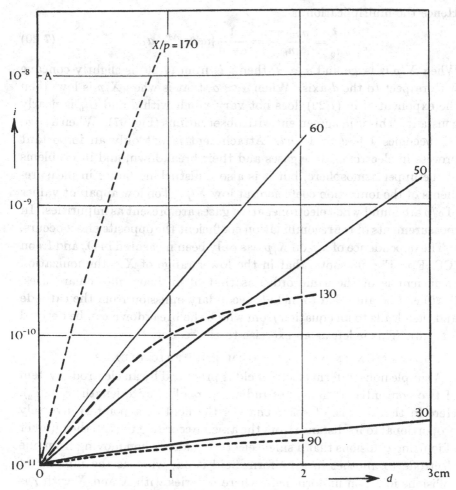

FIG. 97. Observed currents i as a function of the electrode separation d for various values of the reduced field X/p in freon (–––) and O_2 (——). i_0 = constant = 10^{-11} A (148).

the difference in ionization and attachment coefficients:

$$dn_e = (\alpha - a)n_e\, dx. \qquad (7.16)$$

Integration between 0 and d, taking $n_e = n_0$ at $x = 0$, gives

$$n_e/n_0 = e^{(\alpha-a)d}. \qquad (7.17)$$

The increase in number of negative ions dn^- produced along dx is

$$dn^- = an_e\, dx. \qquad (7.18)$$

Integration between 0 and d, using n_e from (7.17), gives

$$\frac{n^-}{n_0} = \frac{a}{\alpha - a}\{e^{(\alpha-a)d} - 1\}. \qquad (7.19)$$

Hence the multiplication is

$$\frac{i}{i_0} = \frac{n_e + n^-}{n_0} = \frac{1}{\alpha - a}\{\alpha e^{(\alpha - a)d} - a\}. \tag{7.20}$$

When X/p is large and $a \ll \alpha$, then i/i_0 from (7.20) is slightly concave with respect to the d-axis. When $\alpha \approx a$, that is when X/p is low, then the exponential in (7.20) does not vary much with d and i/i_0 is nearly constant. This is in agreement with observations (Fig. 97). When $\alpha = a$, i/i_0 becomes $1 + \alpha d$ or $1 + ad$. Attachment is not only an important process in electronegative gases and their breakdown, and in problems of the upper atmosphere, but it is also a disturbing factor in measurements of the ionization coefficient at low X/p. Too low apparent values of α/p are found when electronegative gases are present as impurities. In measurements of the recombination coefficient the opposite effect occurs.

The dependence of α/p on X/p has only been measured in O_2 and freon (CCl_2F_2). Fig. 98 shows that in the lower range of X/p the ionization coefficient is of the same order as that of ordinary molecular gases. (7.20) can be amended to include secondary emission from the cathode and then leads to an equation from which the breakdown can be derived (cf. 7.6). This is left as an exercise to the reader.

2. IONIZATION IN NON-UNIFORM ELECTRIC FIELDS

A simple non-uniform electric field is produced by an electrode system of two concentric cylinders of radius r_1, r_2 of infinite length ($r_1 < r_2$). Here in the absence of space charges the field at a point is inversely proportional to its distance from the axis. According to (7.2) the number of ionizing collisions that a single electron makes when moving from one electrode to another in a uniform field is αd, whereas the number of collisions in a non-uniform field where α varies with X and X with r is given by $\int_{r_1}^{r_2} \alpha_r \, dr$. It can be seen that the integral is independent of the polarity of the electrodes. Since the integral is equal to $\ln(i/i_0)$, we obtain for small V when secondary processes can be neglected, from (7.2) and (7.11) with $X = V/(r \ln[r_2/r_1])$:

$$\ln(i/i_0) = \int_{r_1}^{r_2} \alpha \, dr = \frac{A}{B} \frac{V}{\ln(r_2/r_1)} \times$$

$$\times \left\{ \exp\left(-\frac{Bpr_1 \ln(r_2/r_1)}{V}\right) - \exp\left(-\frac{Bpr_2 \ln(r_2/r_1)}{V}\right) \right\}, \tag{7.21}$$

where i and i_0 refer to currents per unit axial length. V is the applied potential difference and r_1 the radius of the inner cylinder.

If the potential is raised sufficiently then, assuming γ to be constant, breakdown should occur when

$$\gamma \left\{ \exp\left(\int_{r_1}^{r_2} \alpha \, dr \right) - 1 \right\} = 1. \tag{7.22}$$

It must be pointed out, however, that neither (7.21) nor (7.22) agrees with observations. For example, (7.22) suggests that breakdown should

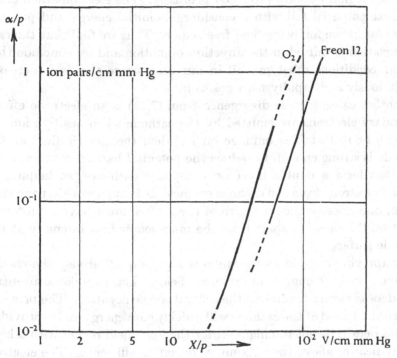

FIG. 98. Reduced electron ionization coefficient α/p as a function of the reduced field X/p for O_2 and freon ($= CCl_2F_2$) (148).

not depend on the polarity; however, in general polarity does influence breakdown. There are several reasons for the observed discrepancy: these can be due to effects in the gas or on the electrodes. In very strong fields which vary from point to point, the electrons in the gas do not acquire quickly enough the speed which is in accord with the field. In order to acquire speeds in equilibrium with the field, the electrons have to collide often enough to transform the energy taken from the field into random energy. This is not possible when the field varies by more than, say, a few per cent over a mean free path. If, for example, the field increases along the direction of motion, then the electrons will always have a random energy at a given point which is lower than the equilibrium value. In other words, the rate of ionization appears to be displaced in the direction of motion. This can be shown by replacing the non-uniform field by a fictitious system. Suppose an electron falls first through a potential difference applied between two grids, less than one mean free path apart, and then enters a field free space followed by similar systems. In the region of potential fall hardly any ionization can occur, but in the field free region ions are produced profusely. The electrons then enter the next potential fall with a considerable initial energy and produce more ions in the following field free region. Thus we find that the rate of ionization is shifted in the direction of motion and we conclude that similar conditions must prevail in our case where the field increases continuously and rapidly with position.

Another cause for the divergence from (7.22) is an electrode effect. Secondary electrons are emitted by the cathode when positive ions or photons or metastables impinge on it, when the electric field at the cathode is strong enough to reduce the potential barrier, or when positive ions form a double layer on composite cathodes, so helping to extract electrons from the cathode surface. It is apparent that the size, shape, and microscopic structure of the surface must have a profound effect on the emission apart from the microscopic field intensity at the cathode surface.

By applying these ideas we can now discuss qualitatively the results obtained with strongly non-uniform fields. Let us take concentric cylinders as electrodes and let the central one be negative. The primary electrons released at the cathode will quickly gain energy and—provided the gas density is not too high—enter the weak field region with a high energy near or above the maximum ionization efficiency. The electron energy distribution is either lacking in slow electrons (which excite atoms rather than ionize) or it has at least a large fraction of fast ones compared with a distribution it would acquire in a uniform electric field produced by the same potential difference. The positive ions also rapidly acquire speed without losing much by impacts when approaching the cathode and hence γ may be larger than in a uniform field.

Reversing the polarity at constant potential difference changes the ionization completely. In the strong field around a positive inner wire the electrons very seldom collide inelastically with gas molecules and hence impinge with large energy upon the wire (any secondary electrons return immediately). The positive ions, on the other hand, reach the large cylindrical cathode with low speed and release fewer secondaries.

It follows that, with a given initial current per cm axial length, a given discharge current requires a larger potential when the wire is positive. The same arguments hold for the breakdown potentials with different polarities.

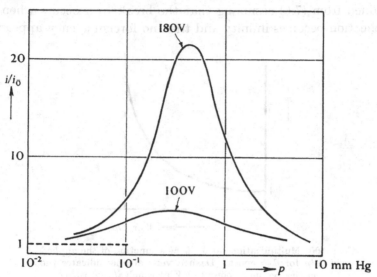

FIG. 99. Multiplication ratio i/i_0 as a function of the pressure p in H_2 for two applied voltages. Ni wire cathode and cylinder Cu anode ($r_1 = 1\cdot53$ mm, $r_2 = 43$ mm) (149). Re-evaluation of the results shows that, e.g., the maximum for 180 V is smaller by a factor 4 (155 b).

In a cylindrical gap with the inner cylinder as the cathode it is observed that for a given potential the reduced current i/i_0 passes a maximum as p is increased (cf. Fig. 92, 92 a). The explanation is that at low p few collisions will occur between electrons and gas molecules and most of the electrons will transfer their energy to the anode; thus i/i_0 will increase with p. At high p the electrons never acquire much energy since X/p is small (except near the cathode) and thus excitation is more likely than ionization; since X/p decreases with increasing p, the ionization falls and so does i/i_0. Fig. 99 shows the reduced current per cm length as a function of the applied potential for H_2 at p between 10^{-2} and 10 mm Hg, which illustrates the various points discussed above.

Uniform fields will become non-uniform when the current and hence the space charges become large. The effects which such fields have on α/p will be discussed in section 4.

3. STARTING POTENTIALS AND BREAKDOWN

(a) *Ordinary gases at low and medium pressure, uniform field.* From Fig. 100 or (7.6) it can be seen that at a certain critical value of X/p, which is found by slowly raising the applied potential difference, multiplication to infinity of the initial current will occur. (Actually the current would only rise to a large finite value which is determined by the resistance in the circuit and the finite power of the source.) The same result is obtained from (7.6) assuming that the breakdown occurs when the multiplication becomes infinity and then no foreign agency appears to

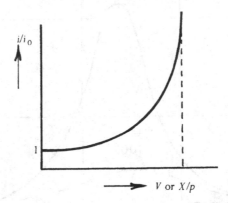

FIG. 100. Multiplication ratio i/i_0 as a function of the applied field X/p for $d = $ const. Dashed vertical line indicates the critical value of potential V at which $i/i_0 \to \infty$.

be necessary for the production of an initial electron; cosmic radiation, radioactive radiation from the surroundings, from the electrodes or walls of the vessel, and so on, are adequate to supply the few electrons which are needed to start multiplication.

In order to find the breakdown field or sparking voltage in terms of the gas pressure, the gas, and the electrode distance, we use

$$\gamma(e^{\alpha d} - 1) = 1, \tag{7.23}$$

which is valid for uniform fields and constant γ, and is likely to hold at medium and low p. p must be large enough to warrant the use of α, that is αd the number of collisions along d ought to be $\gg 1$ and $\lambda \ll d$. On the other hand, p must not be too large as otherwise X/p and α/p become very small: at small X/p excitation processes in the gas become prevalent, secondary emission by ions at the electrodes insignificant, and the theory outlined above inapplicable.

Instead of αd we write $(\alpha/p)(pd)$, and substituting (7.11) for α/p, we obtain for the starting field

$$\frac{X_s}{p} = \frac{B}{C + \ln(pd)} \quad \text{with} \quad C = \ln\left\{\frac{A}{\ln(1 + 1/\gamma)}\right\}, \tag{7.24}$$

and for the starting (or sparking) potential

$$V_s = \frac{B(pd)}{C+\ln(pd)}.$$ (7.25)

Fig. 101 shows V_s as a function of (pd) for air with $\gamma = 10^{-2}$, $A = 15$, $B = 365$ (Table 7.1), which agrees satisfactorily with observations. For large values of (pd) the starting potential V_s according to (7.25) rises

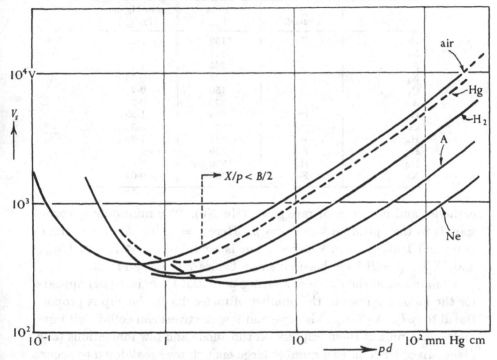

FIG. 101. Sparking potential V_s as a function of the reduced electrode distance pd in several gases (2, 17, 150, 151). For H_2O see (155 a). To the right of the dotted vertical line space charges facilitate multiplication, see (7.31).

nearly linearly with (pd) because the logarithmic term varies slowly. For small values of (pd) the numerator in (7.25) decreases linearly with decreasing (pd), but $\ln(pd)$ decreases faster, with the result that V_s rises when (pd) is lowered. Hence there is a minimum V_s whose value is found from $dV_s/d(pd) = 0$, viz.

$$(pd)_{min} = (2 \cdot 72/A)\ln(1+1/\gamma) \quad \text{and} \quad (V_s)_{min} = B(pd)_{min}.$$ (7.26)

It follows that, for example, the lowest sparking potential is to be expected for gases and cathodes for which B/A is small and γ is large (Fig. 101). Table 7.4 shows the minimum values for a number of gases. The general trend is in agreement with (7.26): for instance, for a given cathode in rare gases, the constant A is often smaller and γ larger than in molecular gases and thus $(pd)_{\min}$ is larger. Again B is small for the rare gases and so is V_s. Values of V_s below 100 V are observed with alkali

TABLE 7.4

Minimum sparking potentials (2, 5, 17, 20)

Gas	Cathode	*Volts* $V_{\min}$	*mm Hg cm* $(pd)_{\min}$
He . .	Fe	150	2·5
Ne . .	,,	244	3
A . .	,,	265	1·5
N_2 . .	,,	275	0·75
O_2 . .	,,	450	0·7
Air . .	,,	330	0·57
H_2 . .	Pt	295	1·25
CO_2 . .	?	420	0·5
Hg . .	W	425	1·8
Hg . .	Fe	520	~ 2
Hg . .	Hg	330	?
Na . .	Fe ?	335	0·04

cathodes and mixtures of rare gases (He–Ne). The minimum V_s corresponds to the optimum value of α/p. Here $\eta = (\alpha/p)/(X/p)$ is a maximum and $1/\eta = X/\alpha$ in V per ion pair is a minimum. Values of $1/\eta_{\min}$ and $(X/p)_{\mathrm{opt}} = B$ have been given in Tables 7.1, 7.2, and 7.3.

A minimum in the curve of starting potential V_s against (pd) appears for the following reason: the number of molecules in the gap is proportional to (pd). At low p, λ is large and few electrons can collide with gas molecules; most of them impinge on the anode and few ionizations take place. In order to have a number large enough for breakdown to occur, V_s has to be the larger the smaller p. At large p, however, λ is small and few electrons acquire sufficient energy over a mean free path to ionize. Hence most of the electrons produce electronic or molecular excitation. Consequently in order to produce enough ionization in the gap, V_s must be large and is higher for larger p. A similar argument would apply for a variation of d.

The question arises why electric breakdown of a gas (even under the most favourable conditions of p and d) does not simply occur when the potential between the electrodes exceeds ionization potential. The answer is that after the event of the first electron the electric field has not only to produce ionization in the gas and to remove the charges to the electrodes, but multiplication of charges in the gas must occur at such a rate that—together with the electrode processes—a large current can finally pass through the gas which no longer requires external means of ionization. With ionization potential applied across the gap each random electron may produce an ion pair but no further multiplication of charges occurs.

(b) *Ordinary gases, non-uniform field, low and medium pressure.* So far cathode and anode have been assumed to be infinite parallel plates. In the case of a long outer cylinder (r_2) with a central wire of radius r_1, we have from (7.15) and the continuity of the electric flux $rX = r_1 X_1$

$$\ln(1+1/\gamma) = \int_{r_1}^{r_2} \alpha \, dr = \frac{A}{B} X_1 r_1 \{e^{-B/(X_1/p)} - e^{-(B/X_1)(p \cdot r_2/r_1)}\}, \quad (7.27)$$

where $X_1 = V_s/[r_1 \ln(r_2/r_1)]$, V_s being the starting potential (155 b). This is a transcendental equation which does not indicate directly the dependence of V_s on p. However, the existence of a minimum in V_s can be easily shown. Neglecting the dependence of γ on X_1 and p, i.e. back-scattering, photo-electric effect, etc., by writing (7.27) in the form

$$\ln(1+1/\gamma) = \frac{A}{B} X_1 r_1 e^{-Bp/X_1} \{1 - e^{-Bp(n-1)/X_1}\}, \quad (7.28)$$

where $n = r_2/r_1$, on expanding the bracket and rearranging,

$$V_s \propto \frac{p}{\ln p}, \quad (7.28\,a)$$

a function which has a minimum for small values of p and is of a form equivalent to (7.25) valid for uniform fields. The minimum value of V_s is found by differentiating (7.28) with respect to p giving

$$V_{s\,min} = \frac{(B/A)\ln(1+1/\gamma)\ln n}{n^{-1/n} - n^{-1}} \quad (7.29)$$

and the value of p_{min} is found by substituting (7.29) in (7.28). Comparison between theory and experiment shows reasonable agreement, with γ of the order 10^{-2} or so. $V_{s\,min}$ from (7.29) is approximately 300 V and p_{min} about 1 mm Hg. Fig. 102 shows experimental curves in molecular and Fig. 103 in rare gases, confirming the above.

Finer details would require, for example, inclusion of the dependence of secondary emission on the reduced field. It is seen that all equations conform to the similarity laws (pr_1). Moreover, the values of V_s when the central wire is positive are always higher than the corresponding values for the negative wire. This is because with a positive central wire back-scattering takes place at the outer cathode cylinder (Chapter 2) which is very appreciable since the field there is relatively small. Thus fewer

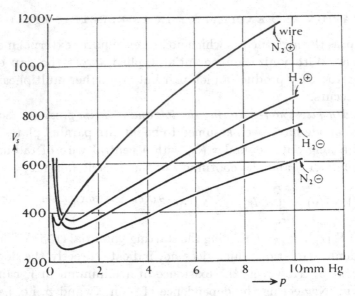

Fig. 102. Sparking potential V_s as a function of the pressure p for concentric cylindrical electrodes ($r_1 = 0.165$ cm, $r_2 = 2.3$ cm) in molecular gases when the wire is positive or negative (152, 153).

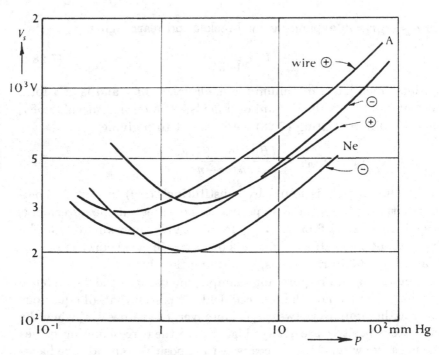

Fig. 103. Starting potential V_s as a function of the pressure p in rare gases ($r_1 = 0.087$ cm, $r_2 = 2.3$ cm) for a positive and a negative wire (2).

electrons are joining the multiplication process and in order to satisfy (7.28) a larger value of V_s is required.

(c) *Ordinary and electronegative gases, high pressure.* There is a wealth of experimental data available in this field which is mainly due to its practical importance (20). We shall discuss here only a few of the points which may be of interest to the applied physicist.

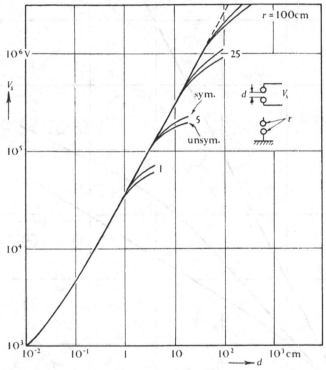

FIG. 104. Sparking potential V_s between two equal spheres of radius r in atmospheric air (s.t.p.) as a function of their distance d for potentials symmetrically applied with respect to earth (sym.) and with one sphere earthed (unsym.) (17). V_s begins to become dependent on r when $d > r$. This may be caused not only by field distortion but also by onset of local corona discharges.

When two spheres of radius r are brought d cm apart, the sparking potential V_s in air at 1 atm is observed to rise (Fig. 104) as d is increased. If the distance d becomes about equal to r, then V_s rises less rapidly with d than before. This means that in the central part the original (nearly) uniform field between the spheres becomes distorted in the sense that the field along the line of centres is larger than V_s/d. Another type of field distortion is that caused by asymmetry. For example, the influence of an earthed sphere is shown in Fig. 104. The result of it is a reduction of V_s provided $d > r$.

The non-uniformity of the field has a still more profound effect on V_s for gaps with point electrodes. Fig. 105 shows that the positive point-negative plane arrangement has a lower sparking potential than either

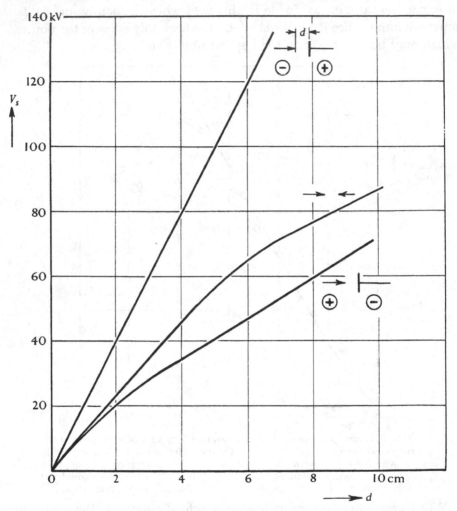

FIG. 105. Sparking potential V_s as a function of their distance d in air (s.t.p., 80 per cent hum.) for various electrode shapes (17). Note remarks to preceding figure.

that with opposite polarity or with two points. The condition of the electrode surface has a great influence on V_s at high p. This observation does not contradict what has been said before on low pressure break-down since at high p corona precedes final breakdown.

The likely explanation of the low value of V_s for a positive point is as follows. Intense ionization is produced near the point which attracts and removes the electrons. The positive ions are repelled and move slowly towards the cathode. However, at the far side of the ion cloud

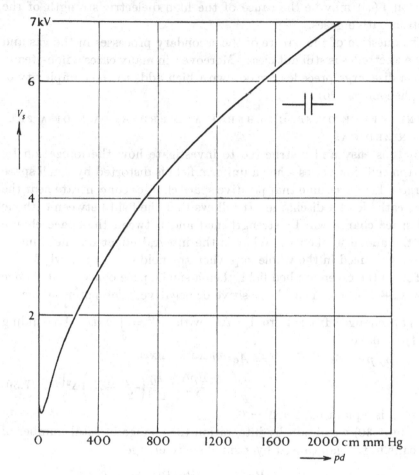

FIG. 106. Sparking potential V_s between parallel brass plates as a function of the reduced electrode separation pd in pure argon (154).

the field and the rate of ionization is increased with the result that a positive filament grows towards the cathode. With a negative point, the positive space charge in front of the cathode reduces the field at its anode side and thus a higher field is needed for breakdown to occur. The explanation for the two-point gap is left to the interested reader.

It can be seen from Fig. 101 that at high p with a good approximation the reduced starting field $X/p \approx 40$ for dry air and plane metal electrodes, whereas in pure argon (Fig. 106) $X/p \approx 4$ or less at large values of pd. The corresponding values for electronegative gases are much higher: for Cl_2 and freon $X/p \approx 100$, for CSF_8 $X/p \approx 200$, for CCl_4 $X/p \approx 230$. Large inelastic losses and large attachment coefficients (section 1 (d)) may be the cause of the high dielectric strength of the electronegative gases.

The question of the nature of the secondary processes in the gas and at the electrodes is still not clear. Moreover, in many cases a non-steady corona discharge precedes breakdown which adds to the complexity of the phenomena (20).

4. INFLUENCE OF SPACE CHARGE AND BREAKDOWN OF A PRE-IONIZED GAS

(a) It is easy and instructive to investigate how the ionization by electron collision varies when a uniform field is distorted by small space charges. Let us assume that positive space charges concentrate near the plane cathode of a discharge. It follows that the field between cathode and space charge will be strengthened and between the space charge and the anode weakened. What is the integral effect on the number of ions produced in the whole gap when the field changes slowly?

If X_0 is the given applied field, then a small space charge will change it to $X_0+\Delta$, where Δ can be positive or negative. The potential across the gap changes from V to $V+\Delta V$ with $\Delta V = \int_0^d \Delta\, dx$. Expanding (7.11) we have

$$\alpha_x/p = Ae^{-Bp/(X_0+\Delta)} = Ae^{-(Bp/X_0)\{1/(1+\Delta/X_0)\}}$$
$$= \frac{\alpha_0}{p}\left\{1+\frac{Bp\Delta}{X_0^2}+\frac{Bp}{X_0^4}\left(\frac{Bp}{2}-X_0\right)\Delta^2\right\}, \quad (7.30)$$

where α_0 is equal to α_x for $\Delta = 0$.

From (7.30) we obtain by integration the change in total number of ion pairs in the gap caused by a small space charge:

$$\int_0^d \alpha_x\, dx - \alpha_0 d = \alpha_0 \frac{Bp}{X_0^2}\Delta V + \alpha_0 \frac{Bp}{X_0^4}\left(\frac{Bp}{2}-X_0\right)\int_0^d \Delta^2\, dx. \quad (7.31)$$

Since the last integral is always positive (Δ^2), the second term will be positive when $X_0/p < \frac12 B$ and negative when $X_0/p > \frac12 B$. Hence with a constant V, or $\Delta V = 0$ and the first case, the ionization in a space-charge distorted field is larger than in a uniform one for a given potential $X_0 d$ and hence space charges facilitate breakdown. The critical value $\frac12 B$ corresponds to the point of inflexion in the curve Fig. 92 which is of the form $y = e^{-1/x}$ (7.11). By differentiating twice and equating to zero we obtain $x_c = \frac12$; thence $(X_0/p)_c = \frac12 B$ and $(\alpha/p)_c = A/e^2 \approx 0.135$ A.

The net result is that the total ionization in a gap increases through space charges when X/p is smaller than a value $\frac{1}{2}B$ and vice versa. For example, in N_2 we find from Table 7.1 that $\frac{1}{2}B$ is 170 V/cm mm Hg. Since for large p (7.24) X/p decreases as pd rises, it follows that for the region well to the right of the minimum V_s (Fig. 101) space-charge distortion helps to increase the current and furthers instability. At very low pd the reverse takes place.

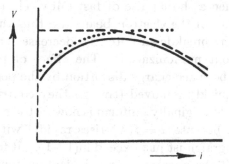

FIG. 107. Potential V of a dark discharge as a function of the current i. Effect of space charge distortion (dashed) and of initial ionization (dotted).

Also from (7.31) the change in potential ΔV can be found which makes the total ionization in the absence of space charges equal to that obtained in their presence. Thus putting the left-hand side to zero, we have

$$\Delta V = (X_0 - Bp/2)\frac{1}{X_0^2} \int_0^d \Delta^2 \, dx. \qquad (7.32)$$

But $\Delta \propto \rho$, the space-charge density, and $\rho \propto j$; hence the relation between current density and voltage follows from (7.32):

$$V = V_0 \pm cj^2 \qquad (7.33)$$

according to whether the bracketed factor in (7.32) is negative or positive. When $j > 10^{-6}$ A/cm^2, then $\Delta V/V_0 > 10^{-4}$ in air at 1 atm. For $\Delta V < 0$ we find the voltage current characteristic of the dark discharge due to space-charge effects to follow the dashed curve in Fig. 107, whereas the dotted curves represent the influence of initial ionization. The solid curve is then the actual dependence.

(b) The fact that a lower field is required to bring about a transition from a feebly ionized gas into a self-sustaining discharge than to obtain conventional breakdown has been known for a long time (Fig. 4). The detailed mechanism, however, is not well understood. It is also known that spurious electric breakdown between electrodes at atmospheric pressure occurs at voltages that are only a fraction of the standard values. Often this has been traced to the presence of hot ionized gas, metal vapour in the gas, or, for example, nearby corona discharges preceding breakdown. It transpires that this problem is fairly complex and that an exact definition of the initial conditions is necessary.

Few simple cases are known. It has been shown recently (158c) that a plane parallel gap with a constant voltage applied breaks down at a lower voltage when a short pulse of fast (MeV) electrons passes the gas. As the intensity of the electron beam and thus the initial ionization is increased, the breakdown potential decreases gradually to $\approx \frac{1}{5}$ of the value without pre-ionization. The effect of pre-ionization is mainly thought to be space-charge distortion by the positive ions since the electrons are quickly removed (though they contribute to ionization). Hence the field originally uniform is now distorted. For the same V, $\int \alpha \, dx$ is larger, because $\alpha = f(X)$ rises rapidly with X. Since for breakdown the integral must just exceed $\ln(1 + 1/\gamma)$, it follows that with pre-ionization V_s will decrease as the pre-ionization density rises.

5. OBSERVATIONS OF BREAKDOWN AND TIME LAGS AND THE MECHANISM OF THE SPARK AT HIGH PRESSURE

(i) We have seen in the preceding section that as the gas pressure is raised the value of X/p at which breakdown occurs decreases. This means that whatever the distribution of electron energy the number of electrons with energies above ionization potential will decrease and thus photons produced in the gas will become relatively more important. In other words, the ratio of photons per cm of path to ion pairs per cm measured in field direction will rise. This fact is the basis of all recent arguments attempting to give a physical picture of the breakdown process, particularly at higher pressures.

Before speculating about the likely mechanism let us first give a short summary of the main experimental evidence on which the conclusions are based.

(a) Measurements of the sparking potential $V_s = f(\text{gas}, pd)$. (There are few observations on V_s at higher temperatures.)

(b) Spectroscopic observations of the light emitted from the discharge as a whole as a function of the discharge parameters and measurement of the diameter of the discharge channel.

(c) Cathode-ray oscillograms showing either the collapse of voltage across the gap as a function of time or in a few cases the rise of current flowing through it.

(d) Study of the statistical fluctuations associated with the break-down.

(e) Cloud-chamber photographs of the spark.

(f) Kerr cell studies of the spark.

(g) Investigations of the intensities of spectral lines as a function of time, position of the emitting centre, gas, etc., and of the thermal and excitation equilibria.

(h) Measurement of the electric field as a function of position and time from the line broadening due to the Stark effect.

(i) Sound measurements accompanying formation of spark discharges.

We shall deal here only with a few selected topics and refer the reader to the bibliography (20) for further study.

If the voltage across a gap is increased from zero at a very low rate, the onset of breakdown is found to be subject to statistical fluctuations. The cause of these fluctuations is the likelihood of appearance of a first electron near the cathode in the gap. Instead of waiting for a cosmic electron, etc., initial electrons can be injected into the gas by irradiating the cathode with ultraviolet light, X-rays, α-particles, etc., or by inserting small amounts of artificially radioactive material. For example, a quantity of Co^{60} equivalent to about $\frac{1}{2}$ mg radium metal suffices; it emits γ-rays and β-rays, with a half-life of about 6 years (that is the time after which the intensity is half of the initial one) (22, 28).

If, therefore, a certain potential, higher than the static breakdown voltage, is suddenly applied to a gap and the time delay between application and breakdown is determined for a large number of experiments, this delay follows the law of change. Thus if n_t is the number of trials which has not led to a breakdown of the gap after time t and n_0 the total numbers of trials, then

$$\frac{n_t}{n_0} = e^{-kt}, \tag{7.34}$$

where k is a parameter which depends on the initial ionization and the fraction of voltage in excess of the static breakdown value $1/k = \tau_s$ is the statistical time lag. The larger the rate at which initial electrons are produced and the larger the excess voltage ('overvoltage') the larger k. Fig. 108 shows that this law is obeyed. With increasing excess voltage and constant ionization, k approaches a constant value, because every electron then leads to breakdown. Exceptions are found, for example, with point electrodes when field emission sets in or when surfaces have adsorbed gas or oxide layers.

Apart from this statistical time lag it is usual to speak about the formative lag τ_f of a discharge which was assumed to be negligible in the above introduction. This is the time required to build up the discharge from the instant the first electron appears until the discharge has

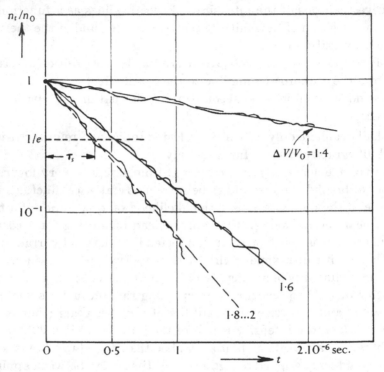

Fig. 108. Fraction n_t/n_0 of trials which have not led to breakdown, as a function of the time t of application of the field for various excess voltages $\Delta V/V_0$. V_0 is the static breakdown voltage for which $\tau_s \to \infty$.

reached equilibrium value. There is of course considerable uncertainty about the beginning and end of the process: the definition of the start depends on the sensitivity of the measuring device and of the end on the mathematical definition of equilibrium and the circuit. However, approximate figures can be found from oscillograms which enable one to check the theory of the growth of discharges. Fig. 109 shows the growth of current in argon for varying pressure. The maximum rate of rise obviously increases with p and possibly (pd); the formative time τ_f is of the order 10^{-5} sec and the lag τ_s larger than τ_f but of the same order. Fig. 110 gives the first cathode-ray oscillogram of the breakdown of air in which the gap voltage was measured as a function of time. Here at atmospheric pressure the formative time is of order 10^{-8} sec whereas the statistical time is about 10^{-7} sec. Though it is in general true that under the same conditions the higher p is made the lower are both time lags, it is important to remember that the dependence on the

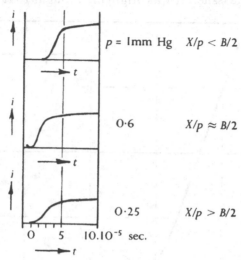

$p =$ 1mm Hg $X/p < B/2$

0·6 $X/p \approx B/2$

0·25 $X/p > B/2$

0   5   10.10^{-5} sec.

FIG. 109. Current i as a function of time t in an argon discharge between plane Ni electrodes 2·3 cm apart. Final current: order mA (17). The rate of rise increases as p rises. Compare the results with (7.31).

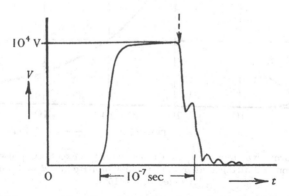

10^4 V

10^{-7} sec

FIG. 110. Voltage as a function of time for plane electrodes at 1 atm, 0·2 cm apart in air, when 50 per cent excess voltage is applied (156).

excess voltage is equally strong. Fig. 111 shows τ_f in argon as a function of $\Delta V/V_s$. It is seen that when $\Delta V/V_s$ is reduced to a few per cent, τ_f increases to $\sim 10^{-4}$ sec which is well within the region of τ_f at low pd but high $\Delta V/V_s$. In fact, at higher pressures τ_f does not depend on pd at all but varies strongly with d and little with p. It appears that photo-electric emission at the cathode is the secondary process (158 e).

(ii) The time has not yet come to give a fairly balanced view of the various mechanisms of breakdown which apply to different gases over certain ranges of pressure and distance; it would seem still less satisfactory to evade the issue. It was originally thought that a first electron

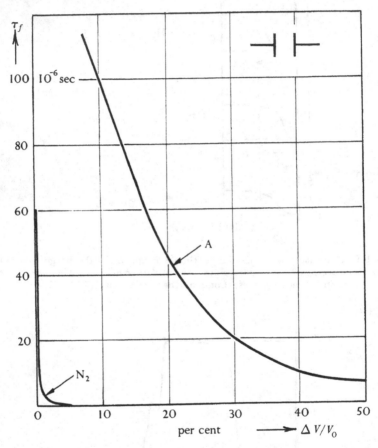

FIG. 111. Formative time lag τ_f as a function of the excess (impulse) voltage $\Delta V/V_0$ in argon and nitrogen at $p = 150$ to 500 mm Hg and $d = 1$ cm between plane electrodes (155).

starting at the cathode and multiplied in a uniform electric field would produce a large number of ions which would then return to the cathode. Provided the field was of the magnitude to induce breakdown, this number would be large enough to release on the average one further electron. Under these circumstances the lag τ_f would be of the order of the transit time of positive ions across the gap d, that is at 760 mm Hg of order 10^{-5} sec (for air $k^+ \approx 1$ cm/sec per V/cm, $X/p \approx 40$, $d = 1$ cm). In fact, as reported earlier, lags of order 10^{-8} sec were found at atmospheric pressure. For this and other reasons such as the observed independence of the cathode material, etc., this picture was abandoned.

Later Townsend's multiplication process was modified by including photo-ionization in the gas in addition to ionization by electron collision. It was argued that since ions could not arrive at the cathode in time to produce a sufficient number of secondary electrons, light quanta (produced by electrons colliding with molecules) may be absorbed by those molecules of the gas mixture which have a low ionization potential. Thus this component of the gas would become ionized in a very short time. In addition space charges would help to develop the electron avalanche because the ions would remain essentially stationary while the electrons are moving forward. This positive space charge would set up a radial electric field which might reach values of the order of the longitudinal field. The radial electric field, the importance of which was uncertain (157), would give rise to radial avalanches starting from photo-electrons in the gas which would strengthen the main avalanche. In this way the time factors involved would be reduced to transit times of electrons (10^{-7} sec or less) since the time of flight of photons corresponds to the velocity of light. However, difficulties still remained in dealing with very long gaps (63).

There have been recent attempts to measure the longitudinal spread of luminosity in a gap and to relate it finally to the spatial variation of the electric field and its change with time (13). The transition mechanism from avalanches to the more condensed form of discharge remains still obscure. An obvious step in the direction of explaining the rapid multiplication and rise in current is to include secondary processes in the gas such as ionization of excited species which would enable slower electrons in the distribution to play an active part besides charge transport (30 b). Once a finite length of path has a high density of ionization (which because of time cannot diffuse) the electric field along this path length must drop to a very low value. As a result of it the field in the remaining gap will rise and the multiplication process will speed up. However, there is no satisfactory treatment which deals with this problem.

A factor which was at first not sufficiently appreciated was the influence of the excess voltage on τ_f. Another circumstance was here the misinterpretation of cloud-chamber photographs of sparks *in statu nascendi*. Luminous streaks (called streamers) appear in the photographs which gave besides the speed of growth the length and the diameter of the discharge. These results were thought to apply to the first stage of the discharge. Photographs of the first stage, however, could not have been obtained by this technique; they describe, therefore, the spark in a later stage of development (158).

The advancement in the technique of breakdown measurements with very small excess voltages has made it possible to test the validity of Townsend's theory once again. When $\Delta V/V_s$ is small (Fig. 111), τ_f is so large that even ions would have a certain chance to cross small gaps, though it seems more likely that photo-electric emission from the cathode is the main secondary effect. Besides, with certain surface layers at the cathode a kind of (low)-field emission may be operative (5). With larger excess voltages the formative time is so small that only a photo-effect at the cathode can account for it. In gas mixtures like air photo-ionization in the gas cannot be ruled out. Several effects may occur: the component with lower ionization energy like O_2 may be ionized, slow electrons may become attached, forming negative ions, at least temporarily, and chemical changes like dissociation and formation of NO with a very low ionization energy and of O_3 can take place. The chemical effects would of course be only of interest here if they were able to develop within a time $< 10^{-6}$ sec at 1 atm and if the concentration of the products obtained were sufficiently high.

The other experiment which confirms the classical theory of electron multiplication in a gas at high pressure is the measurement of the multiplication factor i/i_0 in the steady state for low values of X/p just below the breakdown (Fig. 90). It shows that for plane parallel electrodes at atmospheric pressures no processes other than electron collision in the electric field (α) and secondary emission at the cathode (γ) are needed to describe the approach to breakdown and thus it becomes unnecessary under these circumstances to include field distortion by space charges, photo-ionization in the gas, or other electrode processes. Whether or not this is also true for the transient state when large excess voltages are suddenly applied to the gap or for breakdown in alternating electric fields remains to be seen.

Another way of testing the theory of electron multiplication is to measure the rise of current in a discharge which is started in the region near the minimum sparking potential. It has been shown in section 4 that the effect of space charges is to increase or reduce the rate of growth according to whether the discharge is working below or above a critical value of X/p, the Stoletow constant. Fig. 109 showed current oscillograms of a discharge in argon between nickel electrodes 2·3 cm apart at three pressures when a potential difference of 360 V is suddenly applied.

This corresponds to an excess voltage of $\approx$ 50 per cent. At 0·25 mm Hg where $X/p > \frac{1}{2}B$ (equation 7.31) the space charge slows down the rate of growth of current. At 1 mm Hg, however, with $X/p < \frac{1}{2}B$ a fast rise is obtained, as is expected from the theory.

So far it has been assumed that two large plane electrodes are used for the breakdown experiment. If the electrodes are of different shape, for example a wire at the axis of a cylinder, the sparking potential V_s is very different for the two polarities, the difference depending on the nature of the gas (Fig. 103) and the pressure. This is due to the fact that at high p the spark breakdown is preceded by a corona discharge around the wire electrode which is very similar to a glow discharge.

The effect of the polarity on V_s at high p is pronounced in gaps with point electrodes: the positive point shows the smallest sparking potential (Fig. 105). Again, first a corona discharge (Chapter 8) develops at the positive electrode as the potential is slowly increased. The electrons which ionize the gas in the high field around the point are quickly removed into the point electrode and the stationary positive space charges grow into the gap. Any additional ionization produced will increase the field which attracts more electrons coming from the direction of the cathode, with the result that a positive space-charge filament grows towards the cathode. When the point is the negative electrode, the corona discharge surrounds it with positive ions which reduce the field externally, and thus an electron starting at the cathode will only have a short path in the strong field, and a long one in a reduced field, while all the electrons produced further away in the gas are ionizing in weaker fields. This is probably one reason why at higher pressure positive points show low and negative ones high sparking potentials (Fig. 105). These sparking potentials are thus potentials at which the transition occurs from one discharge, the corona, into another, namely the spark (20). The starting potentials of the corona discharge is, of course, much lower.

Finally it should be mentioned that the lateral growth of discharges has been studied, in particular spark columns—which have been shown to differ very little from the positive column of arc discharges—and also glow and corona (counter) discharges initiated at medium pressures. For example, Fig. 112 shows the velocity of lateral spreading of glow discharges (without a positive column) in rare gases which is started at the ends of two parallel nickel strips and propagates along these electrodes. The speed v_l is found to depend on the applied voltage V and the gas, but is independent of p (between 6 and 35 mm Hg) in accord with similarity rules (Appendix 1). The highest speeds observed are well below the random velocities of the gas molecules and thus it is still an open question whether the limiting factor is the diffusion of ions or photons or the rate of production of charges.

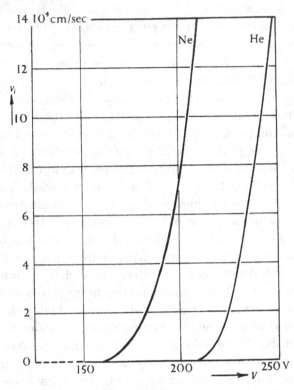

FIG. 112. Speed of lateral spreading v_l in a glow discharge as a function of the applied voltage V (Ni electrodes) (17).

6. EXCITATION AND DISSOCIATION IN AN ELECTRIC FIELD

When a swarm of electrons passes through a gas in a uniform field, electrons of energy above the various excitation potentials will excite the gas according to the respective cross-sections. Although there is a great variety of energy levels particularly in molecular gases, it is possible to lump together groups of these levels from which radiation is emitted. For example, in rare gases metastable or far ultra-violet emitting levels can be grouped together and thus an excitation coefficient ϵ/p for all or part of these processes can be defined in analogy to the ionization coefficient α/p and expressed in excitations per electron per cm of path in the field direction and per mm Hg. In molecular gases certain excitations lead finally to dissociation and for convenience a dissociation coefficient χ/p expressed similarly can be used. The coefficients ϵ/p and χ/p can be found theoretically and have been measured experimentally in some cases.

Fig. 112 a gives the calculated number of excitations to radiating and metastable levels in He, Ne, and A. It can be seen that excitation starts

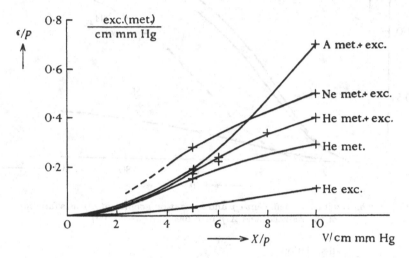

FIG. 112 a. Excitation coefficient ϵ/p as a function of the reduced field X/p in rare gases (145 a).

at fairly low values of X/p when α/p is negligibly small, as is to be expected from the difference between the lowest critical energy and V_i, as well as from the small value of the ionization efficiency. Because of the large metastable cross-sections (Fig. 27 b) their rate of production is large whereas the radiation produced by faster electrons, is always smaller. ϵ/p has also been calculated for X/p up to large values and seems to pass a maximum at $\approx$ 150 V/cm mm Hg (158 b).

In H_2 and N_2 measurements of ϵ/p are given in Fig. 112 b. ϵ/p in H_2 refers to light emitted partly in the far ultraviolet ($\approx$ 1000 Å) and in the near ultraviolet; the former probably predominates (145 a); α/p is also shown for comparison. In N_2 emission from the near ultraviolet ($C \rightarrow B$), the second positive band is more powerful than the far ultraviolet light ($a \rightarrow X$), the Lyman–Birge–Hopfield band. Fig. 112 c shows the measured values of χ/p in H_2 and some estimates of the dissociation coefficient in O_2 which seems to be associated with the reaction $O_2 + e \rightarrow O + O^* + e$, different from the reaction $H_2 + e \rightarrow H + H + e$. The

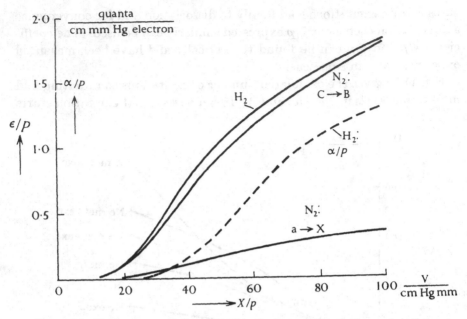

FIG. 112 b. Excitation coefficient ϵ/p and ionization coefficient α/p as a function of the reduced field X/p in molecular gases (209).

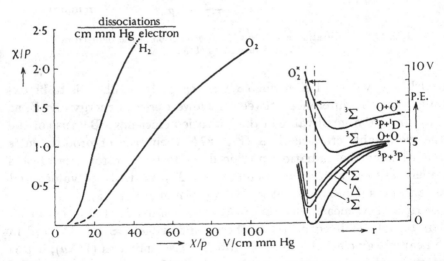

FIG. 112 c. Dissociation coefficient χ as a function of the reduced field X/p in H_2 and O_2, and potential energy P.E. as a function of the interatomic distance r for various electronically excited states of O_2 (145 b, 209).

potential energy diagram of O_2 is also included. A comparison with Fig. 112 b shows that dissociation in H_2 is here more probable than ultraviolet emission (148 c, 149 b).

It is now possible to set up the energy balance of an electron swarm in H_2. The following approximate results (in per cent) have been found:

$$X/p = 40: \quad 70 \text{ diss.} \quad 25 \text{ exc.} \quad 5 \text{ ioniz.} \quad (145\,b)$$

$$X/p = 100: \quad 60 \text{ ,,} \quad 20 \text{ ,,} \quad 20 \text{ ,,}$$

For the effect of intense laser light see Appendix 4.

Dissociation by an electron swarm in an electric field in N_2 is in general negligible. The energy balance in N_2 is not certain; it seems to be similar to that in the rare gases. At $X/p = 100$ about two thirds of the energy are estimated to produce metastably excited molecules, one-quarter produces ordinary excited molecules, i.e. light emission, $< \frac{1}{10}$ is for ionization, and the rest for elastic and vibrational losses.

QUESTIONS

(1) Given a swarm of electrons which moves through a gas of pressure p in a uniform electric field X causing multiplication. Derive from first principles the number of electrons and positive ions as a function of position x, i_0/e being the number of electrons at $x = 0$. Also find the ratio $i^+/i^- = f(x)$.

(2) Calculate in a plane parallel gap the current i as a function of the gas pressure p when X and i_0 are kept constant. Under what conditions can i become smaller than i_0?

(3) Compare the ionization by mono-energetic electrons with that by a swarm. Discuss the parameters which govern the applicability of ionization coefficient concepts and give an estimate of the limits.

(4) Give an account of the principles of the experimental methods from which α/p can be obtained. Indicate the experimental precautions which are necessary. How are secondary processes eliminated?

(5) Given a plane parallel gas-filled condenser at potential difference V and gap width d. Assume a slab of gas (parallel to the plates) of thickness Δ being irradiated such that N_0 ion pairs per cm^3 sec are produced in it. Find i/i_0 for given X/p when the slab is at different distances a from the anode.

(6) Give a qualitative account of the ionization in the gas and the current drawn when the anode of a plane gap contains a radio-active impurity which emits α-particles.

(7) Multiplication of charges by electron collision in an electric field is taken as being proportional to the concentration of molecules present. Derive the multiplication when in addition metastable particles are present which are subsequently ionized by electron collision. Comment on the result.

(8) Some experiments suggest that $\gamma \approx a(X/p)^2$. Derive the equation for the sparking potential V_s of a plane gap and show that the difference in V_s is small when γ varies in this way and compare it with the case when γ is constant.

(9) Derive from Fig. 106 the values of A, B, and γ and compare the results with the constants given in tables.

(10) Write an essay on the formative time-lag and the experimental methods used. Explain the cause of the difference between N_2 and A in Fig. 111.

(11) By examining the expression for the electric field required for breakdown as a function of the secondary emission coefficient, γ, show that at atmospheric pressure in air, breakdown occurs across a gap of 1 cm if γ exceeds a very small value, viz. 10^{-1}–10^{-4}. Show that for $\gamma = 0\cdot06$, $X_s = 30\cdot8$ kV/cm, and find the limits to gap width d, and to γ, which separately change X_s by 10 per cent. What value of X_s would be required for $\gamma = 10^{-5}$? Explain in physical terms why the dependence of X_s on γ and d is small.

(12) A photo-emissive cell with a Cs–O–Ag cathode is filled with argon to a pressure of 1 mm Hg so as to increase the photo-electric current. Draw the current/voltage characteristic of such a device with constant cathode illumination. Assuming plane geometry with an electrode separation of 1 cm and an effective value of γ of $0\cdot1$, find the breakdown voltage of the tube. Comment on the high value of γ.

Why would the value of the breakdown voltage increase if the geometry were cylindrical, assuming a ratio of radii of order 10 and the same gap length?

If the cell is run with a voltage of 90 V applied, find the multiplication (a) neglecting γ, (b) including γ.

It is desirable that such a cell has a region where the photo-current is not appreciably amplified. Find the multiplication for applied voltages of 20 and 30 V.

Also find the maximum light flux (in watts/cm²) if the incident light is monochromatic with an energy of 2 eV/quantum and if the electron current has reached saturation when the applied voltage is increased up to 20 V. Assume a photoelectric efficiency of $0\cdot1$ electron/photon at this wavelength.

(13) A neon lamp of effective radius 1 cm and length 100 cm is filled to a pressure such that it radiates strongly in the red (2 eV quanta). The luminous efficiency is 10 per cent when the current is 500 mA and the applied voltage 1000 V. The cathode area of the cell is 10 cm². How far from the lamp must the cell used in problem 12 be placed so that its current varies by less than 3 per cent as the supply voltage varies from 20 to 30 V?

The circuitry associated with the photocell is capable of detecting $1\,\mu A$. Find the maximum distance from the cell that the neon lamp will be detected with 90 V applied.

(14) Derive an expression for the value of $(r, p)_{min}$ for the breakdown between two concentric cylinders, discuss its dependence on the various parameters, and compare the numerical result with the experiments. Make an estimate of the influence of back-diffusion to account for the polarity effect.

(15) Show that $\alpha = f(p)$ has a maximum for constant V and d using for α (7.11). With A and B from Table 7.1, find the values of α and p at the maximum for N_2.

(16) Give data for the design of a detour tube filled with air to demonstrate the fact that breakdown can occur along a long path (electrode separation) rather than across a short path.

8

GLOW DISCHARGE

1. GENERAL FEATURES

A glow discharge is conveniently described as a discharge in which the cathode emits electrons under the bombardment of particles and light quanta from the gas. The cathode field is essentially determined by positive space charges. Thermal effects are either absent or at least not a necessary condition for maintaining the discharge.

The glow discharge derives its name from a luminous zone which develops near the cathode and is separated from it by a dark space. When a direct-current glow discharge is established in a long cylindrical tube which is filled with a rare gas at a pressure of between 0·1 and 1 mm Hg, the visible light emitted from the discharge is distributed over the length of the tube as shown in Fig. 113. Starting at the cathode there exists sometimes a very narrow dark space (Aston's) close to it followed by a thin relatively feeble sheath luminous layer—the cathode glow—which in turn is followed by the cathode dark space. Aston's dark space and the cathode glow are not always clearly visible. A sharp boundary separates the cathode dark space from the negative glow, which becomes progressively dimmer towards the 'Faraday dark space'. At the positive end of this is the positive column. It is either a region of uniform luminosity or regularly striated. At the positive end of the positive column there is sometimes visible an anode dark space followed by the anode glow close to the anode itself.

It is known that there are a variety of glow discharges and their appearance varies with the nature of the gas, the pressure, the dimension of the vessel, the type, size, separation, and material of the electrodes. However, we shall confine the discussion to a type of discharge shown in Fig. 113 and shall try to explain, first of all qualitatively, the characteristic features of the various zones. Later on the theories of the cathode fall and the positive column will be outlined. We shall not consider striations and discharges in electronegative gases (see 203 a).

When the distance between anode and cathode of a glow discharge is varied it is found that the axial length of the negative zones, i.e. those near the cathode, remain unchanged while the length of the positive column varies. In fact the positive column can be extended to any length provided the voltage for maintaining and starting the discharge is sufficiently large. Further, if a plane cathode is mounted in a large spherical bulb and rotated with respect to a fixed anode, the negative

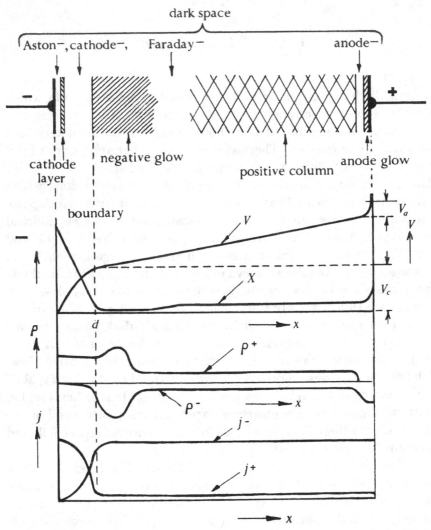

Fig. 113. Spatial distribution of dark and luminous zones, electric field X, space-charge densities ρ^+ and ρ^-, and current densities j^+ and j^- in a glow discharge (schematically).

zones swing round as if they were fixed to the cathode surface while the positive column simply fills the remaining space between the Faraday dark space and the anode. From these observations we can conclude that the motion of the charged particles in the negative zones must be of a beam-like nature whereas the motion in the positive zones appear to be essentially of the random type. This is partly supported by studies of the polarization of the emitted light. Consequently there should be little influence from the walls in the cathode region, i.e. the light emitted and the potential distribution there should not depend on the diameter of the glass cylinder, whereas the positive column should depend on the diameter of the tube. It will be shown later that this is so.

When the gas pressure is increased above, say, 0·1 mm Hg it is seen that the negative zones of a glow discharge contract towards the cathode. In fact, before pressure gauges were introduced which gave continuous readings it was an accepted practice to estimate the gas pressure in a vacuum system by the length of the cathode dark space. Above 100 mm Hg only the Faraday dark space is clearly visible. The positive column always fills the remainder of the gap, but it contracts radially at higher pressures. It is then in no way different from the positive column of an arc discharge carrying the same current, except perhaps for the ends of the column where the gas may contain a certain amount of vapour from the cathode and anode material.

The transport of current through a glow discharge occurs by the axial motion of electrons and positive ions. The flow of current through the cathode zones can be understood by referring to the distribution of the electric field, that is its axial component, as shown in Fig. 113. The field has been found to be large at the cathode, decreasing in intensity towards the negative glow, and, after passing a minimum in the Faraday dark space, it stays constant throughout the positive column and only rises again at the anode.

Consider an electron emitted from the cathode, for example by a positive ion which impinges on it. This electron is first accelerated in a strong field, but initially it executes few ionizing collisions because its energy is not sufficiently far above ionization potential. However, further from the cathode, though the field has become weaker, the electron ionizes more efficiently and strong electron multiplication will take place. Near the boundary between the cathode space and the negative glow the field has become very weak, and thus only the fast electrons which have not lost energy by inelastic collisions will be able to ionize in that region. However, a large number of electrons will cross the boundary and enter the negative glow.

Due to multiplication the number of electrons able to ionize has increased between the cathode and the glow boundary, and a large number of positive ions has been formed representing a strong positive space charge. These positive ions will move through the cathode dark space and impinge on the cathode. Also metastable atoms, fast unexcited atoms (charge transfer), and radiation will fall on the cathode with the result that secondary electrons are emitted from it. In order to have a steady state, each electron which is emitted by the cathode must produce sufficient ionizations and excitations to effect the release of one further electron from the cathode.

This picture also explains the distribution of the emitted light (27). An electron usually starts at the cathode with a very small initial speed of order of 1 eV. It is not able to excite gas molecules unless its energy has reached at least excitation potential, which will not occur within the first few V (5–10 eV) from the cathode corresponding to Aston's dark space. The cathode layer is the region within which the electron acquires an energy corresponding to the maxima of the excitation functions (Chapter 3). Since the maxima for different spectral lines lie at different energies it follows that the lines should be observed in such order that the lowest energy should lie nearest to the cathode. This is actually the case. At higher cathode fall Aston's dark space and cathode layer are replaced by the cathode glow. At larger distances from the cathode most (but not all) electrons have speeds which lie far beyond the maximum of the excitation functions and thus little visible light is emitted from the cathode dark space. At the negative glow boundary the number of slow electrons has become very large and their speed decreases with increasing distance from the cathode; thus in general these spectral lines appear in the reverse order as one approaches the glow (Seeliger's rule) (16).

The electrons entering the glow consist of at least two groups: those which have been produced at or near the cathode and are fast have not suffered losses by collisions in the dark space. The other larger group of electrons which have made many inelastic collisions have been created in the dark space and are therefore slow. Since the slow electrons have energies below the ionization maximum but above or at excitation maximum, they experience many exciting collisions and produce the negative glow. Afterwards their energy becomes so small that recombination with positive ions can take place. This process is likely to occur in and beyond the negative glow since the concentration of ions and electrons in that region is large and the field low. The emission of light due to recombination is, however, in general small. With increasing distance from the boundary fewer fast electrons are found and less visible light is emitted. The field rises slowly, the probability of recombination decreases, and the Faraday dark space develops. This is the anteroom to the positive column, and its properties are probably intermediate between those of the positive column and those of the negative zones. Since the field is found to increase in the direction of the positive column, we shall find first those spectral lines whose maximum excitation probability lies at lower energies.

In the uniform positive column the axial component of the electric field is found to be constant at any point (Fig. 113). It follows that the net space charge is zero or that the concentration of electrons at any point is equal to that of positive ions. Because of the small mobility of positive ions the electrons carry practically the whole discharge current while the positive ions compensate the electron space charge. The field in the positive column is several orders of magnitude smaller than that found in the dark space. This fact as well as the uniform appearance of the positive column indicates that ionization is not obtained from the drift velocity of electrons in the field direction but rather from their large random velocity acquired by numerous elastic collisions in the electric field. We have shown earlier that the random velocity under these conditions is several orders of magnitude higher than the drift velocity (Fig. 93).

At the anode side of the positive column the electrons are attracted by the anode and the positive ions repelled. Consequently a negative space charge is set up in front of the anode. As shown in Fig. 113 this gives rise to an increase of the electric field as well as a rise in potential —the anode fall in potential (160). An electron emerging from the positive column enters the anode fall region with a small initial energy. It is now accelerated towards the anode and after having crossed the anode dark space it has acquired a speed which is sufficient to excite and ionize the gas in front of the anode. The anode is therefore covered with a luminous sheath—the anode glow which is sometimes divided into several luminous spots. About measurements see (159, 160 a).

2. EVOLUTION OF A GLOW DISCHARGE

When a long cylindrical glass tube with two plane electrodes at its ends is filled with a gas of order 1 mm Hg of pressure and the potential difference V between the electrodes is slowly raised, then a small current of order 10^{-12} A or so can be observed to flow through the gas. This current is due to ionization in the gas and at the electrodes or walls by cosmic particles. As V is increased, the ionization by collision in the gas begins and thus the current will rise. At the same time some of the electrons will become attached to the inner glass walls (continuously but only partly neutralized by incoming positive ions), while an equal amount of positive charge will stay in the gas (Fig. 114). In the neighbourhood of the anode the conditions are of course reversed, but not symmetrically. The majority of positive and negative charges, however, will flow to their respective electrodes.

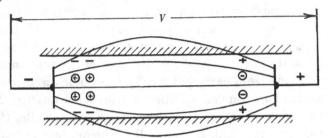

FIG. 114. Starting of a glow discharge in a long cylindrical tube.

As a result of these wall charges a radial electric field develops which restricts the flow of electrons to the walls. By increasing V still more, space charges appear which will distort the axial component of the field. The sequence of events is thus that at low V the field is given by the electrostatic field of the electrodes only, at higher V negative wall charges are set up acting as an electrostatic lens and making the field more uniform near the cathode. At still higher V space charges collect in front of the electrodes, and finally the current becomes large enough to bring about a transition from the dark discharge (page 202) into the more complex form known as a glow discharge (Fig. 115). There is a considerable drop in V and a rise in i. The fall in V suggests that processes have become operative which either facilitate ionization, reduce the losses, or both. The current i, however, is determined by the property of the discharge as well as by the applied voltage and the circuit constants. The transition from electric breakdown into a glow discharge has been studied both theoretically and experimentally in H_2.

3. THE CATHODE FALL REGION

(a) *The normal cathode fall of potential.* Fig. 115 shows that under certain conditions (section 4) cathode fall, dark space, and current density remain constant when the current is raised by several orders of magnitude. This is the normal cathode fall régime. An approximate calculation of the normal cathode fall, the thickness of the dark space, and current density, which gives all the essential features, can be carried out by the following argument.

In the steady state for every electron released from the cathode a certain number of ions, photons, metastables, or neutrals must fall upon the cathode. Let $1/\gamma$ be this number of ions which includes the effects of the other species per secondary electron released and let α be the

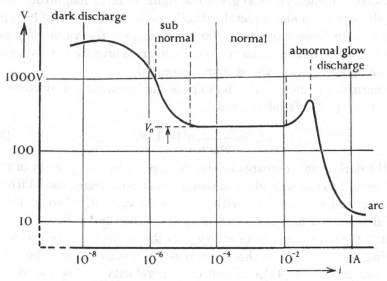

FIG. 115. Lowest maintenance potentials of the three main types of self-sustained discharges. V_n is the normal cathode fall in potential.

number of ion pairs produced in the gas per electron along 1 cm in field direction. Then each electron at $x = 0$ gives rise to $\exp \int_0^d \alpha\, dx$ electrons at $x = d$, that is at the boundary (Fig. 113). The number of ions is one less; this number times γ is equal to the number of secondary electrons to be replaced for the outgoing one, namely 1. Hence

$$\gamma\left\{\exp\left(\int_0^d \alpha\, dx\right) - 1\right\} = 1 \quad \text{or} \quad \int_0^d \alpha\, dx = \ln\left(1 + \frac{1}{\gamma}\right). \qquad (8.1)$$

The integral represents the number of ionizing collisions in d which is approximately

$$\int_0^d \alpha\, dx = \bar{\alpha}d_n \approx \eta V_n, \qquad (8.2)$$

where η is the number of ion pairs per electron per volt potential difference through which it has fallen, a quantity independent of the gas pressure. Combining (8.1) and (8.2) we obtain for the normal cathode fall

$$V_n = (1/\eta)\ln(1 + 1/\gamma). \qquad (8.3)$$

Thus V_n rises with increasing average energy necessary to produce an ion pair in the gas. This depends on the nature of the gas. V_n also decreases as γ, the efficiency with which secondary electrons are released from the cathode, decreases. γ depends on the substance of cathode, the gas, and the geometry. (8.3) gives the right order of magnitude of V_n. For molecular gases and a metal cathode $1/\eta = 50$ V/ion pair (Chapter 7) and $\gamma \approx 10^{-3}$, we have $V_n \approx 300$ V. Note that the value of γ used here is in general not the same as that to describe breakdown; otherwise V_n would be equal to V_s, the starting potential.

The normal thickness of the dark space can be estimated by considering (8.2) and (8.3). It follows that

$$pd_n = \frac{1}{(\bar{\alpha}/p)}\ln(1+1/\gamma). \qquad (8.4)$$

Thus the dark space increases as the average ionization per cm of path decreases and as the secondary emission coefficient decreases. With the same value of γ as above and with $\bar{\alpha}/p = 10$, a value d_n of about 0.7 cm is found for $p = 1$ mm Hg which is again of the right order.

Finally the normal current density j_n at the cathode can be estimated assuming that the current there is carried by positive ions. The space-charge density $\rho = (1/4\pi)X_c/d$ and the ion velocity can be found from the mobility k^+ and the field X_i at the cathode, of order V_n/d_n; hence the normal current density

$$\frac{j_n}{p^2} \approx \frac{1}{\eta^2}\frac{(\bar{\alpha}/p)^3}{\ln 1/\gamma}(k^+p) = (k^+p)V_n^2/(pd_n)^3. \qquad (8.5)$$

With $k^+p = 10^5$ e.s.u., $V_n = 1$ e.s.u., and $pd = 1$ cm, we obtain at $p = 1$: $j_n = 10^5$ e.s.u. $\approx 10^{-4}$ A/cm². This is again of the right order and shows that the current density increases for larger ionization α/p in the gas and larger γ.

As will be shown later, this treatment gives physically the correct picture, but the numerical agreement with observations is largely fortuitous. In fact it is an example which demonstrates that even such a threefold agreement is not a proof that the processes on which the argument was based are those which are operative.

Neither the fact that the normal current density at the cathode nor that the normal thickness of the dark space is approximately constant can be explained in simple terms without invoking dicta like the minimum energy principle. It can be assumed that the rise of the emitting cathode area is controlled by dispersive forces acting radially outwards, viz. electric fields originating from space charges rather than diffusion,

and by inward forces the nature of which is unknown. The distance
between the boundary and the cathode is probably a compromise between the space required for the multiplication process in the dark
space coupled with the excitation and ionization by fast electrons in the
negative glow (164 e, 169 a).

(b) *The abnormal cathode fall of potential.* When the negative glow
covers the cathode, an increase in current produces an increase in cathode
fall (Fig. 115) and in current density and a decrease in thickness of the
dark space. This is the region of the abnormal cathode fall.

If one positive ion impinging upon the cathode at $x = 0$ together
with its associated species liberates γ electrons, then the ion current
densities are related by

$$j_0^- = \gamma j_0^+ \tag{8.6}$$

and the total current density is

$$j = j^+ + j^-. \tag{8.7}$$

Let us assume an electric field which decreases linearly with x as observed down to $X = 0$ and $V = V_c$ at $x = d$, i.e. at the boundary between
cathode dark space and negative glow. If X_0 is the field at the cathode
surface where $V = 0$, then

$$(X/X_0) = 1 - (x/d). \tag{8.8}$$

Neglecting the negative space charge in the dark space, we obtain from
Poisson's equation

$$\frac{1}{4\pi}\frac{dX}{dx} = \frac{1}{4\pi}\frac{X_0}{d} = \rho^+ = j_0^+/v_0^+. \tag{8.9}$$

Since from (8.8) $X_0 d = 2V_c$ we have at the cathode

$$j_0^+ = \frac{v_0^+}{2\pi}\frac{V_c}{d^2}. \tag{8.10}$$

In a previous treatment (4) j_0^+ was replaced by (8.6) and (8.7), v_0^+ by
field and mobility, and d by the equation of the steady state (8.1). This
leads to an implicit expression from which $V_c = f(j/p^2)$ and $d = f(V_c)$ can
be found assuming that the ion mobility holds for strongly non-uniform
fields and that γ is known and constant. However, recent work shows
that the observed ions of high energy (of order V_c) at the cathode can
only be accounted for by taking recourse to charge transfer (Chapter 4).
Further, the dependence of j/p^2 on the extension of the glow in a hollow
cathode (section 4) makes it necessary to include the glow in the theory.
Probably the ultraviolet light quanta from the glow are often responsible
for the secondary emission of electrons from the cathode (161).

If the mean free path for charge transfer is λ_{+0}, then the potential V' at a point λ_{+0} cm from the cathode is given by integrating (8.8):

$$V'/V_c = 2\lambda_{+0}/d - (\lambda_{+0}/d)^2. \tag{8.11}$$

Since the energy of the ions arriving at the cathode is

$$\tfrac{1}{2}M(v_0^+)^2 = eV', \tag{8.12}$$

we have, for $\lambda_{+0} \ll d$ and for ions of mass M, from (8.11) and (8.12)

$$v_0^+ \approx \left(\frac{4e}{M}\frac{\lambda_{+0}}{d}V_c\right)^{\frac{1}{2}}. \tag{8.13}$$

Instead of operating with γ from (8.1), we consider from first principles the electron emission at the cathode. The electron is assumed to be released by positive ions, by radiation from the dark space where electrons excite molecules to higher states, and by excitation light from the glow. Thus the electron density at $x = 0$ and the current density in the glow j_g^- is related by

$$j_0^- = \gamma_i j_0^+ + Dj_0^- + Gj_g^-. \tag{8.14}$$

D is so defined that account is taken of the multiplication of electrons in the dark space, of the solid angle subtended at the cathode, and the average energy necessary to release a photon of energy larger than the work function of the cathode: but later for simplicity D will be neglected. G is defined by

$$G = f_g \gamma_p n_g, \tag{8.15}$$

f_g is the geometric factor, γ_p the photo-electric yield in electrons per quantum incident, and n_g the number of energetic photons per electron in the glow. Let $j_g^- \approx j$ which expresses the fact that the current in the glow is a pure electron current. Actually only a fraction of these electrons are fast ones, but this will be considered later. Taking (8.7) and (8.14), we obtain

$$\frac{j_0^+}{j} = \frac{1-G-D}{1+\gamma_i-D} \approx \frac{1-G}{1+\gamma_i}. \tag{8.16}$$

Obviously G, the secondary emission coefficient due to radiation from the glow, must depend on the parameters of the dark space. Let there be a fraction of electrons with energy $> V_e$ (100 eV for N_2, 50 eV for He). If y is the distance between the boundary and a point in the dark space, all fast electrons coming from y will have energies $> V_e$ when

$$y/d = \sqrt{(V_e/V_c)}. \tag{8.17}$$

But because electrons multiply from the cathode ($y = d$) to y, the number of ionizing collisions of fast electrons entering the glow is

$$\bar{s}d[1 - \sqrt{(V_e/V_c)}].$$

where $\bar{s}$ is the average ionization coefficient in the dark space. If η_p is the number of active quanta per V produced by one electron in the glow and bV_e the average energy of the fast electrons at $y = 0$ and $j_0^-/j_0^- = e^{\bar{s}d}$, we obtain for the secondary emission by photons

$$j_0^- f_g \gamma_p (\eta_p b) V_c \, e^{\bar{s}d[1 - \sqrt{(V_e/V_c)}]} = G j_0^-, \tag{8.18}$$

or
$$G = f_g \gamma_p \, \eta_p \, bV_c \, e^{-\bar{s}d\sqrt{(V_e/V_c)}}. \tag{8.19}$$

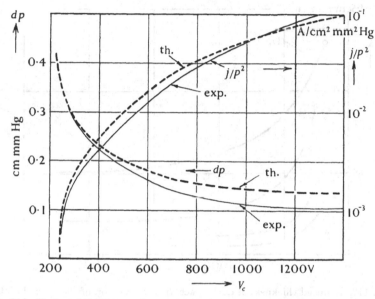

FIG. 116. Reduced thickness of dark space dp and current density j/p^2 as a function of the cathode fall V_c for a glow discharge in N_2 with Fe cathode (165).

The relation between j, V_c, and d follows from (8.10), (8.13), and (8.16):

$$j/p^2 = \left(\frac{e}{M} \, p\lambda_{+0}\right)^{\frac{1}{2}} \frac{1}{\pi} \frac{V_c^{\frac{3}{2}}}{(pd)^{\frac{3}{2}}} \frac{1 + \gamma_i}{1 - G}. \tag{8.20}$$

On the other hand, the condition for maintenance of the discharge is that an electron from the cathode which enters the glow together with the other fast ones formed produces radiation which releases just one photo-electron from the cathode. Hence

$$G e^{\bar{s}d} = 1. \tag{8.21}$$

(8.19) and (8.21) give the relation $pd = f(V_c)$, viz.

$$pd = \frac{1}{(\bar{s}/p)[1 - \sqrt{(V_e/V_c)}]} \ln \frac{1}{f_g \gamma_p \, \eta_p \, bV_c}. \tag{8.22}$$

Also from (8.20), (8.21), and (8.22) $j/p^2 = f(V_c)$ can be obtained. Fig. 116 shows pd as a function of V_c in N_2 from (8.22) and from observations, as well as $j/p^2 = f(V_c)$ from (8.19), (8.20), and (8.22) and from measurements. The agreement is satisfactory.

Various other gases show the expected dependences. For example, He has a small ionization coefficient $\bar{s}/p$ and thus from (8.22) pd should be large, which is known to be so. A and Hg have large values of $\bar{s}/p$ and so pd is small (Fig. 117). The observed values of pd at large V_c are not

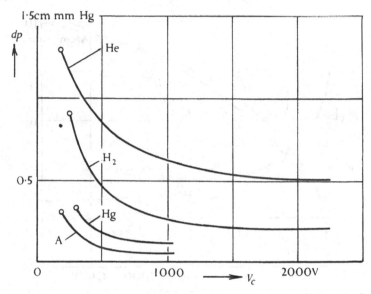

FIG. 117. Reduced thickness of dark space dp as a function of the cathode fall V_c for an Fe cathode in various gases (165).

very reliable because the heat produced in the gas reduces its density; thus the pressure reading gives too large a value. A more accurate method would consist in measuring the gas-density distribution in the dark space and to derive from it the average density. Otherwise the latter can be calculated from the known values of the thermal conductivity of the gas and the energy expended in the dark space (169). A review of the various theories of the cathode fall is given in (169 a). A glow discharge can be maintained at or above atmospheric pressure provided the current is not too large and the cathode remains cold. The current density at the cathode follows the similarity law but the gas temperature change has to be allowed for. At 1 atm in air $j_c \approx 8$ A/cm². Only the Faraday dark space is clearly distinguishable.

(c) *Numerical results.* We shall now give some experimental results and discuss them in the light of the theory indicated above. Table 8.1 contains some data on the normal cathode fall in potential V_n on plane metal cathodes in various gases. V_n is strongly dependent on the purity of the gas and the surface of the cathode. When the cleanliness of the cathode is high, then any impurities of the gas or from the walls (H_2O, grease, etc.) will reach the surface of the cathode which sometimes acts as a getter and the results show large scatter. In general V_n is low for alkali cathodes in rare gases and high for ordinary metals in Hg, O_2, and CO_2. For Cs compound cathodes in Ne about 40 V has been observed, for C in CO over 500 V. A given cathode shows a larger V_n in molecular gases than in rare ones. The values given for Hg vapour and, say, Al are not accurate because of the strong chemical reactions between cathode and gas, and the same applies to O_2, Cl_2, etc. (8.3) shows that, roughly speaking, V_n is the result of two functions: the ionization in the gas (η) and the secondary emission (γ) at the cathode. This may explain why V_n for Cu–He is larger than for K–H_2. By heating above 300° C glass can be used as cathode material; values of V_c for soft glass in H_2 and air are included in Table 8.1.

TABLE 8.1

Cathode fall of potential V_n in volts (17, 28 a, 162)

Cathode: Gas:	He	Ne	A	H_2	N_2	Air	Hg	Gas
Cu . . .	177	220	130	214	208	375	450	CO: 484, CO₂: 460
Zn . . .	143	..	119	184	216	280	..	O₂: 354, CO: 480
Hg . . .	143	..	..	337	226	..	340	
Al . . .	140	120	100	170	180	230	245	Cl₂: 280, O₂: 310
C . . .	..	..	..	280	..	424	475	CO: 525
Mo . .	109	107	103	..	..	..	..	..
W . . .	..	125	..	..	..	..	305	..
Fe . . .	150	150	165	250	215	270	300	O₂: 290, Xe: 306, K: 80, Cs: 340
Ni . . .	160	140	130	271	200	226	275	..
Pt . . .	165	152	130	276	216	277	..	O₂: 364, Cl₂: 275
K . . .	60	68	64	94	170	180	..	K: 80
Glass . .	..	..	..	260	..	310	..	..

The normal current density j_n (Table 8.2) is found to be small in rare gases (except A) and Hg, and large in molecular gases, varying little with the cathode substance. The data are not very accurate, mainly because of impurities at the surface and inhomogeneities. j_n is expected (8.5) to increase for greater k^+ or λ_{+0} (8.20), ionization in the gas ($\bar{\alpha}/p$), and secondary emission. The physical reason for the 'normal' current to be concentrated is probably the radial electric field component which develops at the outer edge of the glow (see above).

TABLE 8.2

Reduced normal current density j_n/p^2 in 10^{-6} A/cm^2 $(mm$ $Hg)^2$ (17, 162)

Cathode:	Gas:	He	Ne	A	H_2	N_2	Air	Hg	Gas
Cu .	.	..	..	..	64	..	240	15	..
Au	.	..	..	..	110	..	570	..	..
Mg	.	3	5	20	..	..	..	..	..
Al .	.	..	..	..	90	..	330	4	..
Fe, Ni	.	2	6	160	72	400	..	8	Kr: 43, Xe: 16
Pt .	.	5	18	150	90	380	550	..	..
Glass	.	..	..	..	~ 80	..	~ 40	..	..

The normal thickness of the dark space d_n (Table 8.3) according to (8.4) and (8.22) increases as the ionization in the gas and the secondary emission both diminish. For example, A and N_2 show large ionization coefficients and thus small d_n; the opposite is true for He. The dependence of d_n on the cathode is less pronounced.

TABLE 8.3

Reduced thickness of cathode dark space $d_n p$ in cm mm Hg (17, 162)

Cathode:	Gas:	He	Ne	A	H_2	N_2	Air	Hg	Gas
Cu .	.	..	..	..	0·8	..	0·23	0·6	..
Mg	.	1·45	0·85	..	0·61	0·35	..	..	O_2: 0·25
Hg	.	..	..	..	0·9	..	..	..	..
Al .	.	1·32	0·64	0·29	0·72	0·31	0·25	0·33	O_2: 0·24
C .	.	..	..	..	0·9	..	..	0·69	..
Fe .	.	1·3	0·72	0·33	0·9	0·42	0·52	0·34	O_2: 0·31, Xe: 0·23
Glass	.	..	..	..	0·8	..	0·3	..	..

The variation of the cathode fall V_c with current density j/p^2 and of dp with V_c is shown in Figs. 118 and 117 respectively for different gases with a cathode of iron. It is seen that when V_c is increased by about one order of magnitude, j_n varies by up to three orders of magnitude. The dark space shrinks with rising V_c to about one-quarter of its normal value and seems to remain constant at higher V_c (cf. end of section (d)).

The energy distribution of positive ions at the cathode has been measured at various gas pressures. It is obvious that at sufficiently low pressure all ions arriving at the cathode will have nearly uniform energy corresponding to the voltage drop in the tube; this has been observed

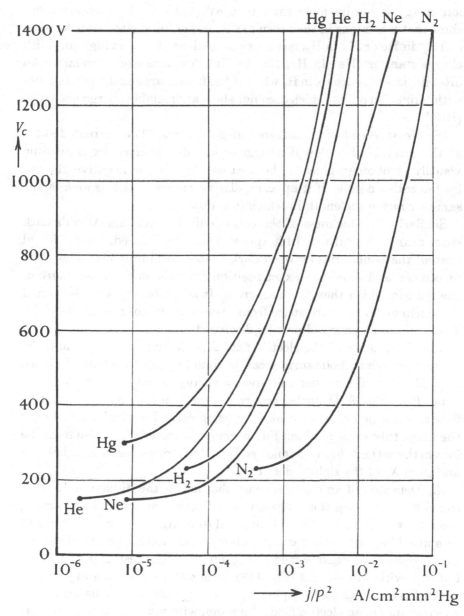

FIG. 118. Abnormal cathode fall V_c dependent on the reduced current density j/p^2 for an Fe cathode in various gases (165).

with canal rays. As p is increased from 10^{-3} mm Hg, collisions between ions and neutrals occur and the maximum of the distribution gradually moves towards lower energies. At p of order 1 mm or more the average ion energy is hardly more than 1–10 eV (164b, c). New observations show that when ions in their own gas do not suffer charge transfer such as He_2^+ in He or H^+ in H_2, fast ions predominate (even at high p). With charge transfer (He^+ in He, Ne^+ in Ne) slow ions are abundant. An intermediate case occurs in H_2 when H_2^+ suffers charge exchange together with formation of H_3^+ which does not show appreciable charge exchange (164d).

(d) *Measurement of the cathode fall parameters.* The current density at the cathode of a glow discharge is usually obtained by measuring visually or photographically the area covered by the negative glow or by the cathode glow, if it appears. This assumes that this area is the same for any wavelength in which it is observed.

Similarly the thickness of the cathode dark space and Aston's dark space near to the cathode (if it appears) can be obtained. It was found that all the boundaries between dark spaces and luminous zones show dispersion, and thus their exact position depends on the wavelength of the light in which they are measured. It is, however, possible to find the thickness of the dark space from measurements of the electric field. These measurements will now be discussed.

The distribution of the electric field $X(x)$ in the dark space and the value X_c at the cathode surface can be found by the following methods.

(i) Mechanical force per unit area = energy density per unit volume, hence $P_c = X_c^2/8\pi$. Disturbances are caused by uncontrolled convective flow of gas which can be avoided by using cathodes which extend over the whole tube cross-section. Furthermore, transfer of impulse from the ions to the cathode has to be allowed for. P_c was measured with a balance and gave X_c of the right order (163).

(ii) Deflexion of an electron beam shot across the discharge. $X(x)$ is found by measuring the deflexion on a fluorescent screen and varying the relative position between beam and discharge. A correction has to be applied because as the beam travels across the dark space it is deflected into regions of decreasing field strength (164). The results give a linear fall of X with x in general (Fig. 113) except at very low p and j (164f).

(iii) Stark effect or the splitting of energy terms and hence of the spectral lines in an electric field. An atom, without a permanent electric moment, becomes an induced electric dipole whose moment M is proportional to the field X at that point. Because of the orbital motions of electrons its angular momentum vector J precesses about the field vector with an angular velocity proportional to $MX \propto X^2$ and thus in the simplest case the splitting is proportional to the square of the field.

The first experimental results, though confirming the linear decrease in X, showed a maximum near the cathode; this has been found to be in error. It was caused by the 'constricted' discharge which developed in the narrow 5-mm tube (section 4), and later results confirm the linear fall in X in agreement with (ii) (164 a).

(iv) Hot probes have been applied to find the space potential and its distribution. The probe current as a function of the probe potential is found to show saturation which is more pronounced as the temperature of the probe is increased. For temperatures at which electron emission becomes appreciable, a sharp kink is observed which is assumed to occur at space potential. The change in gas density is of course a source of error (203 a).

The cathode fall V_c is measured by placing the anode into the negative glow and recording the lowest voltage, or by measuring $X(x)$ and integrating between $x = 0$ and $x = d$, where d is taken as the point at which the extrapolated field curve intersects the abscissa, or by using a probe in the glow for measuring the space potential. Differences in the values obtained are often due to the fact that the methods have not been used simultaneously in the same discharge apparatus.

4. VARIOUS KINDS OF GLOW DISCHARGES

(a) *The normal, subnormal, and abnormal discharge.* When a glow discharge is maintained between two plane electrodes at a pressure of the order of 1 mm Hg, the potential drop across the discharge (without a positive column) as a function of the current is as shown in Fig. 115. There is a constant voltage drop over a current range of 2–3 orders of magnitude (say, 10^{-5}–10^{-2} A). It is found that the current density at the cathode remains hereby sensibly constant, that is the 'covered area' expands proportionally with the current. This is the region of the 'normal' cathode fall in potential. When the current is reduced until the cathode area of the discharge has a diameter of the order of the thickness d_c of the dark space, the cathode fall increases. This is because as the diameter of the discharge becomes small, more charges will diffuse radially outwards and be neutralized. The current density at the cathode decreases and fewer electrons are emitted from the cathode. In order to maintain a steady state a larger cathode fall is required. This is the region of the 'sub-normal' cathode fall where the discharge is often intermittent.

On the other hand at larger currents, when the whole cathode surface appears to be covered by the glow, an increase in current can only be obtained by an increase in current density. A larger cathode emission is necessary, and this again increases the cathode fall. At the same time the dark space shrinks. Small changes in cathode fall are here associated with large changes in the current. This is the region of the 'abnormal' cathode fall already treated in section 3.

(b) *The obstructed glow discharge.* When the anode is moved towards a fixed cathode (i = constant), the anode fall disappears when the anode has reached the glow. By further reducing the distance d the potential begins to rise, before the anode enters the dark space. This can only mean that the ionization becomes more difficult when the discharge is confined to a space whose length is not large enough for developing the dark space and part of the glow. The rise in V sets in when $D < d_n$. This and other evidence suggests that the glow is taking part in the mechanism of the cathode fall in potential and acts most likely as an intense source of light quanta. Fig. 119 shows the potential V_c across such an 'obstructed' discharge at different currents in H_2. When $D \approx 1.5$ cm the anode fall is thought to have vanished; this is accompanied by a drop of ≈ 15 V; however, at $d < d_n$, V_c rises steeply with decreasing pd.

(c) *The constricted glow discharge.* The radius of the tube has practically no influence on the cathode fall of a glow discharge without a positive column, provided the gas pressure is high enough, that is the tube radius is larger than the electron mean free path. When, however, the pressure or radius R is made sufficiently small, the maintenance potential rises steeply. This sets in when the tube radius is of the order of one mean free ionic path (Fig. 120). Such a 'constricted' discharge in air cannot therefore be easily maintained below $pR < 2.10^{-2}$ except by very high voltages. In this case it is likely that fast electrons from the cathode hit the anode and produce X-rays which together with the fast positive ions release secondary electrons from the cathode.

(d) *The spray discharge.* So far we have always assumed that the cathode is a simple conducting surface. If the cathode consists of a thin insulating or semi-conducting substance like C or Al with a layer of Al_2O_3 or glass powder, a discharge develops without a dark space; only the negative glow, the Faraday dark space, and the positive column exist. The potential of this 'spray' discharge which is known to occur in rare and molecular gases is about ten times smaller than that of a conventional glow discharge, namely of order 40 V, and the current densities j/p^2 about 10^3 times larger than j_n/p^2 (Table 8.2) at gas pressures of 10^{-2} to 10^{-1} mm Hg. Ageing of the cathode is very pronounced and this prevents any application of this phenomenon. The absence of the cathode fall has been explained by field emission: positive ions are adsorbed at the surface of the insulating particles and the ensuing field

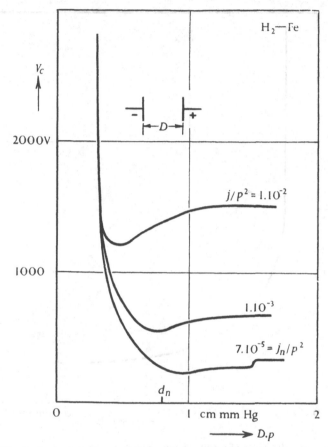

Fig. 119. Cathode fall V_c as a function of the reduced inter-electrode distance Dp for various reduced current densities j/p^2. Applied to an obstructed discharge in H_2 with a Fe cathode (166).

in the crystallites is sufficiently high to extract electrons. At the cathode side of the particles a similar process or else transfer of ions occurs. The light emitted originates mainly in the gas (168).

(e) *The high-pressure glow discharge.* Glow discharges are also obtained at and above atmospheric pressure. When the pressure is increased, the negative zones contract with the result that at 1 atm the Faraday dark space is only just visible to the naked eye. The value of j/p^2 is approximately the same as that for a low-pressure glow discharge provided due allowance is made for the change in gas temperature which can rise above 1000° C in the negative zones (169). In order to avoid the change-over to an arc, the cathode has to be kept cool. The glow discharge is easily distinguishable from an arc, for the potential of the former is

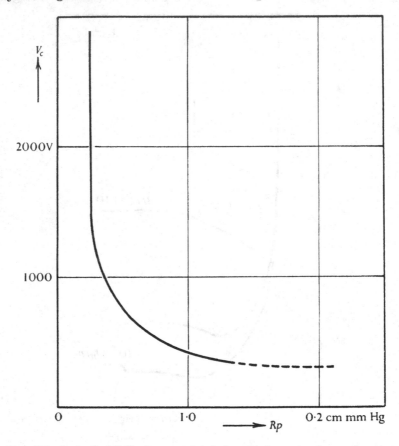

FIG. 120. Cathode fall V_c of a constricted glow discharge in air as a function of the reduced tube radius Rp ($R \approx 1$ cm) (16).

much larger and the light emitted from the cathode region contains the spectrum of the gas and not of the metal vapour of the cathode substance (see section 1). Glow discharges in air between electrodes of resistance material can carry currents of $> 10^3$ A for 10 sec (lightning arrester).

(f) *The hollow cathode discharge.* This effect is best understood from the following experiment. When the distance d between two plane cathodes of a glow discharge is reduced—the anode may be, for example, a ring of large diameter—while the potential is kept constant, the current density is seen to rise according to Fig. 121. At small values of pd, the current densities are 10^2 to 10^3 times larger than the 'normal' ones (Table 8.2).

The rise in j begins when the two separate negative glows are about to coalesce. At the same time the light intensity from the glow increases. One explanation is that at pressures of the order 0·1 to 1 mm Hg ultra-

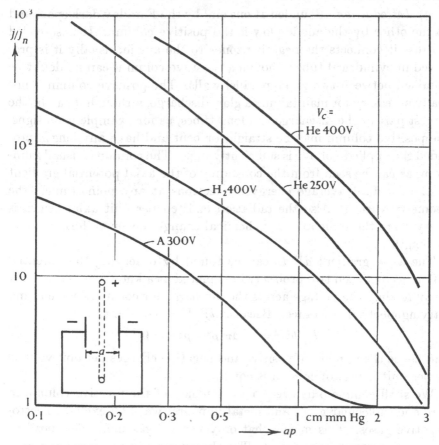

FIG. 121. Hollow cathode glow discharge. Reduced current density j/j_n as a function of the reduced inter-cathode distance ap for various gases and cathode falls V_c using plane Fe cathodes (167).

violet quanta mainly emitted from the glow will release secondary electrons from the cathodes (161). By reducing the inter-cathode distance both dark spaces and glows are compressed. Thus for a given cathode fall an electron entering the glow will lose its energy there in ionizing and exciting collisions since it cannot advance in the retarding field of the other cathode. Another factor is the increase in ionization in the gas by faster electrons from one dark space which enter the other. Instead of two plane cathodes a cylindrical cathode is usually used. The discharge can carry large currents at cathode falls of a few hundred volts. The gas temperature is not unduly large and hence the Doppler width of the emitted lines relatively small.

5. THE POSITIVE COLUMN OF A GLOW DISCHARGE

(a) *Introduction*. Bounded at one end by the Faraday dark space and at the other by the anode glow is the positive column. It is so called because it connects the negative zones to the anode; usually it is produced in cylindrical tubes, though a 'positive column' can no doubt be obtained between two plane parallel walls. The positive column is not really necessary for maintaining a glow discharge, although it can be the largest part of the discharge. In long tubes, as for example neon signs, the positive column may be straight or bent and have any length provided the applied voltage is sufficiently high. The column is usually uniform, as can be seen from the constancy of the axial potential gradient or the constant average energy of electrons at any point (unless the geometry varies). Also the radiation emitted per unit axial length is everywhere the same and any chemical changes occur uniformly along its length.

The axial gradient dV/dx can be found by observing the potential difference between two probes (wires) placed at a known distance apart or by reading the voltage across the discharge for constant current and varying electrode distance. Because $dV/dx =$ constant,

$$d^2V/dx^2 = -4\pi(\rho^+ - \rho^-) = 0,$$

and hence the number of positive and negative charges per unit volume or per unit length of column is equal.

We shall confine ourselves to a discussion of the positive column in rare and molecular gases and metal vapours; we shall exclude electronegative gases. This means that only neutral gas molecules, positive ions, and electrons are present. The theory can of course only apply to a certain range of pressure, tube radius, current, etc. In general it can be expected to hold for $p = 0.1$ to 10 mm Hg, tubes of a radius R of 1 to 10 cm and currents 10^{-4} to 1 A. The current has to be large enough to produce a sufficiently strong concentration of charges and low enough so as to exclude excessive heating or cumulative ionization of the gas. Stationary or moving striations and plasma oscillations are often found in rare as well as in molecular gases; their possible origin and effect will not be discussed here.

(b) *The theory of the positive column*. In the steady uniform column the electric field must have such a value that the number of electrons and ions produced per sec just balances the loss of charge.

Following Schottky's theory we assume many electron collisions along the tube radius, that is $\lambda_e \ll R$. Ionization in the gas occurs only by single collision between fast electrons and gas molecules. The number loss is due to ambipolar diffusion (Chapter 5). Thus both electrons and ions move with the same speed radially outwards. Since they do not recombine in the gas they neutralize their charges on the wall. Hence their concentration is large at the axis and practically zero on the insulated wall. We have therefore in any volume element of the column the same number of positive and negative charges, but this number is different in volume elements which are not at the same distance from the axis.

Also we have to remember that when the positive column is first set up, electrons will diffuse quickly to the wall since the initial positive ion space charge is much too feeble for ambipolar flow. Therefore the wall acquires a negative potential with respect to the axis. This is equivalent to saying that the equipotential surfaces are shells curved convex with respect to the cathode. The field lines originating from negative wall charge end on positive space charges distributed throughout the volume of the column. This positive excess charge and its counterpart on the wall determines the radial field.

The current in the positive column is mainly carried by electrons because of the small mobility and drift velocity of the positive ions. It thus appears that since an equal number of charges of both signs are produced in the column, more electrons than ions would leave unit length of column. The result would be an accumulation of positive charges, increasing with time. However, this is not so. When the transport of charge along the column is considered the situation at its ends has to be included. There is a constant influx of electrons from the Faraday dark space which constitutes the current. They are removed by the anode at the positive end of the column. There is also a constant flow of positive ions down the column as a result of ionization in the anode region driven into the positive column by the field of the anode.

The elementary theory of the positive column gives a good quantitative description of the relations between the axial and radial electric field, the pressure, the tube radius, and the nature of the gas. The dependence on the current can only be discussed qualitatively. We shall derive first the radial distribution of charge concentration by equating ionization rate and diffusion loss; then we shall calculate the electron temperature necessary to maintain the rate of ionization, and finally the axial strength which is needed to produce this electron temperature and to balance the energy losses.

(i) *The radial distribution of charges.* Let the mean free path of electrons $\lambda_e \ll R$, so that diffusion laws apply (Chapter 5). We have

$$N^+ \approx N^- = N \tag{8.23}$$

and

$$\frac{dN^+}{dr} \approx \frac{dN^-}{dr} = \frac{dN}{dr}. \tag{8.24}$$

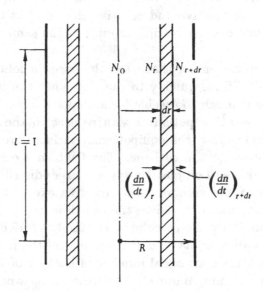

Fig. 122. To derive the radial distribution of charges in a long cylindrical positive column.

From Fig. 122 the number of ion pairs entering the volume element dr radially per unit length of the cylinder is

$$(dn/dt)_r = -2\pi r 1 D_a (dN/dr)_r. \tag{8.25}$$

The number leaving dr is

$$(dn/dt)_{r+dr} = -2\pi (r+dr) 1 D_a (dN/dr)_{r+dr}, \tag{8.26}$$

where D_a is the coefficient of ambipolar diffusion and N the concentration of ions and electrons. Since for reasons of symmetry and absence of recombination in the gas $dN/dt = 0$ at $r = 0$, we know that its value must increase with r. Thus from (8.25) and (8.26) the number leaving dr by diffusion exceeds the number entering it by

$$dZ_{\text{diff}} = 2\pi r D_a \left[\frac{1}{r} \frac{dN}{dr} + \frac{d^2 N}{dr^2} \right] dr. \tag{8.27}$$

This number loss has to be balanced by ionization in the same element dr. Let each electron make z ionizing collisions per sec; thus we have

$$dZ_{\text{ioniz}} = 2\pi r z N\, dr. \tag{8.28}$$

Equating (8.27) and (8.28), we find the differential equation

$$\frac{d^2 N}{dr^2} + \frac{1}{r}\frac{dN}{dr} + \frac{z}{D_a} N = 0. \tag{8.29}$$

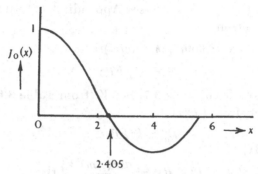

FIG. 123. Zero-order Bessel function $J_0(x)$ with real argument.

The solution is a Bessel function of zero order and real argument

$$N_r/N_0 = J_0[r\sqrt{(z/D_a)}] = J_0(x), \tag{8.30}$$

which is shown in Fig. 123 to be an oscillating function with a variable period. Since the concentration is a positive quantity and $N_r = N_0$ at $r = 0$, the first part of the curve has to be applied here. As we assumed that neutralization occurs on the wall only, we can write to a first approximation,
$$N_R/N_0 \approx 0 \quad \text{at} \quad x = R.$$

Hence from (8.30) $x_R = R\sqrt{(z/D_a)} = 2\cdot405,$ (8.31)

the first zero point of J_0, and substituting it in (8.30), we obtain the distribution
$$N_r/N_0 = J_0(2\cdot4r/R). \tag{8.32}$$

The concentration of charges in a column thus varies with r in a nearly parabolic manner. This has been confirmed experimentally in various gases by measuring N_r with cold probes (170, 171), by measuring the current distribution flowing to an anode which consisted of several concentric electrodes (172), and also by measuring the quantity of light emitted by sections of the column at different r (172). For example, in a Ne discharge at $p \approx 5$ mm Hg N_0 is about 10^{11} electrons and ions per cm^3 at $i \approx 50$ mA and $R = 2$ cm. If the positive column is bounded by two parallel walls, a cosine distribution of charge is found. The proof is left as an exercise for the reader.

(ii) *The electron temperature.* The next step is to find the mean energy of electrons. These are assumed to have a Maxwellian distribution. It is thus legitimate to talk about the temperature T_e of the electron gas. Both D_a and z, the ambipolar diffusion coefficient and the rate of ionization respectively, depend on T_e, p, and the nature of the gas. Also the number losses of charges will increase as R decreases and so we must find a relation between these quantities.

We start with (8.31) and write (see Appendix 3 (11.9)) for the rate of ionization per electron

$$z \approx 600(2/\pi)^{\frac{1}{2}}ap(e/m)^{\frac{1}{2}}V_i^{\frac{3}{2}}x^{-\frac{1}{2}}e^{-x}, \tag{8.33}$$

where $$x = eV_i/kT_e;$$

a in volts^{-1} follows from Table 3.7, but V_i (from Table 3.6) and e/m are in e.s.u. Further, from (5.21)

$$D_a \approx k^+kT_e/e. \tag{8.34}$$

Hence from (8.31)

$$z/D_a = (2 \cdot 4/R)^2 = \frac{ap^2(e/m)^{\frac{1}{2}}V_i^{\frac{1}{2}}}{(k^+p)}x^{\frac{1}{2}}e^{-x}, \tag{8.35}$$

or $$e^x/x^{\frac{1}{2}} = 1 \cdot 2 . 10^7(cpR)^2, \tag{8.36}$$

where $c = \left(\frac{aV_i^{\frac{1}{2}}}{k^+p}\right)^{\frac{1}{2}}$ in V, cm, and mm Hg and pR in mm Hg cm. (8.36) is the relation between T_e (or x) and (pR) for all gases. The nature of the gas is contained in the mobility of the positive ions, the efficiency of ionization (the coefficient a), and the ionization potential.

(8.36) is represented in Fig. 124 showing $T_e/V_i = f(pR)$. Since the number of charges lost is smaller for larger pR, a lower T_e is needed to maintain the balance of charges. It must be emphasized that the theory will fail at very low pR when the extended space-charge sheath on the wall is included and at large p when the positive column becomes constricted so that it does not fill the tube. Moreover, the theory excludes recombination in the gas and ionization in stages.

At large p it becomes increasingly difficult to measure T_e accurately and also the purity of the gas (fraction of electro-negative molecules) becomes a major practical problem; electrons become attached to impurity molecules (O_2, vapour from tap grease, H_2O, etc.) and recombine rapidly with positive ions.

A numerical example will serve to illustrate the use of Fig. 124. A tube of $R = 1$ cm is filled with Ne at $p = 1$ mm Hg. Since $cpR = 6 . 10^{-3}$ we find $T_e/V_i = 1600°$ per volt, and with $V_i = 21 \cdot 6$ V we have $T_e = 35\,000°$ K. Since 1 V corresponds to $7740°$ K $(eV = \frac{3}{2}kT_e)$, the electron temperature is said to be 4·5 V. This is borne out by observations. For $pR = 10$, $T_e = 21\,000°$ K or 2·8 V.

The observed values of T_e are between a few tenths and about 10 eV. The agreement between theory and experiment is good in the rare gases but it is not satisfactory in vapours, mainly because ionization is likely to occur in steps through excited atoms. Finally, according to the

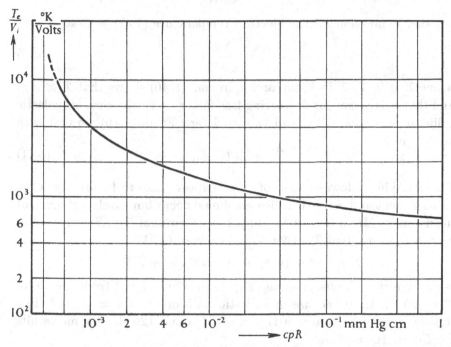

FIG. 124. Electron temperature T_e (divided by the ionization potential V_i) as a function of the reduced tube radius Rp from theory. Approximate values for c: He 4.10^{-3}, Ne 6.10^{-3}, A 4.10^{-2}, Hg 7.10^{-2}, H_2 1.10^{-2}, N_2 4.10^{-2}. These figures refer to atomic ions. This graph cannot be applied for very small values of pR where the theory fails. (8.36) is a double valued function!

elementary theory, T_e does not vary with r; this too has been confirmed by probe measurements (2).

(iii) *The (axial) potential gradient.* The gradient X in the positive column is given by balancing the energy the electrons gain from the electric field and the energy they lose by collisions. If an electron loses in one collision on the average a fraction κ of its energy, the energy gained per sec is given by the electric force times the drift velocity v_d, whereas the energy lost in elastic collisions is the product of κ, the mean electron energy, and the collision frequency. Thus

$$eXv_d = \kappa(m\bar{c}^2/2)(\bar{c}/\lambda_e) = \kappa \tfrac{3}{2}kT_e\bar{c}/\lambda_e. \qquad (8.37)$$

Also from (4.22) the ratio of drift and random velocity is

$$v_d/\bar{c} \approx (\tfrac{1}{2}\kappa)^{\frac{1}{2}}, \tag{8.38}$$

and combining these equations we have, apart from a factor of order 1 (actually 0·47) with T_e in °K and the other quantities in e.s.u.,

$$T_e \approx \frac{X\lambda_e}{\sqrt{\kappa}}\left(\frac{e}{k}\right). \tag{8.39}$$

Another expression for the electron temperature (1 eV = 7740° K) is

$$T_e = \frac{X\lambda_e}{\sqrt{(2\kappa)}}, \tag{8.40}$$

where T_e is in V, X in V/cm, and λ_e in cm. (8.40) shows that $T_e \propto X/p$ and that it is larger for decreasing loss factor κ. In the case of inelastic collisions $\kappa \gg (2m/M)$ as will be seen later. From (8.40) we find with $\lambda_e = \lambda_1/p$

$$\frac{X}{p} = T_e\sqrt{(2\kappa)}/\lambda_1. \tag{8.41}$$

For example, at lower values of T_e we expect there to be few fast electrons present and inelastic collisions should occur but rarely. Hence, for neon $\kappa \approx 2m/M \approx 5\cdot 6 . 10^{-5}$. For $pR = 10$ cm mm Hg ($R = 2$, $p = 5$) $\lambda_1 \approx 0\cdot 1$ cm, we find $T_e = 2\cdot 8$ V. Thus from (8.41)

$$X/p \approx 2\cdot 8 . 10^{-2} . 10 \approx 0\cdot 3 \text{ V/cm mm Hg},$$

which agrees with observations (Fig. 125). For $pR = 1$ from the above $T_e = 4\cdot 5$ V. Here because of inelastic collisions $\bar{\kappa} = \kappa \approx 2 . 10^{-3}$ (Fig. 63) and $X/p = 3$ V/cm mm Hg. Figs. 125 and 126 show some results for Ne, A, H_2, and N_2.

All curves have in common that after a steep fall X/p tends to approach a lower limiting value or to decrease very slowly with rising pR. This would agree with (8.41) as T_e decreases slowly and λ_e and κ are nearly constant. Further, we find that X/p in molecular gases is much higher than in atomic gases. This is to be expected; though T_e is lower in molecular gases because of the numerous inelastic collisions between electrons and molecules, κ is larger than $2m/M$ by orders of magnitude (Fig. 63) and the net result is that X/p in molecular gases is larger than in rare gases by a factor 10 or more.

X/p does not depend on the current in molecular gases provided due correction is made for density changes (p designates the pressure which is equivalent to the gas density). In rare gases and metal vapours, X/p decreases with increasing i (Fig. 125) because the electrons probably ionize by colliding with excited atoms. Their number may be only a small fraction of the atoms in the ground state, but the number of electrons having energies equal to or larger than $V_i - V_{\text{exc}}$ is very much larger. Moreover, the collision cross-section for ionization of the excited atoms is usually larger than for the unexcited species by a factor 10 or so. For large i, X varies with p as shown in Fig. 127.

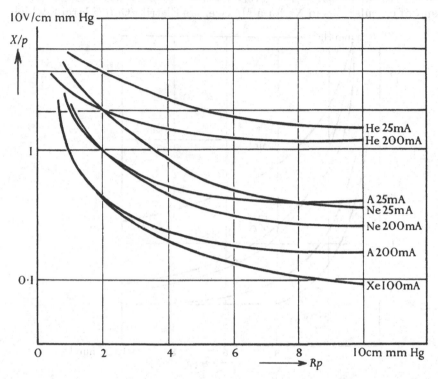

FIG. 125. Measured values of the reduced field X/p in the positive column of various rare gases as a function of the reduced tube radius Rp for different currents (176, 177). The curve for Hg at $p \leqslant 0.3$ mm Hg coincides approximately with that for A.

There is a more direct way of computing the potential gradient in a column, namely by determining the elastic (L_e) and inelastic losses (L_i) and equating them to the energy taken from the field (173). We have for the energy loss per electron per sec

$$L_e = p \int_0^\infty \kappa_{\text{el}} \frac{mv^2}{2} \frac{v}{\lambda_1} f(v)\, dv, \qquad (8.42)$$

$$L_i = p \int_0^\infty e V_e \, \eta \frac{v}{\lambda_1} f(v)\, dv, \qquad (8.43)$$

and also $$eXv_d = L_e + L_i, \tag{8.44}$$

where $$\eta = q_{\text{inel}}/(q_{\text{el}} + q_{\text{inel}}).$$

With a known velocity distribution $f(v)$, cross-section $1/\lambda_e$ (Chapter 3), ionization and excitation function η, critical potentials V_e, and drift velocity (Chapter 4), $X/p = f(T_e)$ can be found. Fig. 128 shows the result of calculations in Ne for a Maxwellian distribution (i large), which

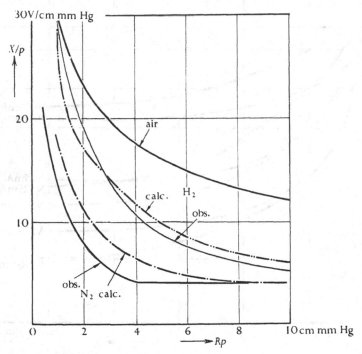

Fig. 126. Comparison between calculated and observed values of the reduced field X/p in various molecular gases as a function of the reduced tube radius Rp (178).

agrees with observations in positive columns and also theoretical and experimental results for electron swarms (i low). One reason for the divergence seems to be that the current density in beams is orders of magnitude smaller than in positive columns where the inelastic losses are numerous, thus reducing T_e for a given X/p. A Maxwellian distribution may not develop at low currents and particularly not in argon where the present theory gives a wrong answer. The ultimate reason for the non-Maxwellian distribution may be the variation of λ_e with the energy (Fig. 21). A large number of slow electrons have a large λ_e and these are quickly lost to the wall, thus X/p rises. The concept of ambipolar diffusion seems to be no longer applicable here.

Fig. 129 gives the observed variation of T_e with the reduced field for very large values of X/p in molecular gases and Fig. 130 for various gases at low X/p; the currents are here very low ($< 10^{-8}$ A) as used in swarm experiments (Chapter 7).

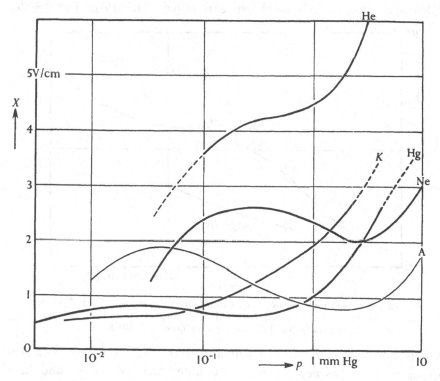

FIG. 127. Measured field X in the positive column of gases and vapours as a function of the pressure p for 0·3 A and 1-cm tube radius (179).

(iv) *The radial potential distribution.* From (5.19), the equations of ambipolar flow, the radial field is obtained:

$$X_r = \frac{1}{N}\frac{dN}{dr}\frac{D^- - D^+}{k^+ + k^-} \approx \frac{1}{N}\frac{dN}{dr}\frac{D^-}{k^-} \approx -\frac{1}{N}\frac{dN}{dr}\frac{kT_e}{e}. \qquad (8.45)$$

When $r \to R$, $N_r \to 0$ and thus $X_r \to \infty$, a result which is due to the approximations made.

By integration of (8.45), the potential V_r is found at a point r from the axis. N_r is the concentration. Putting $N_r = N_0$ and $V = 0$ at $r = 0$, we have

$$-V_r = \frac{kT_e}{e}\ln\frac{N_0}{N_r}. \qquad (8.46)$$

It can be rewritten in the form

$$N_r/N_0 = e^{-e|V_r|/kT_e}, \qquad (8.47)$$

showing that the concentrations of charges follows the Boltzmann distribution. Thus from (8.46) for any point off the axis, V_r is negative with respect to the axis as shown earlier (section 5(b)). For $r \to R$,

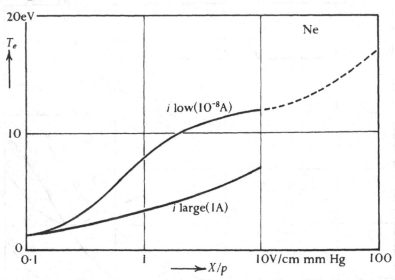

Fig. 128. Electron temperature T_e as a function of the reduced field X/p in an electron swarm in Ne (low current) and in a positive column in Ne (large current) (173). 1 eV corresponds here to 7740° K.

$N_r \to 0$, and $-V \to \infty$. This result and that from (8.45) are not surprising since it has been originally assumed that $N^+ = N^-$ and thus $dX_r/dr = 0$ at any point.

A more accurate treatment leads to $N_R/N_0 \approx 1\cdot7\lambda_e/R$. It takes account of the fact that there is a wall layer with electrons on the wall and positive ions close to it in which ambipolar diffusion ceases. This sheath takes control of the radial motion of charges at low p and the above treatment no longer applies. The potential difference V_b between axis and boundary of the sheath is

$$-V_b \approx \frac{kT_e}{e}\ln\left(\frac{R}{1\cdot7\lambda_e}\right). \qquad (8.48)$$

With $R = 1$ cm, $p = 1$, $\lambda_e = 0\cdot1$ cm, $T_e = 4\cdot5$ eV, as used previously, we obtain a potential of $-V_b = 5\cdot4$ volts. To V_b we have to add the potential across the sheath. However, even this approach is unsatisfactory and new basically different treatments have been attempted recently (181 a).

The function of the sheath is as follows: in the steady state equal numbers of ions and electrons must arrive at the wall. However, in the absence of a wall potential (field) $N_e v_e/4$ electrons and $N^+ v^+/4$ ions impinge on 1 cm² of the wall per sec and since $N_e \approx N^+$ but $v_e \gg v^+$ the number of electrons and ions would differ. In order to have equal num-

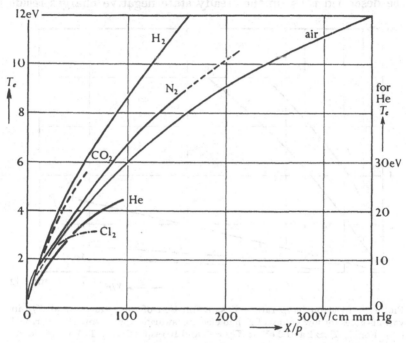

FIG. 129. Electron temperature T_e as a function of the reduced field X/p in various molecular gases at large values of X/p in swarm experiments (low current) (180). 1 eV corresponds here to 7740° K; He curve calculated (158 b); use right scale.

bers arriving, a sheath potential develops such that it retards the electrons and speeds up the ions. If V_s is the potential across the sheath,

$$(N_e v_e/4)_b \, e^{-eV_s/kT_e} = (N^+ v^+/4)_b,$$

where b designates the boundary between plasma and sheath. Since $N_e \approx N^+$ and $v^+ \approx v_{gas}$

$$V_s = \frac{kT_e}{e}\ln\frac{v_e}{v_{gas}} = \frac{kT_e}{2e}\ln\frac{T_e}{T_{gas}}\frac{m^+}{m^-}. \tag{8.49}$$

With the same data as above, we find $V_s \approx 22$ V and hence the wall potential is $V_R \approx -27$ V. This agrees roughly with probe measurements. Actually $v^+ > v_{gas}$ in the sheath and thus V_R is somewhat smaller. The concept of a sheath is likely to fail at high p.

To the critical reader the elementary theory of the positive column presented here appears neither satisfactory nor rigorous. This is partly due to the indiscriminate use of the concept of ambipolar diffusion and partly due to the choice of the boundary conditions on the wall, when zero charge densities are assumed. Qualitatively the physical picture can be described thus: in the steady state negative charges reside on

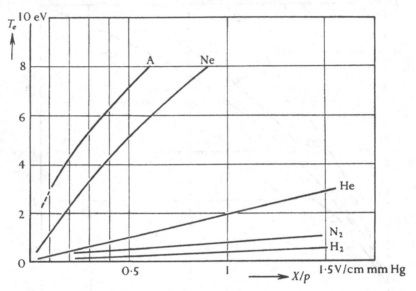

FIG. 130. Electron temperature T_e as a function of the reduced field X/p in various gases for low values of X/p in electron swarm experiments (low current) (10). For H_2, $T_e \approx 1$ and 2 eV at $X/p = 5$ and 10 respectively. 1 eV corresponds to 7740° K.

the wall. When the positive column is set up, electrons arrive on the wall before the ions, and the electron surface charge is finally such that the number of ions and electrons arriving balance. Because of the negative potential of the wall (with respect to any point in the gas at the same equatorial plane) only those electrons can reach the wall which have sufficiently large velocities, while the others are turned back. The positive ions are gaining speed when approaching the wall. The charged wall therefore acts as a flux selector. At low p it becomes a velocity selector for electrons. The radial electric field due to the wall charge can affect the drift velocity or the random velocity of ions or electrons or both. However, in order to obtain simple mathematical expressions, it is necessary to concentrate on specific limiting cases.

At low p the electrons will be mainly affected by the wall charge which leads to a Boltzmann distribution in the 'sheath' (8.47). This crude model can be improved by using instead of a sharp boundary, which defines a sheath region between 'plasma' and wall, a region in the plasma where the ions 'feel' the electric field. As a result of this, the ions acquire a directed initial velocity when crossing the 'boundary' and are accelerated in the 'sheath', so that their final directed velocity component v^+ is approximately $\frac{1}{2}m^+(v^+)^2 = kT_e$. The wall potential is determined by T_e because the larger T_e (or the mean energy of the electrons) the larger is in general the relative number of fast electrons reaching the wall. At high pressure, the wall field will act both on the positive ions and electrons which move through the gas with speeds determined by their respective mobilities. However, because of the numerous collisions, kinetic-theory relations apply, and fluxes must be equal to the product of the respective concentrations and random velocities ($\frac{1}{4}N\bar{v}$). This is equivalent to saying that the Boltzmann factor $(\exp(-eV/kT_e))$ approaches unity as p is increased since $V = X_r . \lambda_e$. Since the electrons have always larger random velocities than the ions the respective charge concentrations must be different in the same ratio. This again shows that the assumptions in the approximate theory are erroneous.

The concept of ambipolar diffusion is a device to avoid the use of the radial electric field. Its radial dependence given in (8.45) does not hold rigorously, particularly near the wall, where the gradient would be infinite. The reason is that the difference between ion and electron concentrations is finite and so is the charge density of electrons on the wall.

(v) *Extension of the theory of the positive column.* The treatment given for the cylindrical positive column is only applicable to a certain range of p, R, j and describes the conditions in molecular gases rather than in rare gases and vapours (1).

At low p when $\lambda_i \geqslant R$, ions and electrons wherever produced move to the wall, colliding rarely with gas molecules. A strong negative charge builds up on the wall. The larger part of the electrons are deflected back into the gas by this charge with the positive ions following it. Though collisions between electrons and molecules are rare, they are numerous enough to provide firstly the necessary number of ionizations—equal to the number of charges lost to the wall—and secondly the Maxwellian energy distribution of electrons assumed to exist at low p. The energy exchange between electrons is then likely to occur by electron–electron interaction, via collisions of the second kind or plasma oscillations. The mean electron energy can here be obtained by equating the rate of loss of charge to the rate of ionization z_i (2). If the positive ions are in equilibrium with the electrons and $\frac{1}{2}m^+v_i^2 \approx kT_e$, the rate of loss of charge is v_i/R per second and thus

$$z_{loss} \approx \frac{1}{R}\left(\frac{kT_e}{m^+}\right)^{\frac{1}{2}} \approx 4apV_i\left(\frac{kT_e}{2\pi m}\right)^{\frac{1}{2}}e^{-eV_i/kT_e} = z_i,$$

or
$$kT_e/eV_i \approx 1 \Big/ \ln\left\{aV_i pR\left(\frac{m^+}{m}\right)^{\frac{1}{2}}\right\}. \tag{8.50}$$

For Ne $aV_i = 1\cdot1$, $m^+/m = 4.10^4$, and taking $pR = 0\cdot1$, we have $kT_e \approx 7$ eV. This is to be compared with $kT_e = 20$ eV from Fig. 124 which holds for $\lambda_i \ll R$. Thus the left-hand part of Fig. 124 rises less steeply than anticipated. Both treatments lead to $T_e \propto 1/\ln(pR)$. (8.50) fails at very low p.

When j is small ($j < 10^{-4}$ A/cm^2), the positive and negative space charges become unequally distributed, and since the electron losses to the wall are likely to increase, the longitudinal component X of the field rises, increasingly so for less j. At very feeble currents radial space-charge distribution controls the free fall of ions and electrons to the wall.

At large R both T_e and X/p approach lower limiting values. For example, in H$_2$O, O$_2$, N$_2$, H$_2$, He, and Ne, X/p tends to about 50, 13, 3, 2, $0\cdot1$, and 10^{-3} V/cm mm Hg respectively, the corresponding values of T_e can be taken from Fig. 124. At large p the theory fails because T_e decreases with p and so does ΔV, the wall potential with respect to the axis, according to (8.46). At the same time at the axis the higher current density leads to a reduction of the gas density δ so that X/δ is larger at the axis than further out. Electrons and ions thus move radially outwards with reduced speed into a denser gas where recombination is favoured and the column no longer extends to the wall—it contracts radially.

A magnetic field H applied parallel to the axis of the column reduces the radial flow of electrons crossing the boundary of the sheath—that region between column and wall which is controlled by the negative wall charge—and produces a tangential flow of electrons. Since the number of electrons arriving at the wall decreases with increasing H, fewer positive ions must reach the wall, equilibrium being assumed. Hence the wall becomes less negative and so, besides ΔV, T_e and X are lowered (138 b).

A transverse magnetic field presses the column against the wall which increases the losses there more than it reduces the losses on the opposite side. The result is an increase in ΔV, T_e, and X/p (174). Finally, at very high currents (order 100 A) and low p ($\lambda_i > R$) the self-magnetic field of the current becomes significant. This is the case when the potential energy of the electron in the radial electric field ($e\Delta V$) is of the order of the magnetic potential energy $\left(\dfrac{e}{c}\displaystyle\int_0^R Hv_d\,dr\right)$, where v_d is the drift velocity. With R of order 1 cm, $H \approx 10^2$ Oe, $e/c \approx 10^{-20}$ e.m.u., $v_d \approx 10^7$ cm/sec the magnetic term is $\approx 10^{-11}$ ergs compared with $\Delta V = 1$ to 5 V or 10^{-12} to 10^{-11} ergs for the electrostatic energy (175, 181).

(vi) *Contraction of a positive column.* When a positive column is set up in a rare or molecular gas and the gas pressure increased, while $i = $ constant, it is found that the column contracts, i.e. the visible light, originally emitted from all regions, becomes gradually confined to the central region of the tube. At the same time the longitudinal electric field decreases. The mechanism of the effect is not clear (181 a, b). However, it seems as if constriction is about to set in or is already developed when the electron temperature is higher at the axis than near the wall, and the time an ion or electron requires to diffuse from the centre of the tube to the wall is of the order of the time required to recombine in the gas. That is,

$$\tau_D \approx \tau_{\text{rec}},$$

or, since $\tau_D \approx R^2/D_a$ and $\tau_{\text{rec}} \approx 1/(N_e\rho_e)$, the column becomes constricted when

$$N_e p \geqslant (D_a p)/R^2\rho_e. \tag{8.50 a}$$

Using the known data it follows that the observed values of p at which for a given current constriction sets in agree in order of magnitude with (8.50 a). Moreover, the lighter rare gases require a higher critical pressure, and because of $N_e\,p$ the contraction may occur fairly suddenly, as borne out by experiment. With electro-negative gases a lower critical pressure and current is expected, since $\rho_i > \rho_e$.

Positive columns have been shown to contract when an a.c. or high-frequency electric field is applied to the positive column. In the case of high frequency, sleeve electrodes are usually fixed around the glass tube carrying the discharge. By applying a potential difference between them the column contracts, for example, to less than half the diameter of the tube with nitrogen at pressure of the order 10 mm Hg. The explanation is that because of the unsymmetry of the probe curve (Appendix 5) and the absence of current flow, the inner wall acquires a large negative potential with respect to the axis which prevents the ambipolar flow of charge to reach the wall; this enhances volume recombination by retarding the faster electrons.

6. THE CORONA DISCHARGE

This type of gas discharge has received its name from the crownlike shape of a glow which was first observed on points or along wires in air raised to high potentials with respect to their surroundings. The corona occurs at any pressure and in any gas but is only pronounced when p is relatively high. It appears on lightning conductors and masts of ships (St. Elmo's fire) when densely charged clouds are overhead or discharging. It appears on high-voltage transmission lines and is probably closely related to discharges in point and wire counters. It is used extensively to produce electric charges in gases, particularly in electro filters, where the charges becoming attached to dust, smoke, and other particles are then extracted by an electric field, and in ozonizers for sterilizing water (Chapter 2). Physically it is very much the same as a glow discharge in a highly non-uniform electric field. All those convinced of the complexity of discharges may note that from new evidence available, a discharge around a negative point electrode in air shows all the characteristic features of a common glow discharge.

What distinguishes a direct-current corona from the glow discharge? One can regard the corona as a glow discharge without positive or negative zones, according to whether the point or wire is negative or positive.

Take, for example, a very long cylinder of radius r_1, with a thin wire of radius r_0 at the centre, and find the current of the corona discharge. Applying Gauss's law, which states that the difference between the electric flux ϕ entering a volume element of unit axial length and the flux leaving it is 4π times the total charge enclosed, we have

$$\phi_r - \phi_{r+dr} = K\,2\pi r X - K 2\pi\Big(X + \frac{\partial X}{\partial r}dr\Big)(r+dr) = 4\pi[\rho 2\pi r\,dr]$$

or

$$\frac{1}{r}\frac{\partial}{\partial r}[Xr] = 4\pi\rho/K, \qquad (8.51)$$

whence X is the field at any point r, ρ the net space charge there, and K the dielectric constant. This is Poisson's equation for a cylinder with any distribution of $\rho\ (=\rho^+ - \rho^-)$.

When $\rho = 0$, (8.51) becomes Laplace's equation

$$\frac{1}{r}\frac{\partial(Xr)}{\partial r} = 0 \qquad (8.52)$$

and hence

$$X = -\frac{A}{r} \quad \text{with} \quad A = V_1/\ln(r_1/r_0). \qquad (8.53)$$

Finally, integrating $X = -dV/dr$, we have

$$V = A \ln(r/r_0), \tag{8.54}$$

whence V_1 and zero are the potentials at r_1 and r_0 respectively. We take $K = 1$, which holds for charge concentrations encountered in such discharges. Under what circumstances has (8.52) to be used instead of (8.51)? The answer is: when the space charges become of the same order of magnitude as the surface charges on the electrodes. Let ρ be the constant space-charge density (all per unit length) and C the capacity, then this means that approximately

$$\rho.\text{volume} \approx V_1 C$$

and

$$\rho \pi r_1^2.1 \approx V_1 2\pi/\ln(r_1/r_0). \tag{8.55}$$

With $V_1 = 3000$ V $= 10$ e.s.u., $r_0 = 2.10^{-2}$ cm, $r_1 = 1$ cm, $\ln(r_1/r_0) \approx 4$, a marked influence of space charge is to be expected when $\rho \approx 5$ e.s.u. or $N \approx 10^{10}$ charges/cm^3. Hence space-charge effects set in when N exceeds 10^8 to 10^9 per cm^3.

The field distribution follows from (2.31) and (8.52) by integration

$$X^2 = (2i/k)+(r_0/r)^2[X_0^2-(2i/k)], \tag{8.56}$$

which for $r \gg r_0$ simplifies into $X = (2i/k)^{\frac{1}{2}}$ independent of r. Since $i \propto \rho r X$, we obtain for large r, $\rho \propto 1/r$.

The relation between V and i is obtained by integration of (8.56). Neglecting r_0 on the right-hand side this gives

$$V = \int_{r_0}^{r_1} \sqrt{\left\{\frac{2i}{k}+\left(\frac{r_0 X_0}{r}\right)^2\right\}}\, dr = C_1 \int_{\alpha_0}^{\alpha_1} \sqrt{\{1+(1/\alpha)^2\}}\, d\alpha, \tag{8.57}$$

where $\alpha = r(2i/k)^{\frac{1}{2}}/C_1$ and $C_1 = X_0 r_0$. Thus

$$V/C_1 = \sqrt{(1+\alpha_1^2)}-\sqrt{(1+\alpha_0^2)}-\ln\left[\frac{\alpha_0}{\alpha_1}\frac{1+\sqrt{(1+\alpha_1^2)}}{1+\sqrt{(1+\alpha_0^2)}}\right] \approx \alpha_1^2/2-\ln(\alpha_0/\alpha_1), \tag{8.58}$$

and since $X_0 r_0 = V_0/\ln(r_1/r_0)$ (V_0 now the starting potential),

$$i = (V-V_0)V_0 k/[r_1^2\ln(r_1/r_0)]. \tag{8.59}$$

The current is therefore proportional to the excess voltage and the ion mobility, i.e. inversely proportional to the gas density. This relation describes the observations satisfactorily, provided the excess voltage is not too high. When $V > V_0$ empirically $i \propto V(V-V_0)$ (Fig. 12). The starting potential V_0 of the corona is known from experiment; in air at 1 atm the field X_0 is of order 3.10^3, 7.10^3, and 3.10^4 V/cm for $r_0 = 10$, 10^{-1}, and 10^{-3} cm respectively and r_1 about 10 to 100 times larger than

r_0. In the absence of a theory of breakdown in non-uniform fields it is thus not possible to calculate V_0 and X_0 from atomic data. Furthermore, the accuracy of measurement is restricted by the fact that the inner wire surface is usually not sufficiently uniform to produce an evenly distributed corona; a spot structure is visible particularly at higher pressure and in most gases. This explains why V_0 depends on the roughness of the surface and probably to a certain degree on its chemical composition. V_0 also depends on the polarity.

In a corona the electric energy is converted mainly into heat in the gas in that the ions impart their momentum to molecules; only a small fraction of the energy is converted into chemical energy and light; losses by recombination are probably small. The alternating-current corona is complicated by the fact that the space charge is driven into the space by the field of the wire. When the field changes its direction this space charge is attracted by the wire, but at the same time ions of opposite polarity are repelled. Thus in the steady state there is a considerable phase difference between the field on the surface of the inner wire and the velocity of the ions which results in a corresponding phase difference between the peak value of the applied field and the (usually distorted) corona current proper. The alternating-current corona is of considerable practical interest since the losses of power on transmission lines caused by such discharges can be very large (21).

7. Pressure effects and electrophoresis

When ions and electrons move through a gas under the influence of an electric field, they collide with gas molecules and impart momentum to them. The result is a pressure difference in the gas or a flow of gas or both. Let ions of one sign move in the x-direction of the field E, ρ being the net space charge density. Since more lines of force act on the front of the volume element of charge than on the rear, this means that the mechanical force (per unit area) p changes along dx and thus

$$-dp/dx + \rho \cdot E = 0 \qquad (8.60)$$

if mobility is the controlling factor and gas flow is absent. With Poisson's equation we find that the pressure difference between two points 1 and 2 is proportional to $(E_1^2 - E_2^2)$ (17).

In the positive column of an atomic gas, positive ions move towards the cathode as well as to the wall. Since the latter has a negative charge, the ions transfer all their momentum to the wall which they have gained over the last mean free path λ_i. The electrons, however, move towards the anode and the wall, the latter in a retarding field. Hence, in a layer of gas of thickness λ_i a larger momentum is transferred from the electrons to the gas in anode direction than from the ions to the gas and this causes a pressure difference in the column which is proportional to its length and the current, increasing strongly as the tube radius decreases. Experiments confirm that the pressure is highest at the anode and its value agrees approximately with theory (182) in which the pressure difference due to the electric field is balanced by the viscous flow of the gas (184 c).

If a small amount of metal vapour with low V_i is added to a rare gas (V_i high) the degree of ionization of the vapour in the positive column will be high. However, collisions between ions or electrons and the atoms of the vapour will be rare and hence positive ions of the vapour will be driven by the field towards the cathode where they accumulate after being neutralized. In equilibrium the drift by cataphoresis is balanced by diffusion of metal vapours from the cathode region into the gas. At the same time ions of the vapour move to the wall and similarly a higher concentration of vapour develops near the wall. Experiments confirm the above conclusions; the theory has not been tested thoroughly (183, 184).

If the degree of ionization of a gas becomes sufficiently large, the total pressure at a point can rise considerably above that of the feebly ionized gas. This is particularly so when the electron pressure $p_e = N_e k T_e$ is large. Some experiments seem to show that Dalton's law holds (184 a), which states that
$$P = p_{\text{gas}} + p_{\text{ion}} + p_e. \tag{8.61}$$
Also a large local value of p_e often gives rise to a large local emission of light.

QUESTIONS

(1) Give an account of the primary and secondary collision processes in the fall region of an ordinary low-pressure glow discharge. Include an argument which shows that the actual thickness of the dark space is in an equilibrium state by applying the method of virtual displacement. Suggest a possible mechanism for the equilibrium in current density at the cathode, i.e. of the spot size.

(2) What optical and electrical evidence is available on the electron energy and drift velocity in the dark space at the negative glow boundary? When does the electron space charge make itself felt?

(3) Write a review of the work done on the velocity distribution of positive ions at the cathode of a glow discharge and discuss their variation with gas pressure and the nature of the gas.

(4) Give an account of the chemical changes which occur in the different regions of a glow discharge and include examples.

(5) The thickness of the dark space of a normal glow discharge in N_2 at 1 mm Hg and 20° C is $d_n = 0.4$ cm. If the normal cathode fall $V_c = 200$ V, find the current density j_c at 1 mm Hg by two different methods. What are the values of d and j_c at 760 mm Hg assuming 800° C in the dark space, if similarity rules hold. What is the reason for the higher gas temperature in the dark space? Make an approximate calculation of its magnitude.

(6) Use the elementary equations of the cathode fall to compare numerically V_c, pd_n, and j/p^2 for an Fe cathode in H_2 and Ne. Use (8.20) to discuss $V_c = f(j/p^2)$ for the abnormal glow in these gases.

(7) Outline briefly the possible causes of the increase in current density when a second cathode (hollow cathode) is used in a glow discharge.

(8) Fig. 119 shows that for $(Dp) < (d_n p)$, V_c rises sharply. Remember that the steady state requires a certain multiplication in the dark space to produce secondary electrons at the cathode. Find the dependence between V_c and D analytically or by a graphical method for a constant value of j/p^2 when γ_i varies according to Fig. 47 and for linear spatial variation of the field X. Compare the result with the starting potential at low pd, and comment.

(9) The theory of the positive column (8.41) shows that $T_e \propto X/p$ for $\kappa = $ const., i.e. elastic losses. Find the corresponding relation when $\kappa = 10^{-4}[1+(X/p)^2]$ and compare this result with the conventional one.

(10) Calculate X in the positive column in He at 1 mm Hg, tube radius 1 cm and 0.3 A (allowing for gas heating), using (8.42) to (8.44) and the inelastic cross-sections. Compare the result with observations and suggest possible explanations for the rise of X with p (Fig. 127) at $p > 1$ mm Hg in He and other gases. Discuss why, at low p, X varies only slightly with p and why it rises steeply above a certain pressure.

(11) Show that ambipolar flow of charge in the radial electric field of a positive column of a glow discharge is not associated with an energy loss (treat electron and ion current separately). How is the wall charge acquired? Explain its origin and sign, and how it disappears when the discharge ceases. Find the order of magnitude of the wall charge σ for $V_w = -10$ V when $R = 1$ cm.

(12) Determine the lower limiting value of cpR at which a solution of (8.36) can be obtained and comment.

9

ARC DISCHARGE

1. INTRODUCTION

If two carbon or metal electrodes are brought into contact and then separated in a circuit in which more than a few amperes flow, a self-sustained arc discharge occurs. Though this type of discharge appears commonly in atmospheric air, it is also observed at lower or higher pressures and in various gases and vapours. The short arc in air is by no means a form suitable for the study of its essential features. This is because violent chemical and thermal processes take place in the gas as well as at the electrode surfaces. The arc exists in a turbulent mixture of gases or vapours which come partly from the electrodes and partly from the surrounding atmosphere and from their reaction products. In addition, the discharge column anchored between the spots at the electrodes varies in diameter and composition over its length. Thus short arcs which have been studied most extensively in the past do not lend themselves readily to simple deductions. It will be shown later how these difficulties may be overcome.

A convenient definition of an arc is given by comparing its cathode region with that of a glow discharge: the cathode of the glow has a fall of potential of about 100 to 400 V and a low current density. Except at high pressures and currents, the positive column always fills the tube. Thermal effects do not contribute to the working of the cathode of a glow discharge, and the light emitted from the region near the cathode has the spectrum of the gas. The arc cathode, however, has a fall of potential of order 10 V, a very high current density, thermal effects are essential to its working, and the light emitted has the spectrum of the vapour of the cathode material. The positive column is constricted near the two electrodes. Thus if with a Cu cathode in N_2 (or air) at 1 atmosphere a potential difference of 350 V is observed, if the current density at the cathode is of order 1 A/cm², and if the light emitted from the cathode region is due to N_2 bands, then a glow discharge is present. But if (with electrode distances of $\leqslant$ 1 cm) the potential difference is of order 30 V, the current density at the cathode spot 10^4 to 10^6 A/cm², and the light emitted from the cathode region shows mainly the green Cu lines (arc and spark spectrum), then an arc discharge is present.

The main electric properties of an arc can be easily summarized. For a given value of the electrode distance d, the arc voltage E decreases as the current i rises (Fig. 131), and these curves lie higher for larger d. With C electrodes, E suddenly drops at h. The accompanying hissing is an anode phenomenon. For a given current the voltage rises with d as shown in Fig. 132. The linear rise reveals that the positive column has

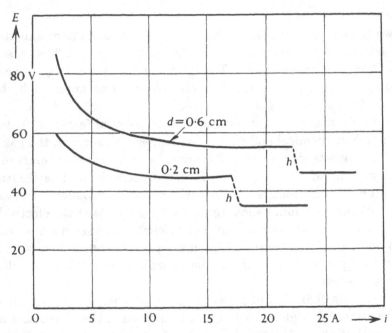

FIG. 131. Arc voltage E as a function of the current i of an arc between two carbon electrodes in air at 1 atm for two arc lengths d (37). The hissing arc appears at h.

become uniform and that its longitudinal electric field is constant. The uniformity can also be tested by measuring the emitted light (per unit length of the column) which is found to be constant. The regions near the electrodes, however, have to be excluded. Their inclusion leads to complex empirical formulae which have been used mainly by engineers and hold only in a very restricted range. It is convenient to write for the total voltage

$$E = e_c + e_a + e_p + (e_1 + e_2), \qquad (9.1)$$

where c, a, p stand for cathode fall, anode fall, and positive column, and e_1, e_2 are the voltage drops in the two transition regions which represent the non-uniform parts of the positive column; their value depends on d and i. Also for long arcs

$$e_p = X d_p, \qquad (9.2)$$

where X is the electric field in the column of length d_p which depends on i, the gas and its pressure.

Another aspect is provided by the measurement of the energy distribution. In short arcs between carbon electrodes with 5 A in air calorimetric observations show that about 42 per cent of the total electric energy

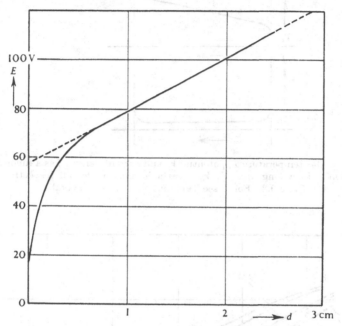

FIG. 132. Arc voltage E as a function of the electrode separation d of a carbon arc in air at 7 A.

flows to the cathode and $\sim$ 37 per cent to the anode; the rest is lost by convection, radiation, chemical changes, etc. When the C cathode is replaced by Cu, 45 per cent of the energy enters the cathode (this may be too high because of the radiation from the anode spot). Thus about 80 per cent of the energy flows into both electrodes. The balance is expected to be different when, for example, N_2 instead of air is used because of the absence of oxidation of the C anode. Fig. 132 a gives the spatial distribution of the gas temperature, the potential, and the current density in a long arc. In long arcs only 10 per cent of the anode fall energy is found to be lost by heat conduction in the anode (185, 186).

2. THE POSITIVE COLUMN

Fig. 133 gives the variation of the field with the current derived from measurements of $E = f(d)$ (Fig. 132). X is large in H_2 and H_2O because

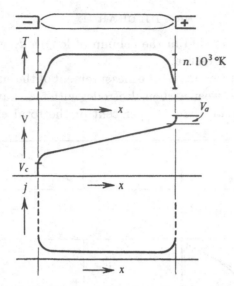

FIG. 132 a. Gas temperature T, potential V, and current density j as a function of position x in a long arc. V_c, V_a = cathode and anode fall respectively. Cf. Table 9.2. For T see Table 9.1. 'n' means several.

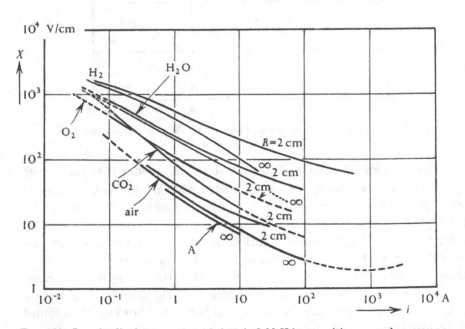

FIG. 133. Longitudinal component of electric field X in a positive arc column as a function of the current i in various gases at 1 atm for different tube radius R (stabilized columns). Values for N_2 slightly larger than those for air (187, 188, 189, 190).

of their large heat conductivity and dissociation losses and it decreases as i rises, probably because the gas temperature T_g increases with the current (see later). Since the current density is given by the concentration of charges and the drift velocity

$$j = eN_e v_d \propto N_e X \tag{9.3}$$

we have $X \propto j/N_e$. In general as i rises j falls, but N_e increases with rising T_g; thus as i rises X decreases (Fig. 133). At very large i, X rises slowly with increasing i. At larger i, X depends on the tube radius R.

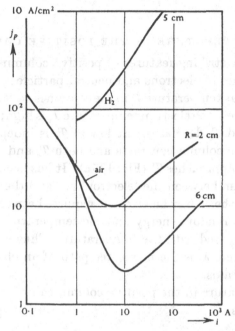

FIG. 134. Current density j_p in the positive column as a function of current i in air and hydrogen at 1 atm for different tube radius R (187, 189).

Fig. 134 shows a few results of photographic measurements of the diameter of the positive column in air and H_2 and the derived dependence of j on i. The observed rise of j at larger i depends on the radius of the tubes in which the columns are enclosed, but it is not certain whether the absolute value of the current density so found is correct since the photographic plates or films record the visible and the near ultraviolet and thus may include light from the flame surrounding the ionized region. The actual conducting path may be wider or narrower. Also, j rises as the speed of the gas flow and its heat losses (thermal conductivity) are increased.

Long (5–50 cm) straight positive columns on which the above results are based can be obtained by three methods: either by injecting tangentially air through jets at one end of the arc tube or by rotating the tube around the arc. In both cases the positive column is kept at the axis because the centrifugal forces displacing the colder gas towards the wall annul any deviation of the hot gas of the column from the axis. In another method a liquid (e.g. water) flows along the walls of the tube containing the arc so that any deflexion of the positive column causes the water to evaporate more violently and to return the arc to the axis (187).

3. The gas temperature in the positive column

One of the most striking features of positive columns is the ratio of the average energies of electrons and neutral particles. Fig. 135 shows that at low p the gas temperature T_g in the column is equal to the ambient temperature, whereas the electron temperature T_e is high; with increasing p, T_g increases and T_e decreases. At low p, T_e is independent of r, at high p the positive column contracts and both T_e and T_g depend on r. Also at large p, T_e approaches T_g (Fig. 135 a). It has been suggested that thermal equilibrium between the 'electron' gas and the neutral gas can be expected when the energy an electron acquires between two collisions is smaller than its random energy and the temperature changes in the gas are small compared with the temperature. Thus $eX\lambda_e < \frac{3}{2}kT_g$ and $\lambda_e \, dT_g/dr \ll T_g$, where λ_e is the mean free path of an electron and r the distance from the axis.

The gas temperature in the positive column can be measured by the following methods:

(a) by measuring the gas density using X-rays, α-particles or sound, and applying corrections to the ideal gas law which allow for dissociation, etc.;

(b) by measuring the intensity of molecular bands;

(c) from the intensity ratio of two spectral lines;

(d) from the intensity of γ-radiation of a radioactive isotope (Hg^{203}) which is proportional to the density of atoms;

(e) by means of an interference refractometer or by applying the 'Schlieren' method which gives the refractive index or its gradient respectively and thus the gas density.

(a) By means of two filters, soft X-rays (λ order 10 Å) from a narrow wavelength range are produced and passed as a beam along the axis

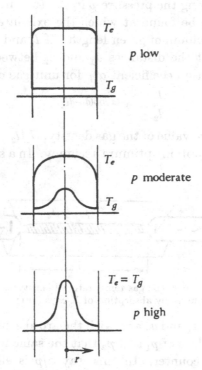

Fig. 135. Radial distribution of electron and gas temperatures
T_e and T_g at various pressures.

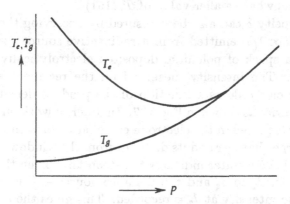

Fig. 135 a. Schematic diagram showing the dependence of the electron and gas
temperatures (T_e, T_g) on the gas pressure (density) p.

of a long positive column (Fig. 136). They are absorbed in the gas
(Chapter 3) to a degree which depends on the number of atoms per unit
volume. The intensity at the exit is measured for various lengths d of

the column. By varying the pressure p in the tube in the absence of an arc, the pressure can be found at which the exit intensity is the same as that with an arc column of given length. If I_1 and I_2 are the relative values of intensity at the distances d_1 and d_2 between the electrodes, then with the absorption coefficient a/p for unit gas density we have

$$\frac{I_1}{I_2} = e^{-(a/p)p(d_1-d_2)}, \qquad (9.4)$$

where p is the effective value of the gas density. I_1/I_2 is usually chosen to be about $2 \cdot 7 (= e)$ to obtain optimum accuracy. In a second experiment

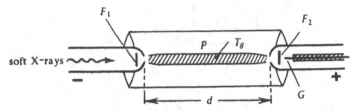

FIG. 136. Measurement of the gas temperature (density) in a long positive column by absorption of X-rays (191).

with the same values d_1 and d_2 and with the arc absent, the gas pressure is reduced to such values of p_1 and p_2 that the same intensities I_1 and I_2 are obtained in the counter. In this way a/p is eliminated and the equivalent density p of the gas and its temperature $T_g = T_{\text{room}} \cdot p/p_{\text{room}}$ is found. A correction is to be made allowing for dissociation, etc., which leads to a somewhat smaller value of T_g (191).

The gas density δ can also be measured by observing the range R of α-particles ($R \propto 1/\delta$) emitted from a radioactive source consisting, for example, of a speck of polonium deposited electrolytically on a small silver sphere. The intensity measured on the receiver—a counter—inside an arc electrode as a function of the product density (p) times distance (d) varies as shown in Fig. 137. In order to work on the point A the α-emitter is placed in the electrode chamber filled with the same gas as the discharge chamber and its distance from the window W (Fig. 137) is varied until the counter indicates the steep fall. Then the arc length is reduced from d_1 to d_2 and the distance : source—W increased by Δ until the same intensity at d_2 is recorded. This gives the average equivalent gas pressure in the arc column. The temperature in the column follows from

$$T_g = T_0 \Delta/(d_1-d_2), \qquad (9.5)$$

where T_0 is the temperature of the gas between the source and W in the water-cooled electrode chamber. T_g has to be corrected as above. If the counter potential is only applied during a fraction of a cycle of an a.c. arc, the instantaneous values of the gas temperature can be measured (192).

The gas temperature can also be found by sending a sound wave along the axis of the positive column and measuring its velocity (193). Again, to eliminate disturbances caused in the regions near the cathode and

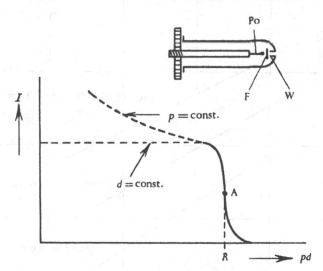

Fig. 137. Measurement of the gas temperature in a positive column by measuring the range R of α-particles. Intensity I of the α-rays as a function of the mass of gas (pd) traversed. Above: cross-section through the arc electrode with polonium source. Its distance from the window W is variable. The curve $p = $ const. shows that I decreases as d is increased because the solid angle subtended by by the source at the detector decreases.

anode, measurements with different arc lengths are required. The velocity v_s of a sound wave is given by

$$v_s = \sqrt{\left(\kappa\,\frac{p}{\delta}\right)} \propto \sqrt{\left(\frac{\kappa RT}{M}\right)}, \qquad (9.6)$$

where $\kappa = c_p/c_v$, δ and p the density and pressure of the gas respectively, M the molecular weight, and T the absolute temperature; (9.6) is valid provided the ideal gas law holds. If dissociation occurs another value c_p/c_v has to be used. Since the distance the sound wave travels is known and its transit time is measured to obtain v_s, T can be found from (9.6).

(b) Spectroscopic methods can be applied to measure the gas temperature. We assume that those molecules emitting rotational band spectra are in thermal equilibrium with the unexcited molecules, and that self-absorption is negligible. The intensity of the emitted lines in a pure rotation spectrum varies as the thermal distribution of the rotational states which is proportional to the Boltzmann factor $e^{-E/kT}$; since each state is composed of $(2J+1)$ levels (J = total angular momentum quantum number), the frequency of its occurrence, the statistical weight, is $(2J+1)$ times greater than that of the state with $J = 0$, the

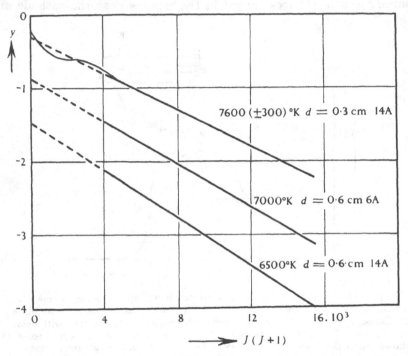

FIG. 138. Intensity y of the rotational lines as a function of the rotational quantum number J (total angular momentum) for a carbon arc in air at different current i and length d.

number of molecules and hence the intensity of a rotational line is more accurately

$$I \propto \nu^4(J'+J''+1)e^{-C[J'(J'+1)/kT]}, \qquad (9.7)$$

where $\nu = 1/\lambda$ is the wave number of the emitted line, J' and J'' the angular momenta of the upper and lower states respectively, and $C = h^2/8\pi^2\theta$, where θ is the known moment of inertia of the molecule. Thus, when $y = kC^{-1}\ln[I/(J'+J''+1)]$ is plotted against $J'(J'+1)$ a straight line should result with a slope proportional to $1/T$. Fig. 138 shows the intensity of the lines of the CN bands of carbon arcs in air at 6 and 14 A, in particular of the higher terms of the R-branch, which confirms (9.7). From the slope a temperature of 6500° K is found. The deviation at low values $J'(J'+1)$ is due to self-absorption. It has to be borne in mind that the exponential term can be written as $e^{-eV_r/kT}$ and that $eV_r \ll kT$ and $\nu \approx$ constant, since eV_r is always of order 10^{-2} to 10^{-3} eV whereas here kT is of order 1 eV. It has also been found that the 'rotation' temperature T is equal to the gas temperature. This method also gives the radial distribution of T in a cylindrical arc column (194).

The results obtained by the various methods are listed below:

TABLE 9.1
Measured temperatures in the positive column of arcs (17, 191, 193, 194)

Gas/electrode	Pressure	Arc length (cm)	$i(A)$	Method	Position	T_g (°K)
Air C	1 atm	0·8	~2	CN band	axis	5900±300
,, ,,	,,	0·3–1·8	1–12	,,	,,	6200–7800
,, C+Al₂O₃	,,	1·5–2	6–7	AlO	flame	2800–3800
,, C+Ca	,,	few cm	180	Ca⁺	axis	~12 000
H₂ W	,,	..	4	H_α–H_ξ	,,	4900
,, W	,,	..	8–9	,,	,,	5300–6300
,, C	,,	..	few	C₂	,,	5000
N₂ Cu	,,	15	2	X-rays) α-rays)	,,	5300±300
Air Cu	,,	,,	2	,,	,,	4700±450
,, ,,	,,	..	14	H_α	,,	5470±10
H₂O ..	,,	(jet)	500	..	,,	35 000
,, ..	,,	,,	1500	..	,,	50 000
Air Cu	,,	..	≈10	sound	,,	6100

4. THE ENERGY AND THERMAL BALANCE IN THE POSITIVE COLUMN

In general at atmospheric pressure and moderate currents only a few per cent of the energy in the column of molecular and rare gases is converted into radiation or chemical energy; except for vapours like Na, Hg, etc., the larger part of the energy heats the gas and is carried away by convection (197).

The influence of convection on arc columns can be demonstrated by eliminating the effect of gravity which induces convection. This is done by enclosing the arc in a box filled with air which is allowed to fall freely while the field in the positive column, its current density (diameter), and the emitted light are recorded. The results are as expected: the heat convection losses disappear and the gradient in the positive column becomes smaller. The constricting effect of the axial gas flow ceases, the diameter of the column grows, and the current density and light emitted per unit length decrease (Fig. 139). (A decrease in gas temperature is to be expected too.) Such experiments have been done with arcs in stagnant air as well as with mercury arcs enclosed in narrow tubes (196).

In the absence of radiation and convection, losses are mainly due to heat conductivity. However, since the temperatures in the column vary radially between 5000° and 50 000° K at the axis and room temperature

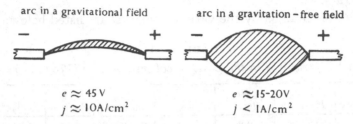

arc in a gravitational field   arc in a gravitation – free field

$e \approx 45\,V$
$j \approx 10A/cm^2$

$e \approx 15\text{-}20V$
$j < 1A/cm^2$

Fig. 139. Horizontal arc between carbon electrodes in air (8 A, 3 cm long) in the gravitational field of the earth and in a gravitational-free field (195). e = arc voltage, j = current density in the positive column.

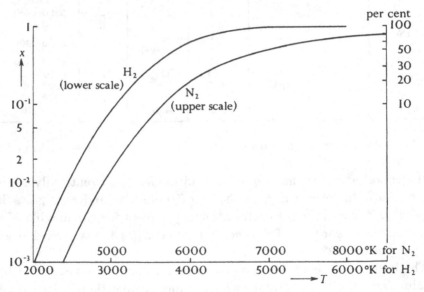

Fig. 140. Degree of dissociation x as a function of temperature T in thermodynamic equilibrium for H_2 and N_2 at 1 atm.

at the edges account has to be taken of dissociation, turbulence, and diffusion as well as of chemical reactions. Thus at large currents or high pressures H_2, N_2, O_2, etc., are partly dissociated into atoms (Fig. 140) and recombined either on the walls surrounding the column or in the colder portion of the gas: NO, OH, and other molecules and radicals are produced in higher vibrational states near the axis and diffuse outwards, partly dissociating on their way. In the steady state a diffusion and dissociation equilibrium is set up. The net result is that in certain temperature regions the apparent conductivity is increased. The change of the apparent heat conductivity Λ with temperature for H_2 and N_2 is shown in Fig. 140 a. At low and high values of T, Λ corresponds to the conductivity of H_2 and H respectively, but there is an intermediate

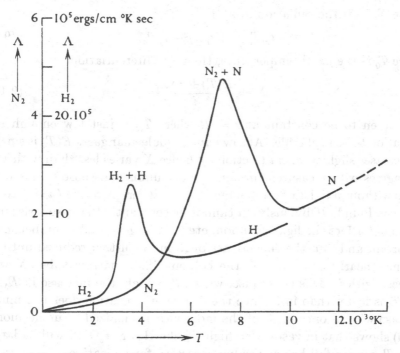

FIG. 140 a. Thermal conductivity Λ of hydrogen and nitrogen as a function of the temperature T. Influence of dissociation. The peak for O_2 lies at about $4000°$ K and is probably twice as high as for N_2.

range of T where dissociation gives rise to a maximum. At very much higher T, Λ should rise again when transitions of electrons set in. This 'thermal excitation' is observed spectroscopically in an electrically heated carbon tube (King's furnace) with elements of low excitation potential. The dissociation and excitation peaks have been suggested to explain radial variations in a positive column of thermal and electric conductivity and other properties (194 a, b).

The purpose of setting up the energy balance in the positive column is to find relations between electric field X, current i, radius R of the conducting cylinder—the seat of thermal ionization—the gas and its pressure. However, though the problem appears relatively simple, it

offers certain formal difficulties. For this reason only approximate
solutions are given here.

Assuming a long column with radial losses by radiation and heat
conduction the energy balance reads

$$iX = E(T) - 2\pi R \Lambda \left(\frac{dT}{dr}\right)_R, \qquad (9.8)$$

where $E(T)$ is the radiation loss. Let

$$T_r/T_0 = \{(r/R)^2 + 1\}^{-1}, \qquad (9.9)$$

where T_0 is the axial temperature, then by differentiation

$$X = \frac{E(T) + \pi \Lambda T_0}{i}. \qquad (9.10)$$

Λ is taken to be constant at $r = R$ where T_R is just low enough for
ionization to be negligible. At low i and in molecular gases $E(T)$ is small.
T_0 increases slightly with i (section 3), hence X varies less than with $1/i$;
this agrees with measurements on open columns and those enclosed in
tubes without axial air flow for moderate i, where $X \propto i^{-n}$ ($n = \frac{2}{3}$ to $\frac{3}{4}$)
has been found. (Obviously Xi cannot be constant, otherwise a rise in i
would not affect the light emission, etc.) At larger i, radiation becomes
important and thus the dependence of X on i is further reduced until X
becomes nearly constant. If the column fills a narrow tube, X will
increase with i. This is because, where T_0 is high, the increase in $E(T)$
and T_0 is faster than in i, since the degree of ionization does not much
increase with T once most of the molecules are ionized. Furthermore,
(9.10) shows that in gases with high Λ, like H_2 or H_2O, X will be large
since T_0 cannot fall below a value necessary for ionization.

With vapours which emit strong resonance lines, $E(T)$ is the important term because V_i is in general low and thus a smaller value of T_0
is needed to provide ionization. Taking a long tube (radius R) in which
a volatile substance is completely evaporated, we have approximately

$$iX = E(T). \qquad (9.11)$$

If radiation is mainly due to a resonance line which is excited by electron
collisions, then the number of quanta emitted per sec in unit length of
tube is proportional to the electron density, and hence to i, and thus

$$X \approx \text{const}, \qquad (9.12)$$

which agrees with observations over a reasonably broad range. For a
given amount of vapour the number of atoms per cm³ in the ground
state is $N_0 \propto 1/R^2$, and with a given mass m of vapour in a sealed tube

we have $N_0 \propto m/R^2$. The electron concentration $N_e \propto N_0^{\frac{1}{2}}$ (3.57). From the continuity equation

$$i \propto jR^2 \propto N_e R^2 v_d \propto N_e R^2 X/p \propto \frac{N_0^{\frac{1}{2}} X R^2 f(T)}{N_0} \propto X R^3 f(T), \quad (9.13)$$

and thus

$$X \propto \frac{i}{R^2 f(T)}. \quad (9.14)$$

This means that for constant power Xi per cm tube, $X \propto R^{-\frac{3}{2}}$. This result is consistent with (9.14), since $f(T)$ rises with i.

So far it has been assumed that the degree of ionization in the positive column at high pressure is given by the gas temperature only, which is thought to be equal to the temperature of the electrons and ions. With the available data this assumption can be tested as follows: for a long positive column in N_2 the gas temperature at the axis has been measured (section 3) and simultaneously the electric field and the average density. Hence in the equation

$$j = eN_e v_d = eN_e k^- X = Xf(T) \quad (9.15)$$

j, X, and T are known while $f(T)$ is taken to be given by (3.57), the equation of thermal ionization in equilibrium. It was found that (9.15) holds for various molecular gases. In gas mixtures (air) formation of molecules with low V_i (NO with 9·5 eV) have to be considered.

In the case of Hg, instead of applying (9.15) another check can be made. Let the input per cm of column be $X.i$; the radiation loss is assumed to come from one fictitious pair of levels between 5 and 8 eV giving 6000 Å quanta, other losses being neglected. Let the transition probability for all lines be $\Sigma A = 10^9$. Then

$$iX = \Sigma A h\nu N_0 e^{-11700 V_a/T}. \quad (9.16)$$

If the density of vapour $N_0 \approx 10^{18}$ atoms/cm³, $Xi = 30$ watts/cm, $h\nu = 4.10^{-19}$ W sec, $V_a = 8$ eV, then $T \sim 5000°$ K. This is of the right order as found from experiments (197).

5. Observations on Arc Cathodes

Self-sustained arcs can be conveniently divided into two groups: those whose cathodes vaporize readily at temperatures too low to give thermionic emission of electrons—the so-called cold cathodes—and those whose cathodes have temperatures which give copious thermionic emission without appreciable vaporization. One of the most puzzling problems of the arc discharge is the functioning of the cathode of the cold arc. Cathodes of Cu, Ag, liquid Hg, and many other metals are examples of this type. It can be stated from the outset that no final solution of this problem has been found yet. Thermionic cathodes like C, W, the rare earths, etc., are fairly well understood. Strangely enough, the cathode fall in potential is in both cases of the same order of magnitude, but perhaps a little higher for substances of the second group. The cathode substances also fall into two groups when the restriking time is investigated: a circuit containing a carbon arc can be interrupted for up to about 1 sec at applied voltages of several hundred V and will

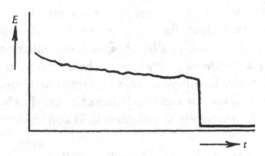

FIG. 141. Arc voltage E between electrodes which approach one another until they touch. The ordinate of the discontinuity is the sum of cathode and anode fall. The abscissa represents time or distance between the electrodes.

restrike without change of the electrode separation, while a Cu arc when interrupted for 10^{-3} sec or less will not restrike. Hg arcs at low pressure have restriking times of less than 10^{-8} sec.

Let us deal first with some elementary observations. An approximate determination of the sum of cathode and anode fall can be made by starting an arc discharge in a gas between identical electrodes several mm apart and gradually closing the gap while measuring with an oscillograph the potential difference across the arc. Fig. 141 shows such an oscillogram.

Initially the potential is about 20 V, decreasing slowly with time and distance, and there is finally a sudden drop down to a few tenths of a volt when the two electrodes touch; the remaining potential is due to the contact resistance. The smallest potential difference before the contacts close minus the anode fall V_a (see later) is taken as V_c, the cathode fall in potential of the arc. By using different materials for anode and cathode V_a and V_c can be separated approximately. Values for different substances in various gases are given in Table 9.2. They lie between the ionization potential of the gas and that of the cathode vapour. This seems to indicate that the presence of the vapour is essential. On the other hand, it has been shown that a cold arc cathode spot can be driven around magnetically or mechanically with speeds of order 10^4–10^5

cm/sec and this was taken to prove that with good conductors like Cu or Ag there is not sufficient energy or time available to melt and evaporate the cathode material. However, it has to be remembered that with positive ions impinging on a surface with low speed no actual melting but direct evaporation of a very thin surface layer could take place. Each ion carries with it an energy of between 1·5 and 2 eV_i of which only a fraction is used to liberate electrons. This intense evaporation may be accompanied by strong back-scattering in its own vapour. Thus from measurements with fast moving spots the conclusion cannot be drawn that the cathode vapour is absent, but only confined to a restricted volume. Spectroscopic work confirms the presence of vapour in the arc spot.

In this connexion the observations of the cathode spot temperatures are of interest. They have been obtained by pyrometry, or by measuring the emitted radiation. The results, which in some cases are a rough guide only, are collected in Table 9.2. As far as carbon is concerned, the value of the cathode temperature T_c together with the thermionic constants A and φ (page 90) gives a current density j_c which is of the same magnitude as that found in Table 9.2. The electron emission on a C

TABLE 9.2

Cathode and anode temperature, current density, and fall of potential at p = 1 atm except for Hg (17, 198, 199)

Electrodes	Gas	i (A)	T_c (°K)	T_a (°K)	j_c	j_a (A/cm²)	V_c (V)	V_a (V)
C—C	air	1–10	3500	4200	470	65	9–11	11–12
C—C	N_2	4–10	3500	4000	500	70	..	..
Cu—Cu	air, N_2	~ 5	< 2200	2400	~ 10^6	~ 10^3	8–9	2–6
Fe—Fe	air	~ 5	2400	2600	high	..	8–12	2–6
Ni—Ni	,,	~ 5	2400	2400	..	..	..	..
W—W	,,	~ 5	3000	4200 ?	..	..	..	..
Al—Al	,,	~ 5	3400	3400	..	..	..	..
Al—Al	N_2	~ 5	~ 2500	~ 2500	..	..	..	..
Zn—Zn	air	~ 5	3000	3000	..	..	..	..
Zn—Zn	N_2	~ 5	low	low	..	..	..	..
Hg—C	Hg	> 2	~ 600	..	10^6	..	7–10	0–10

For Cu, Ag, Al, Ni, Fe, Ti arcs in air, T_a = 2650, 2300, 3300, 3650, 3600, 4750° K and j_a = 10^4 to 10^6 A/cm² have been observed recently (197 a). These temperatures seem to be fictitious (197 b).

cathode thus occurs thermionically. The value of T_c for cold arcs, however, give values of j_c (eq. 3.59) which are about 4 orders of magnitude lower than those observed. In addition there is considerable doubt as to whether these values refer to the surface temperature of the cathode or a dense vapour cushion close to it. In the case of Hg the existence of the latter has been established; the surface temperature of the Hg cathode spot is only about 360° C. Another curious fact is that anode temperatures of metal arcs seem not much different from cathode temperatures (Table 9.2).

Finally we consider the current densities in the electrode spots. Their values though well established for carbon electrodes are not very accurately known as far as metal arcs are concerned. They are derived from photographic observations (198) or from marks left on the electrodes (199). Because of the fast random movement of the spot the 'active' area is not too well defined and is probably much smaller than the optically emitting area taken from photographs. Another point is that the shape is not necessarily circular as is usually assumed, and that sometimes several spots seem to exist simultaneously.

6. THEORIES OF THE CATHODE OF COLD ARCS

A number of suggestions have been made to explain the working of the cathode of metal arcs. Ruling out conventional thermionic emission, the most frequent suggestion is electron emission under the influence of a strong electric field. This is supposed to be due to an intense space charge of positive ions situated about one mean path from the cathode. A variation of this suggestion is the combined effect of electric field and temperature; see e.g. (186 b).

If the constricted part of the positive column above the cathode spot is taken as an ideal plasma and the cathode regarded as a negative probe inserted into it, the thickness of the cathode fall space is obtained from (2.15) to be

$$d = \left(\frac{9}{32\pi} \frac{k^+ V_c^2}{j_c}\right)^{\frac{1}{3}}. \tag{9.17}$$

With $V_c = 10 \text{ V} = \frac{1}{30}$ e.s.u., $j_c = 10^5 \text{ A/cm}^2 = 3.10^{14}$ e.s.u., $k^+ p = 3.10^2$ e.s.u. and $p = 100$ mm Hg, we have $d \approx 10^{-4}$ cm or several mean free electronic paths. The field at the cathode is then $X_c = 2V_c/d = 2.10^5$ V/cm, and from (3.66) the current density due to field emission is found to be negligibly small. Another difficulty is that because of the low V_c the available electric field can only act over small distances in the metal lattice.

Arguments have been advanced to show that local fields due to surface roughness are actually much higher, or that the constants used apply to pure metals whereas the emitting areas consist of compounds which emit more readily. However, no convincing evidence for or against the field theory has been supplied. Another suggestion is that thermal ionization in the hot gas above the cathode spot produces the required charge densities. The cathode attracts the positive ions which carry the whole current in the fall region. The nature of the electron emission from the cathode, however, is not clearly understood from such considerations (see later).

There is another difficulty arising from the heat loss from the spot by conduction into the cathode material. Experiments with water-cooled cathodes of different thicknesses show that the potential difference and thus the cathode fall input is practically independent of that conduction loss. It follows that this loss is hardly influenced by changes in the temperature distribution at distances large compared with the size of the spot. On the other hand, assuming a Cu arc in N_2 with $j_c = 10^5$ A/cm², $T_c = 2000°$ K, $i = 10$ A, the heat loss by conduction is that of a circular area of temperature T_c which is part of a semi-infinite medium, zero temperature being taken at infinity. For constant heat conductivity Λ and radius R of the area, the heat flow

$$H = 4R\Lambda T_c \qquad (9.18)$$

(for a hemispherical surface $2\pi R$ instead of $4R$). Since $i = j_c R^2\pi$, the volt equivalent of the conduction loss is $H/i = 4\Lambda T_c/(\pi j_c i)^{\frac{1}{2}}$ volts, which is about 6 V, that is comparable with the arc voltage (10 V) provided j_c is taken of the order 10^6 A/cm². (Also, heat is produced in the 'constriction resistance' which is caused by the concentration of current flowing to the small area; it is equal to $i^2\rho/4R$, where ρ is the specific resistance of the cathode material. The volt equivalent is thus $i\rho/4R$ and is very small.) With the old low values of j_c of order 10^3 to 10^4 A/cm² the volt equivalent of the conduction loss was of order 100 V and hence more than 10 times the observed value of V_c. Consequently it was necessary to assume thermal resistances between the spot and the cathode substance in form of oxides, etc., or other artifices. In support of this view cathode spots are known to extinguish less easily when the surface of the cathode is, for example, oxidized. This has been explained by the lower work function of the oxidized surface (which is not always true). However, it is also possible that these specks of impurities do steady evaporation and avoid formation and ejection of droplets of the metal. Such droplets act as recombining surfaces and should increase the losses and thus can interrupt the arc. A way out of this impasse is indicated later.

Another way to inquire into the possible function of the cathode of a cold arc is to set up its energy balance. Here one has to define the coordinates carefully; the energy flux has to be calculated for a certain position in the system. Taking this to be the cathode surface, the energy flowing in is made up of the potential and kinetic energy of the positive ions, radiation, and heat conduction from the hot gas of the column which includes impact of excited atoms; the energy flow outwards (cooling) is due to emission of electrons—provided it is not field emission—and evaporation of atoms and the conduction losses into the metal and the gas. From the values given in Tables 9.1 and 9.2 we note that—in contrast to earlier investigations—the temperature along the axis of any 'thermal' positive column is distributed as given in Fig. 132 a, showing that T is lower at the electrodes than in the positive column and thus heat (V_{gas}) flows from the hot gas to the electrodes. If a ions enter the cathode while $(1-a)$ electrons leave it, the energy rate at the cathode per A current, that is the volt equivalents of energy, can be combined in the following energy balance at the cathode:

$$\{a(V_i+V_c-\varphi)+V_{gas}\}-\{(1-a)\varphi+V_{evap}+V_{rad}\} = V_{cond}. \qquad (9.19)$$

From this equation, for example, V_c can be found when a is known or vice versa. The first term is the gain where a is the ion current as a fraction of the total current, and the second is the energy lost by the cathode where φ is the work function. V_{cond} is the heat conduction loss in the cathode. The absolute value of these terms are only partly known, but a rough estimate—which is not free from a personal factor—would be for the Cu arc (heat loss area > current area):

$$\{a(7\cdot7+4\cdot5-4\cdot5)+0\cdot5\}-\{(1-a)4\cdot5+0\cdot1+0\} \approx 5. \qquad (9.20)$$

For the kinetic energy of ions, instead of V_c about $\frac{1}{2}V_c$ is assumed since a large number will probably collide with atoms in the fall space. The net result is that $a \sim 0\cdot8$ and so a larger part of the current is carried by positive ions. Since V_{rad}, V_{evap}, and V_{cond}, etc., are not well known, the information which one can obtain at present from the energy balance is very dubious. (9.20) holds provided the ions are singly charged; the presence of doubly charged ions would make a even smaller. Further, it is assumed here that the ions after having struck the cathode have lost all their energy except for the small amount which corresponds to the cathode temperature.

A small heat loss by conduction can only be understood by abolishing the classical evaporation concept and introducing 'stepwise ablation' under the impact of excited species. This is consistent with the modern theory of evaporation. Here an atom gains sufficient energy to escape from the lattice by absorbing several energy quanta (199 d).

Finally, the emission of electrons from cold arc cathodes has been thought to arise from collisions by excited atoms (199 a). The luminous spot is assumed to be a region of dense vapour very near to the cathode ($< 10^{-3}$ cm) which contains excited atoms in large concentrations. Since the distance between the spot and the cathode is very small, excited atoms scattered back towards the cathode cannot lose their excitation energy by radiation and so impinge on the cathode where they release electrons with a high yield (page 99). The electrons acquire sufficient energy in the cathode fall to excite atoms and its value can thus be well below ionization potential. The small size of the spot has been explained by the concentrating effect of the self-magnetic field on the electrons which is opposed by radial diffusion of charges. The high vapour density at the cathode is the result of momentum transfer from ions onto the atoms of the vapour.

It is noteworthy that a spot is only necessary at the cathode of a 'cold' arc whereas at the anode no spot is needed. A special case is the 'unipolar' arc, observed as a transient in a thermo-nuclear device (199 c). The cathode spot is formed on a metal wall adjoining a fully ionized gas (plasma) carrying very large currents. The 'plasma' acts then as anode and charge flow from the plasma to the metal wall completes the circuit.

Under certain conditions, a thermionically emitting arc cathode can be changed into a 'cold' arc cathode though the cathode material as well as the gas remains the same. This was first observed with W cathodes but more extensive studies have been made with pure C electrodes in different gases. Fig. 141 a shows the cathode current density $j_c = f(p)$ in N_2. As p is reduced the spot area grows, i.e. j_c falls, and at a critical value of p (somewhat dependent on i) j_c rises to extremely large values. At the same time the spot, originally at rest, begins to move rapidly, the total arc voltage drops by a few V and strong ablation of the cathode occurs, observed by blackening of the surrounding tube. The arc cathode in this 'vapour mode' appears to be similar to that on Cu or Hg, i.e. electron emission occurs through excited carbon atoms. Thus, as p exceeds a certain value, the N_2 molecules collide with C* at a rate sufficiently large to reduce their concentration by 'quenching' the excited states. The potential energy of C* is carried away by N_2^* which does not contribute to electron emission. The quenching cross-sections calculated from the critical pressure are of the order of the optical values and seem to confirm this picture. Also no thermionic arc but only vapour arcs are observed in C–Ne or C–A, since the potential energy of C* is too low to be transferable to rare-gas atoms with high excitation energies (199 e).

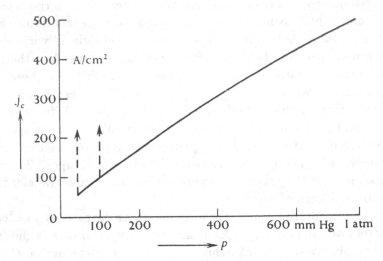

Fig. 141 a. Current density j_c at the cathode of a carbon arc as a function of the gas pressure p; dashed: range of transitions to the vapour arc (the higher i the lower p at transition). Gas: N_2 or air; $i \leqslant 8$ A; $d = 0.35$ cm (199 e).

7. THE ANODE OF AN ARC

In general the electric field of the anode draws electrons from the positive column and repels the positive ions. Consequently a negative space charge is set up in front of it, the field rises, and an anode fall develops. Usually the anode fall in potential of a metal arc is smaller than that of a glow discharge and the current densities at the anode are many orders of magnitude larger. The temperature of the anode spot is high at large gas pressures.

As the pressure in a discharge is increased its positive column contracts gradually (Fig. 135) and the current density rises until evaporation of the anode sets in (see below). The anode end of the column forms a spot filling the region in front of the anode with metal vapour. Chemical reactions between the gas and the hot anode are thought to produce the peculiar hissing noise (section 1).

That electrons and not negative ions impinge in general upon the anode of an arc was shown very early by a crude measurement of e/m of the particles (199 b). Let V be the arc voltage and i the current, then $n = i/e$ is the number of charges arriving at the anode per sec. Assuming that the electric energy is converted into kinetic energy of the particles $Vi = nmv^2/2$, where v is their average velocity. The linear momentum transferred to the anode $P = nmv = mvi/e$, and combining the three expressions one obtains $e/m = 2Vi^2/P^2$ and $v = (e/m)P/i$. P was measured by observing the pressure of a stream of gas directed against a Hg anode which gave the same depression as the arc. e/m was found to be of the order 10^{17} e.s.u. and v of the order 10^8 cm/sec. This shows that electrons strike the anode. It is left to the reader to explain how in spite of the numerous erroneous assumptions (e.g. V being the total arc voltage) the right answer was obtained.

The energy balance at the anode is as follows. The energy gain is given by the sum of the kinetic and potential energy of electrons and by the excitation, chemical reaction, and thermal energy of the gas atoms, whereas the losses are due to evaporation of the anode material, radiation, and conduction from the anode spot into the electrode. Thus per A electron current, we obtain with V_a from Table 9.2 and with V_{cond} from calorimetric observations (section 1):

$$\{V_a + \varphi + V_{therm}\} - \{V_{vap} + V_{rad}\} = V_{cond}, \tag{9.21}$$

and inserting approximate figures for the Cu anode, we have

$$\{2 + 4 \cdot 5 + 0 \cdot 5\} - \{1 + 0 \cdot 1\} \approx 6.$$

The work function φ is energy gained because each electron which arrives at the anode neutralizes a positive metal ion when it enters the lattice. Again V_{cond} is not accurately known.

The energy balance for the anode of a carbon arc is still simpler when only V_{rad} is considered (the low conduction loss is partly balanced by the heat of reaction):

$$j_a(V_a + \varphi) = a_0 \sigma T^4. \tag{9.22}$$

With $j_a \approx 70$ A/cm^2 and $\varphi = 4 \cdot 5$ eV, $T_a = 4100°$ K, $\sigma = 5 \cdot 8 . 10^{-12}$ W/cm^2 (°K)^4, and the emissivity $a_0 = 0 \cdot 75$ for C we find $V_a \sim 12$ V.

The second problem is that of the current density at the anode. What determines the size of the anode spot, and why is the area of the spot at a higher pressure not equal to that of the positive column? One of the main reasons is that the area of cross-section of the positive column is essentially determined by radial diffusion processes and recombination of electrons and ions in the denser outer zones of the gas, whereas heat conduction of the metal of the electrodes and vaporization determine the area of the spot. The ionization and excitation potentials of the electrode vapour are usually lower than the corresponding values for the gas, and the density of the vapour near the spot is higher than that of the gas. Consider a positive column extending with constant diameter right to the anode surface initially at room temperature. The electrons will

heat the anode and produce there intense evaporation at the centre
where the current density is largest. At the same time the near regions
of the column will receive a vapour stream which decreases in intensity
for points lying further off the axis. Ionization is therefore favoured
nearer the axis and as a result of this the positive column contracts.
Thus it seems that the primary cause for the formation of the spot is
the radial current distribution in the column, but the final spot size
seems to be the effect of the subsequent emission of vapour which causes
the column to contract radially (199f).

8. ARC DISCHARGES WITH EXTERNALLY HEATED CATHODES

Take a large vessel containing a hot cathode (W filament or coated
emitter) and an anode (Ni, Mo, or W) filled with a gas at $p = 0\cdot1$ mm Hg
or more. When the cathode is at such a temperature that electron emission is observed (for W: $T = 2500°$ K, $j = 0\cdot3$ A/cm^2), then a 'plasma'
is formed between cathode and anode. This is a region of very dense
equal space charges which appears instead of the positive column when
the walls are far removed from the axis. This plasma shows very small
losses. The potential drop is low and so is the anode fall in potential.
Thus the main voltage drop across the discharge is due to the cathode
fall, the cause of which will now be discussed in more detail.

The cathode, considered to be plane (Fig. 142), can be regarded as
a probe which is kept at a negative potential with respect to the plasma
but which emits electrons carrying the discharge current. Since the ions
which are attracted by the cathode do not have to release secondary
electrons, the electric field at the cathode surface—neglecting the small
potential trough—is approximately zero. In a similar way to a space-
charge sheath surrounding a probe, the field at the plasma side of the
cathode fall region vanishes. Field, potential, and space-charge distribution in a hot cathode arc discharge are shown in Fig. 142. The fall space
d on this picture develops because the plasma electrons are repelled by
the cathode, thus leaving a space in which only positive ions from the
gas and electrons from the cathode move. When the pressure is low and
d of order one electron mean free path, then ions and electrons move
through d as in a vacuum. Calculation of the space-charge-restricted
current for ions and electrons moving simultaneously gives

$$j_e = \frac{1\cdot85}{9\pi}\left(\frac{2e}{m}\right)^{\frac{1}{2}}\frac{V_c^{\frac{3}{2}}}{d^2}, \tag{9.23}$$

$$j^+/j_e = (m/M)^{\frac{1}{2}}. \tag{9.24}$$

From (9.24) it follows that $j^+ \leqslant 2$ per cent of j_e for any gas. This set of equations gives the ratio of the current densities for a particular ion (gas). For a known emission current j_e and cathode fall V_c, p can be found. The complete solution of the problem requires another independent equation.

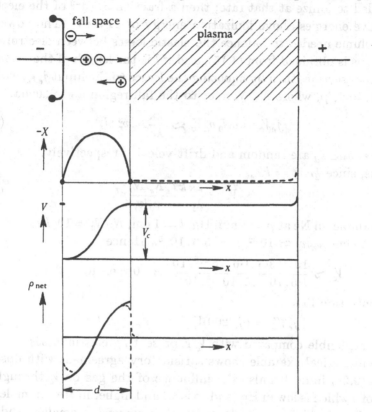

FIG. 142. Field X, potential V, and net space charge ρ of a hot cathode arc discharges as a function of position x for a large plane cathode. ρ_{net} dashed: probable actual distribution.

Since it was assumed that the plasma is everywhere at the same potential, its only source of energy to overcome losses is the kinetic energy of 'primary' electrons entering the plasma from the cathode end. These electrons which arrive at the boundary with nearly uniform energy will give rise to an energy distribution in the plasma partly by interaction between electrons, atoms, and ions and partly by collisions with excited atoms which can accelerate some of them to energies well above ionization potential. At larger currents collisions between excited atoms can lead to ionization. The exact nature of the conversion of a narrow into a wide energy distribution is not yet known. If we assume the plasma loses ions moving to the cathode, a lower limit to V_c can be estimated. For Ne, (9.24) gives $(m/M)^{-\frac{1}{2}} \approx 200$ and hence every 200th electron must ionize once on the average. With a Maxwellian distribution and ionization by single electron collision we find that an average energy of $kT_e = 3 \cdot 5 \, \text{eV}$ is needed to ionize at that rate; then a fraction 5.10^{-3} of the electrons will have energies above ionization energy of $21 \cdot 5 \, \text{eV}$. The power per unit volume needed to balance the elastic losses between electrons and Ne atoms is obtained from (4.15), and if l is the length of the plasma of unit cross-section, then by equating the losses to the input $j_e V_c$, we have with $\lambda_e = \lambda_1/p$, where N_0, N_e refer to the fall region and plasma,

$$j_e V_c = eN_0 v_d V_c \approx \frac{m\kappa}{\lambda_1} N_e v_r^3 \, pl, \tag{9.25}$$

where v_r and v_d are random and drift velocity respectively.

Thus, since $\frac{1}{2}mv_r^2 = kT_e$,

$$V_c \approx \frac{2\kappa}{\lambda_1} \frac{kT_e}{e} \frac{N_e/N_0}{v_d/v_r} pl. \tag{9.26}$$

For example, in Ne at $p = 1 \, \text{mm Hg}$, $l = 1 \, \text{cm}$, $N_e/N_0 = 10$, $kT_e = 3 \cdot 5 \, \text{eV}$, $\lambda_1 = 0 \cdot 15 \, \text{cm}$, $v_d/v_r \approx 10^{-2}$, $\kappa = 5 \cdot 6.10^{-5}$. Hence

$$V_c \approx \frac{10^{-4}.3 \cdot 5.1 \cdot 6.10^{-12}.10}{1 \cdot 5.10^{-1}.5.10^{-10}.10^{-2}} \approx 10^{-2} \, \text{e.s.u.} = 3 \, \text{V}.$$

The ionization loss is

$$j_e V_c' = (j_e/200)V_i \quad \text{or} \quad V_c' = 0 \cdot 1 \, \text{V}$$

and is negligible compared with V_c representing elastic losses.

The numerical example shows satisfactory agreement with observations. (9.26) indicates also the influence of the gas on V_c through the value of κ which is low in Hg (and so is λ_1) and higher in He. In molecular gases κ is orders of magnitude larger, T_e is somewhat smaller, and thus V_c larger than in rare gases. Also at larger currents, because of heating, the gas density in the plasma is reduced though the pressure may be high. Ionization does not occur by single collision only, as has been assumed above. (9.26) is not strictly applicable to cases in which $\lambda_e < d$. The value of d is found from (9.23) to be $\approx 10^{-4} \, \text{cm}$ when $j = 1 \, \text{A/cm}^2$; hence the theory is applicable up to pressures of about $10 \, \text{mm Hg}$.

There is now no doubt that such 'low-voltage arcs' can be maintained at a voltage much smaller than the lowest excitation potential; what is not always emphasized is that in order to start this discharge a much higher potential is necessary. It has also been confirmed that in order to maintain the discharge plasma oscillations are not necessary though these arc discharges tend to work in a discontinuous or oscillatory

manner, particularly at low currents. However, there is no final answer
to the question of whether the primary electrons find in the plasma an
artifice which without extracting too much energy is able to transform
the more or less uniform electron energy into an energy distribution
which is needed to satisfy ion production in the gas (2, 200).

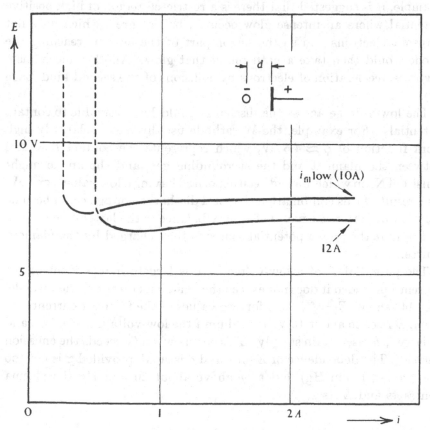

FIG. 143. Variation of voltage E with current i of a low-voltage arc in a spherical
vessel ($R = 5$ cm) filled with argon of a few mm Hg between an oxide-coated
cathode and a nickel anode 1 cm apart, for different cathode temperatures
(201).

In fact it has been suggested, as a result of certain probe measurements, that there is a strong positive space charge accumulated in front
of the cathode, so intense that the space potential there is considerably
higher than the discharge voltage and at least higher than the lowest
excitation potential. How this space charge develops and how the electrons have random energies sufficient to overcome the retardation in the
negative field between space charge and anode is still an open question.

The current to the anode in this case can only flow through this region if the drift velocity due to diffusion exceeds the drift in the negative field, and thus considerable concentration gradients and electron temperatures have to be assumed to make this possible. The experimental evidence supporting this picture is not regarded as quite convincing. For example, it is suggested that there is a restricted region of high positive potential where an intense glow occurs ('ball of fire') which does not transmit electrons and so the major part of the current reaching the anode would then take a path round that glow. Another mechanism involves acceleration of electrons by collisions of the second kind (page 77).

The low voltage across the discharge could be partly due to contact potentials. For example, the W cathode usually has a relatively high work function of $\varphi \approx 4 \cdot 5$ eV, which represents the contact potential between the filament and the surrounding gas, and the anode might consist of Ni on which Ba had been deposited having a low value $\varphi \approx 1$ eV. As a result of this combination the total discharge voltage could become very low or even negative. The energy balance in the latter case is satisfied in that the anode potential source is then charged by the filament source.

The potential E of a low-voltage arc is largely independent of the current i provided it does not exceed the emission current of the cathode. Fig. 143 shows $E = f(i)$ in A for two values of the filament current. At low i, E rises to about 15 V, at medium i the low-voltage arc exists, and at large i, E rises again steeply. This occurs when i exceeds the emission current. The dependence of E on p and d is small provided p is not too low (say < 1 mm Hg), but at p above about 20 mm Hg the plasma contracts and E rises.

QUESTIONS

(1) Outline the essential differences between a glow discharge and a self-sustained arc discharge and emphasize the part of the discharge which determines its principal property. Do discharges exist where one region of the former joins another region of the latter? Give details of the evidence.

(2) Calculate the transition of a glow discharge in N_2 into a thermionic arc assuming a spherical cathode of W which loses heat by black-body radiation only. Neglect anode fall and positive column drop and assume that the electron emission from the cathode is given by $\gamma = 4 \cdot 10^{-2}$, and by (3.59), and use a constant average gas density corresponding to 2000° K. Find the cathode fall $V_c = f(i)$, and the dark space $d = f(i)$ for $p = 100$ mm Hg. Compare the results with experiments and comment.

(3) Give the experimental results obtained from the energy balance in self-sustained arcs and explain to what extent it is possible to set it up for the different regions. What conclusions can be drawn from it with regard to the mechanism?

(4) Outline the character of the light emitted from the various regions of an arc between carbon electrodes in atmospheric air and compare it with that of a copper arc. Give reasons for the stability of the former and explain in physical terms how a copper arc can be stabilized. Comment on the unsatisfactory nature of the conventional stability proof which makes use of the points of intersection of the curves $V - iR$ and $V_a = f(i)$, V and V_a being the total and the arc voltage respectively.

(5) What do you infer from the variations of the current density j_c at the cathode of a carbon arc with the ambient gas pressure when the gas is nitrogen (air) or a rare gas? Give reasons for the observed discontinuity in j_c and your explanation of the variation of the critical gas pressure with the current.

(6) Write an essay on the role of excited particles in a self-sustained arc discharge with a metal cathode and show why in many cases the amount of escaping radiation is restricted.

(7) What types of processes can be envisaged which result in a dependence of the longitudinal electric field in the positive column of an arc discharge as it is given in Fig. 133?

(8) Give a critical account of the methods of measuring the neutral gas temperature in the positive column of an arc and give reasons for the difference between electron and gas temperature at various gas densities. Are electronically excited states of molecules in thermal equilibrium with either temperature and what factors determine the temperature of the excited states?

(9) By what processes are particles issued from the cathode of a 'cold' arc and what experimental evidence exists which supports your view? Why are thermionic arc cathodes not observed with Hg, Cu, and other metals?

(10) Calculate the approximate value of the axial electric field in the positive column of a 10-A arc in atmospheric N_2 when the effective thermal conductivity in the central region $\Lambda = 5.10^4$ erg/cm °K sec ($\approx 5500°$ K) and when the radius limiting this conducting region corresponds to a temperature difference of $500°$ K which defines the total energy loss per cm.

Also estimate the heat convection loss L in the outer region, where $T \approx 600°$ K, the axial gas flow velocity $V = 50$ cm/sec, $r = 0.6$ cm, and the energy per unit mass can be found from simple gas laws. Assume for convection loss $L = \rho' . V dE/dr$, where E is the gas enthalpy and ρ' the mass of gas per cm of length.

THE SIMILARITY RULES

T w o discharges are said to be similar if they are maintained between electrodes of the same material, in the same gas, and show equal potential difference at the same discharge current although the geometry of the two gaps is different.

If the linear dimension of discharge 1 is n times that of 2, the following conditions are to be fulfilled:

$$\lambda_1 = n\lambda_2 \text{ (mean free path for any group of particles)} \qquad (10.1)$$

$$\delta_1 = \delta_2/n \text{ } (\delta = \text{gas density}) \qquad (10.2)$$

$$X_1 = X_2/n \text{ } (X = \text{fields at corresponding points}) \qquad (10.3)$$

$$\lambda_1 X_1 = \lambda_2 X_2 \text{ (hence drift and random velocities at corresopnding} \qquad (10.4)$$
$$\text{points are equal)}$$

$$\sigma_1 = \sigma_2/n \text{ (from (10.3), since } 4\pi\sigma = X) \qquad (10.5)$$

$$\rho_1 = \rho_2/n^2 \text{ } (\rho = \text{space-charge density} = eN) \qquad (10.6)$$

$$j_1 = j_2/n^2 \text{ } (j = \text{current density}) \qquad (10.7)$$

$$dt_1 = n\,dt_2 \text{ } (dt = \text{time interval, follows from constancy of velocity} \qquad (10.8)$$
$$\text{and distance)}$$

The ratios of gas volume, total charge, degree of ionization, frequency, rate of change of current, etc., can be derived from these relations (17, 203, 203 a).

In order to show that in similar discharges only certain fundamental processes are allowed, it is necessary to test whether the rate processes transform according to the above rules, viz.

$$\left(\frac{\partial N}{\partial t}\right)_1 = \frac{1}{n^3}\left(\frac{\partial N}{\partial t}\right)_2. \qquad (10.9)$$

(a) *Ionization by single collision.* From (3.41) we have

$$\frac{dN}{dt} = ap(V - V_i)\frac{dx}{dt}N_e, \qquad (10.10)$$

where all factors are invariant except p and N_e. (pN_e) transforms according to (10.2) and (10.6) with $1/n^3$. This process is thus permissible.

(b) *Stepwise ionization.* It is assumed that electrons produce excited atoms of concentration N^* and average life τ:

$$N^* = \tau\frac{dN^*}{dt}; \qquad (10.11)$$

electrons also ionize N^* at a rate

$$\frac{dN}{dt} = abN_e N^* = abN_e\tau\frac{dN^*}{dt}, \qquad (10.12)$$

where a and b are density-independent constants; one is the probability of ionization of excited atoms, the other is the collision rate constant. For normal excited states and usual collision rates τ is constant and $N_e\,dN^*/dt$ transforms from (10.6) and (10.11) as $1/n^5$. For metastables destroyed on the wall or by collisions with other atoms $\tau_1 = n\tau_2$, and thus the result is $1/n^4$. Both processes are therefore forbidden. Collisions of the second kind are forbidden, the Penning effect is allowed. The proof is left to the reader. A revised and enlarged treatment is found in (203 a).

(c) *Motion of charges by diffusion and in electric fields.* In a steady state the continuity equation is $-\partial N/\partial t = \text{div}(Nv)$, or in one coordinate

$$\frac{\partial N}{\partial t} = -\frac{\partial(Nv_x)}{\partial x}. \tag{10.13}$$

Since v is invariant and N/x transforms with $1/n^3$ (and so do the other components), diffusion and motion in an electric field are allowed processes because (10.13) applies whatever the cause of the drift velocity v.

(d) *Electron attachment.* The rate of change of concentration is proportional to the attachment coefficient a and to the total collision rate:

$$\frac{dN}{dt} \propto aN_e \bar{v}/\lambda_e. \tag{10.14}$$

a and $\bar{v}$ depend on the mean electron energy and N_e/λ_e transforms with $1/n^3$. Attachment is an allowed process and so is detachment.

(e) *Ion–ion recombination.* Since $dN/dt = -\rho N^2$ it follows that for high p, where $\rho \propto 1/p$, recombination is allowed but forbidden at low p, where $\rho \propto p$. From Chapter 6 the reader can easily derive the answer for the various types of recombination including that for electrons and ions.

(f) *Charge transfer.* When fast ions collide with molecules and transfer their charge but not their kinetic energy, the rate of production of fast molecules (or slow ions) is proportional to $N^+\delta$ or $(1/a^2)(1/a)$. Hence neutralization charge transfer is allowed. The reverse process is forbidden (see 30).

(g) *Surface and wall processes.* The surface charge varies in similar discharges as

$$\frac{d\sigma_1}{dt} = \frac{1}{n^2}\frac{d\sigma_2}{dt}. \tag{10.15}$$

Most secondary emission processes and neutralization on the wall conform with this requirement (202). Photoelectric and field emission are forbidden (203a).

The similarity rules serve a dual purpose: they predict the behaviour of a discharge when its parameters are changed and they facilitate sometimes discrimination between fundamental processes.

THE PRINCIPLE OF DETAILED BALANCING

THIS principle, first enunciated by Klein and Rosseland, can be expressed thus: in equilibrium the total number of particles leaving a certain quantum state per sec equals the number arriving in that state per sec, and the number leaving by a particular path equals the number arriving by the reverse path. By this any system containing particles in different states provides a means of preventing the population of a certain species from increasing indefinitely. This principle of microscopic reversibility, as it is also called, has been applied first to thermodynamic equilibria. Here are a few examples:

fast electron+atom → excited atom+slow electron (collision of the first kind).

Applying the above principle, the reverse process must exist, viz.

excited atom+electron → fast electron+atom (collision of the second kind).

Or, since a fast atom can excite a slow one, there is the reverse process

excited atom A+atom B → excited atom B+atom A.

The excitation energy can change its form: electronic excitation can be converted into vibration or sometimes into translation energy, etc. In a collision of the second kind, potential energy of some kind is converted into another form of energy, but not radiation.

An experimental confirmation of collisions of the second kind has been carried out in mercury vapour, in which excited atoms collided with slow electrons (205 a). The measuring principle is similar to that of Franck and Hertz: electrons from a hot cathode are accelerated to a nearby grid after which they move through a field-free space (of length $\approx \lambda_e$) and then through a retarding field to a collector. When the vapour in a field-free space is illuminated with resonance light 2537 Å, the retarding potential shows that the fastest electrons have 7·2 eV; but without illumination they have only 2·5 eV. The difference of these energies is 4·7 eV, which agrees with the excitation energy of the atoms. It is concluded that the slow electrons from the cathode picked up resonance energy in a collision.

The main interest is of course to determine the probability of this process. Two results of a quantum-mechanical treatment will be quoted: the probability of a second-kind collision is larger the smaller the energy difference between the two states. For example, the cross-section for energy transfer from an excited Hg atom ($^3P_1$) on to a Na atom in its ground state forming a Na atom ($^2S_{\frac{1}{2}}$) is very large because the two levels at $\sim$ 4·9 eV are only 20 mV apart. Further, in that type of collision the angular momentum of electron spin is conserved (Wigner's rule), and thus if in a collision the total resultant spin of both atoms remains unchanged, the cross-section for energy transfer is large. Metastable Kr ($S = 1$) with normal Hg ($S = 0$) leads to Hg ^3D and not Hg ^1D, because in the former case $\sum S = 1$, in the latter $\sum S = 0$. All this has been confirmed experimentally. Another example is the collision between metastable Ne atoms ($^3P_2$, 16·6 eV) and A atoms present as impurity, resulting in ionization of A (15·8 eV), known as the Penning effect.

An example of the dependence of the cross-sections for collisions of the first and second kind on the electron energy is shown in Fig. 144. Q_1 is the cross-section for electrons colliding with Hg atoms in the resonance state $^3P_1$ to raise them to the metastable state $^3P_2$ and Q_2 for collisions between $^3P_2$ atoms and electrons resulting in $^3P_1$ atoms, assuming thermal equilibrium. The critical potential for excitation is about 0·6 V, but there is no critical potential of course for the 'down' process; however, only slow electrons have a good chance of carrying away the excess energy. Hence the peak at 0·2 eV (204, 205).

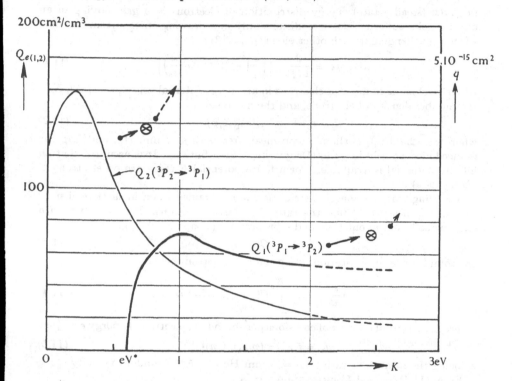

FIG. 144. Cross-section for collisions of the first (Q_1) and second kind (Q_2) as a function of the electron energy K for excited Hg atoms (205).

Returning to the transfer of excitation energy from atoms to electrons, it can be seen that, since the electron spin is always conserved and 'energy resonance' is not required, the cross-section for this process must be very large (probably of order 1). It is, however, necessary that the electron collides with the excited atom during its life, which for metastable atoms is 10^{-3} sec or longer; also that the metastable does not collide earlier with another atom or molecule which 'quenches' it.

Calculation of Q_2: Assume an atom A is hit by an electron of energy eV exciting it to A^*; the electron now having energy $e(V - V^*)$. Theory (97 a) shows that

$$Q_2(V - V^*)/Q_1(V) = (g_1/g_2)\{V/(V - V^*)\},$$

where $g = 2J + 1$ and $eV^* = E_2 - E_1$, the energy level difference. Note that for a cross-section of the first kind a cross-section of the second kind for an electron of energy different from the initial is obtained.

APPENDIX 3

THE ELECTRON ENERGY DISTRIBUTIONS

(a) THE (steady-state) energy distribution of electrons in a gas moving in an electric field, suffering elastic collisions only and drawing energy from the field without exchanging it with other electrons, is (2)

$$N(\epsilon)\,d\epsilon = \frac{2}{\Gamma(\frac{3}{4})}\,N_e\left[\left(\frac{\epsilon}{\epsilon_0}\right)^{\frac{1}{2}}e^{-0.55(\epsilon/\epsilon_0)^2}\,d\left(\frac{\epsilon}{\epsilon_0}\right)\right], \tag{11.1}$$

where $N(\epsilon)$ is the number of electrons in an energy interval $d\epsilon$, $\Gamma(\frac{3}{4}) = 1.23$, N_e is the number density of electrons, and the mean energy

$$\epsilon_0 = (2/3\kappa)^{\frac{1}{2}}e\lambda_1(X/p), \tag{11.2}$$

where $\kappa = 2m/M$, λ_1 is the electron mean free path at 1 mm Hg, and X/p the reduced electric field in e.s.u. (Druyvesteyn distribution). It is assumed that an infinite potential is available. For a finite potential V there are no electrons of energy $> eV$.

Assuming that this energy distribution is maintained when inelastic collisions set in, it is possible to find the rate of inelastic collisions. For example, with exciting collisions of one type and a probability (Chapter 3 A.3) given by

$$P = C(\epsilon - \epsilon_{ex}) \tag{11.3}$$

the total number of excitations per sec for N_e electrons/cm³ is

$$\left(\frac{dN}{dt}\right)_{ex} = \int_{\epsilon_{ex}}^{\infty} C(\epsilon - \epsilon_{ex})f_c N(\epsilon)\,d\epsilon, \tag{11.4}$$

where f_c is the total number of collisions per second per electron of energy $\epsilon = \frac{1}{2}m\bar{v}$ and is given by (3.19)

$$f_c = \bar{v}/\lambda_e = (p/\lambda_1)(2/m)^{\frac{1}{2}}\epsilon^{\frac{1}{2}}, \tag{11.5}$$

λ_1 being the mean free path at $p = 1$ mm Hg and $\bar{v}$ the random velocity.

From (11.4) and (11.5) integration yields

$$\left(\frac{dN}{dt}\right)_{ex} = \frac{1}{2\Gamma(\frac{3}{4})}\frac{N_e p}{\lambda_1}\left(\frac{2}{m}\right)^{\frac{1}{2}}C\frac{\epsilon_0^{5/2}}{\epsilon_{ex}}e^{-(\epsilon_{ex}/\epsilon_0)^2}, \tag{11.6}$$

where ϵ_0 is given by (11.2) and C is given by (11.3). Comparing this result with that based on the Maxwellian distribution it will be seen that for the same value of ϵ_0 a smaller excitation rate dN/dt is found here. This is because the exponential term in (11.6) contains the square of the ratio of energies and thus fewer fast electrons are available.

The same result (11.6) is obtained for the rate of ionization provided C represents the ionization probability per unit excess energy. However, since ionization cannot occur without excitation, the energy distribution may not be of the form assumed above. It can be derived by analysis (18).

(b) The energy distribution of electrons in a gas moving in an electric field and interacting strongly is approximately Maxwellian. The velocity distribution is (cf. 3.1)

$$N(v)\,dv = \frac{4N_e}{\sqrt{\pi}}\left(\frac{v}{v_m}\right)^2 e^{-(v/v_m)^2}d\left(\frac{v}{v_m}\right). \tag{11.7}$$

With $\epsilon = mv^2/2$ and $\epsilon_m = mv_m^2/2 = kT_e$, the energy distribution follows:

$$N(\epsilon)\,d\epsilon = \frac{2N_e}{\sqrt{\pi}}\left[\left(\frac{\epsilon}{\epsilon_m}\right)^{\frac{1}{2}} e^{-\epsilon/\epsilon_m} d\left(\frac{\epsilon}{\epsilon_m}\right)\right]. \tag{11.8}$$

Adopting the same procedure as before, we can calculate for example the total number of ionizations per sec for N_e electrons/cm^3 for a gas with ionization energy ϵ_i (potential V_i):

$$
\begin{aligned}
N_e z = \left(\frac{dN}{dt}\right)_i &= \int_{\epsilon_i}^{\infty} C_i(\epsilon - \epsilon_i) f_c N(\epsilon)\,d\epsilon \\
&= \frac{6.10^2}{(e/m)\sqrt{\pi}} u p N_e\left(\frac{2}{m}\epsilon_m\right)^{\frac{3}{2}} e^{-\epsilon_i/\epsilon_m}\left(1 + \frac{\epsilon_i}{2\epsilon_m}\right) \\
&\approx 9.10^7 a p N_e\, e^{-\epsilon_i/\epsilon_m}(kT_e/e)^{\frac{1}{2}} V_i, \tag{11.9}
\end{aligned}
$$

where V_i and (kT_e/e) are in V, p in mm Hg, and a in ions per V. The factor a is given in Table 3.7 and $\epsilon_i/\epsilon_m = eV_i/kT_e$.

For numerical calculations it is useful to remember that the total number of collisions between all electrons (probably velocity v_m) and atoms is of the order (11.5)

$$N_e f_c = N_e v_m/\lambda_e. \tag{11.10}$$

It is often necessary to estimate the fraction of electrons which are able to excite or ionize. Again assuming an electron swarm with energy distributions (11.1) or (11.8), we observe that these are of the form $x^{\frac{1}{2}}e^{-x^2}\,dx$ and $x^{\frac{1}{2}}e^{-x}\,dx$ respectively. Thus integrating these expressions between the limits ϵ_1/ϵ_m or ϵ_1/ϵ_0 and ∞, where ϵ_1 is a critical energy, we obtain (Appendix 6):

$\dfrac{\epsilon_1}{\epsilon_m}$ or $\dfrac{\epsilon_1}{\epsilon_0}$ = 2	3	4	6	8	10	14
$\displaystyle\int_{\epsilon_1/\epsilon_m}^{\infty} [(\ )] \approx 2{\cdot}5.10^{-1}$	1.10^{-1}	4.10^{-2}	7.10^{-3}	1.10^{-3}	2.10^{-4}	3.10^{-6}
$\displaystyle\int_{\epsilon_1/\epsilon_0}^{\infty} [(\ )] \approx 9.10^{-3}$	5.10^{-5}	4.10^{-8}	..	..	..	..

The energy distribution of electrons is not only sensitive to losses but particularly to variations of the cross-sections with energy (Chapter 3). Assuming elastic collisions only to occur, theory shows that electrons with the same mean energy have the largest number of fast electrons (that is electrons exceeding, say, ionization energy) in a gas with constant collision cross-section, fewer fast electrons in He, still fewer in Ne, and least in A (18). No experimental confirmation is yet available. For application of the foregoing to the calculation of α/p see (148 b). The energy distribution is non-Maxwellian particularly in A (fewer fast electrons), He (more fast electrons), and in molecular gases at low X/p. One cause is thought to be the Ramsauer–Townsend effect (206).

APPENDIX 4

PHOTO-IONIZATION BY BLACK BODY RADIATION

THE intensity of radiation E_λ (ergs/cm³ sec) of wavelength λ and the radiative power dS (ergs/cm² sec) in the range λ to $\lambda+d\lambda$ which an ideal black body emits over the solid angle 2π are connected by

$$dS = 2\pi E_\lambda \, d\lambda. \tag{12.1}$$

Planck's law for any temperature T or $kT \ll h\nu$ respectively is with $x = eV_i/kT$, $\nu = c/\lambda$, and $d\lambda = -c\,d\nu/\nu^2$,

$$E_\lambda = \frac{c^2 h}{\lambda^5} \frac{1}{e^x - 1} \approx \frac{h\nu^5}{c^3} e^{-h\nu/kT}. \tag{12.2}$$

Since $dS = h\nu\,dn$, where dn is the number of quanta per cm² and sec of energy $h\nu$, (12.1) and (12.2) give for $h\nu \gg kT$

$$dn = \frac{2\pi}{c^2} e^{-h\nu/kT} \nu^2 \, d\nu. \tag{12.3}$$

This is the number of quanta which cross 1 cm² per sec at any point in space assumed to be filled with a gas in thermal equilibrium at absolute temperature T. If μ is the absorption coefficient (Table 3.11) and if absorption is identical with ionization, the rate of photo-ionization is found by integrating between $\nu_i = eV_i/h$ and $\nu = \infty$, viz.

$$\frac{dN}{dt} = \int_{\nu_i}^{\infty} \mu(\nu)\,dn. \tag{12.4}$$

Using an average absorption coefficient $\bar{\mu}$, we have from (12.3):

$$\frac{dN}{dt} = \bar{\mu}\frac{2\pi}{c^2}\left(\frac{kT}{h}\right)^3 e^{-x}(x^2 + 2x + 2). \tag{12.5}$$

This is the number of ion pairs/cm³ sec produced by black-body radiation in a gas. For example, radiation from a source at 6000° K passing a gas of $V_i = 15\,\text{V}$ at 1 mm Hg and 0° C and $\bar{\mu} \approx 1$ renders 10^{12} ion pairs/cm³ sec provided the optical path is less than 1 cm.

A relation between the optical absorption cross-section q_a and the electron recombination coefficient ρ_e can be obtained by detailed balancing, thermodynamic equilibrium being assumed.

Let isotropic radiation of density ρ_ν (in ergs/sec cm³ per unit frequency range $d\nu$) be present in a gas-filled enclosure at temperature T. With $y = h\nu/kT$

$$\rho_\nu = \frac{8\pi h\nu^3}{c^3} \frac{1}{e^y - 1}, \tag{12.6}$$

where c is the velocity of light. The factor 8π is to be replaced by the factor 2 if ρ_ν refers to unit solid angle. If stimulated emission is neglected $e^y \gg 1$ and $\rho_\nu \propto \nu^3 e^{-y}$. Since $c\rho_\nu\,d\nu/h\nu$ is the number of quanta crossing unit area per sec, the number of ionizing quanta absorbed (per cm³ and sec) by atoms of cross-section q_a, i.e. for light of frequency $\geqslant \nu_i$, is

$$\left(\frac{dZ}{dt}\right)_a = N \int_{\nu}^{\infty} q_a(\nu)c\rho_\nu\,d\nu/h\nu, \tag{12.7}$$

where N is the number of absorbing atoms in the ground state per cm³, $h\nu_i = eV_i$, and the absorption cross-section $q_a = \mu_1/N_1$ (3.12).

The rate of ionization by light must be balanced by electron–ion recombination according to

$$\text{atom} + h\nu \rightleftarrows \text{positive ion} + \text{electron} + \text{K.E.},$$

so that

$$\left(\frac{dN_e}{dt}\right)_{rec} = -\rho_e N^+ N_e = \rho_e N_e^2 = q_e \bar{v}_e N_e^2. \tag{12.8}$$

This implies that atomic ions are assumed to be present only.

From the Lindemann–Saha equation $N_e = f(T)$ can be found, but instead of (3.57) we write (in c.g.s.)

$$\frac{x^2}{1-x^2} N = \frac{2g^+}{g}\left(\frac{2\pi}{h^2}\right)^{\frac{3}{2}} (mkT)^{\frac{3}{2}} e^{-\nu} \approx N_e^2/N, \tag{12.9}$$

where the g's are the statistical weights of the ions and gas atoms and N is given by $p = NkT$, the sum of all partial pressures. In equilibrium, (12.7) equals (12.8) and by eliminating N_e^2 by means of (12.9) the ratio of the average of the cross-sections for recombination and absorption (apart from a factor of the order 1) is obtained, viz.

$$\frac{q_e}{q_a} \approx \frac{g^+}{g}\left(\frac{h\nu_i}{mc\bar{v}_e}\right)^2 \approx \frac{(h\nu_i)^2}{kTmc^2}, \tag{12.10}$$

where $\bar{v}_e$ is the mean electron velocity and q_a a mean absorption cross-section which often does not vary much within the frequency range considered if ν is sufficiently distant from the nearest X-ray absorption edge. (12.10) shows that the larger the electron velocity and the smaller the ionization potential the smaller is q_e for a given value of q_a. For N_e, $eV_i \approx 21$ eV and $kT \approx 1$ eV, $q_a \approx 2.10^{-17}$ cm². Thus $q_e = 6.10^{-23}$ cm² or $\rho_e \approx 4.10^{-15}$ cm³/sec; at room temperature the values are about 30 times larger.

Photoionization of a gas molecule requires $h\nu \geqslant eV_i$ if it occurs in a single interaction process or $h\nu \geqslant eV_{exc}$, if in stages. However, it has been found (158 d), that gas ionization occurs with very intense light beams, even when $h\nu < eV_{exc}$. If the light from a 7000 Å ruby laser, producing a strong pulse for $\approx 10^{-8}$ sec, is focused by a lens on N_2, H_2, He, or A at 0·3 to 10^2 atm, spectral lines of singly ionized molecules, afterglow (50 μsec) and ionization are observed. With a beam energy of $F = 10^{18}$ ergs/cm² sec, the associated electric field, from $F = cX^2/8\pi$, is $X \approx 10^7$ V/cm at 4.10^{14} c/sec. On the other hand 1·8 eV quanta appear to ionize gas atoms of $V_i \approx 15$ to 25 V with the lowest levels between 11 and 19 V.

Thus the single quantum model seems not applicable whilst the electromagnetic wave concept holds. However, it may be argued that the energy density $F/c \approx 10^7$ ergs/cm³ so that the quanta (10^{19}/cm³) are $< 10^{-6}$ cm apart. If their 'size' were $\lambda = 7.10^{-5}$ cm, this would mean that the quanta strongly overlap. Thus the energy is transferred to an electron by a smeared-out train of quanta, and ionization is due to accumulation of energy in an atomic electron. A free electron could acquire large energy (inverse Bremsstrahlung) but numbers and cross-sections involved are low.

THEORY OF PLANE PROBES

(a) Single probe

The plasma of a discharge is that region in which the concentrations of positive ions N^+ and electrons N_e are equal, negative ions assumed to be absent. In a uniform plasma the electron temperature T_e and ion temperature T^+ are independent of position. Let the concentration of neutral particles be large compared with that of the charges and the corresponding gas pressure low enough for $T_e > T_{gas}$. A plane conducting surface immersed in the plasma—the probe—is connected to a potential source V_S and the latter to the anode of the discharge, the anode fall and potential drop in the positive column being assumed independent of the current flowing to the probe. Taking a point P in the plasma near the probe S, just outside the zone of influence of the probe, and neglecting the contact potential between the anode and S, we have

$$V_S = V_P + V_A, \tag{13.1}$$

where V_P is the potential difference between S and P, and V_A the potential difference between the anode A and P.

When V_S is large and its polarity is such that S is negative with respect to P, all the electrons in the plasma are repelled from S but an ion current flows to S. A fall space develops in front of S with an electric field that decreases with distance x from S and becomes zero at d—the boundary B of the plasma. Thus ions cross the boundary by virtue of their thermal velocities, v^+ being the average. Their number per cm² and sec is $n^+ = \frac{1}{4} N^+ v^+$. The positive ion current density at the probe, en^+, is independent of V_P and depends on N^+ and T^+ only. In the absence of collisions in the fall space between ions of mass m^+ and molecules ($d < \lambda_i$ or λ_e), the current density at S is the same as if there were an ion-emitting electrode at B in high vacuum, viz.

$$j^+ = \frac{1}{9\pi} \left(\frac{2e}{m^+} \right)^{\frac{1}{2}} \frac{V_P^{\frac{3}{2}}}{d^2} = \frac{1}{4} e N^+ v^+. \tag{13.2}$$

Since j^+ for a given plasma is constant, we have $d \propto V_P$. Also for $V_P = $ const, $d \propto (N^+)^{-\frac{1}{2}}$. Both relations have been found to be approximately true provided that V_P is sufficiently strongly negative, the gas pressure not too high, and the size of the probe smaller than λ_i and λ_e.

If the probe is too large, the ions collected come from a larger volume of the plasma than that corresponding to the probe area and d. Also the boundary may become blurred. In fact v^+ and T^+ are usually found to be much too large because probe measurements give N^+ too low (see later) and j^+ too high. If collisions in d occurred and the motion of ions could be described by their mobility, then again $j^+ = $ const, but $d \propto V_P^{\frac{3}{2}}$ and $d \propto (N^+)^{-\frac{1}{3}}$. Because of the non-uniform electric field, this case is never realized. However, ionizing collisions in d would make j^+ increase with V_P and ions as well as photons and metastables would fall on the cathode producing secondary electrons, so that S becomes finally the cathode of a glow discharge.

For any V_P the measured current density at S is

$$j = j^+ + j_e,$$ (13.3)

and hence the electron current density j_e can be found by allowing for j^+. The variation of $j_e = f(V_S)$ can be obtained by using smaller negative values of V_P up to, say, $V_P = V_A$, so that a larger fraction of slow electrons of the distribution can reach S. From (13.2) as V_P is reduced d becomes increasingly smaller. Thus the field of S acts on a region through which electrons of temperature T_e fall without colliding with molecules but retarded by a potential V_P giving rise to a current density

$$j_e = en_e = \tfrac{1}{4}eN_e v_e e^{-eV_P/kT_e}.$$ (13.4)

Taking logs and differentiating with respect to V_S, remembering that $V_A = \text{const}$, we have $\ln j_e = \text{const} - eV_P/kT_e$ and

$$\frac{d(\ln j_e)}{dV_S} = -\frac{e}{kT_e},$$ (13.5)

showing that the slope of the semi-log plot $\ln j_e = f(V_S)$ gives either T_e or the volt-equivalent of T_e, 1 V corresponding to 11600° K.

Let $V_S = V_0$ for $j = 0$ and $j^+ + j_e = 0$. V_0 is negative with respect to P and the number of electrons and ions arriving is equal. Hence, as in (8.49),

$$V_0 = \frac{kT_e}{e} \ln \frac{v_e}{v^+} = \frac{kT_e}{2e} \ln\left(\frac{m^+}{m}\frac{T_e}{T^+}\right).$$ (13.6)

When $V_S = V_A$, ideally the relation (13.4) should suddenly change into a less steeply sloping function, the kink indicating that S is at plasma potential ($V_P = 0$). For larger positive potentials S acts as an anode, attracting electrons from further regions of the plasma until finally the whole discharge current flows into S. For details regarding probe measurements see (6a). Probes for the investigation of the energy distribution of electrons are found in (203a).

The concept of the sheath around a probe is, of course, limited. Physically the boundary between the sheath and the plasma is, in general, not well defined and thus any extension of a probe theory, for example, to higher p, is of doubtful value if the original fundamental assumptions are adhered to. This remark applies for example to the idea that the sheath thickness (if it exists) equals one 'Debye length'. The presence of excited particles has been ignored for convenience. It is known that simple results have been obtained under conditions where 'the probe theory' was invalid.

(b) Double probe

A double probe consists of two equal electrodes, usually wires, which are a few cm apart and immersed in the plasma (Fig. 145). To find its electron temperature, the probe current i is measured as a function of the voltage applied between 1 and 2, varied between $\pm V_a$. Since the measuring circuit is floating, its insulation must be such that leakage currents do not exceed $\approx 10^{-9}$ A (206a).

Consider first an isolated single probe. Electron and ion currents of equal magnitude but opposite sign arrive at its surface from the plasma so that the probe current $i = 0$. Since the electrons move faster than the ions the probe must acquire a negative potential V_p to maintain this state and thus only the 'faster' electrons can reach the probe. With two identical probes and no potential differences in the plasma the current in the external probe circuit is zero for $V_a = 0$. (Any p.d. in the plasma or asymmetry in the probes or sheaths will act like an external voltage and $i \neq 0$—a useful test for exploration; by changing the direction of the discharge current or superimposing an a.c. current, the cause of the unsymmetry may be deduced.) When $V_a \neq 0$ the external probe current will always be finite:

$$i = i_1 = i_2 \quad \text{and} \quad V_a = V_1 + V_2, \tag{13.7}$$

where V_1 and V_2 are the changes in potential at 1 and 2 irrespective of the wall potential V_p. Since the two probe characteristics are assumed to be equal, the changes in plasma potential V_1 and V_2 must take such values that the above condition is satisfied.

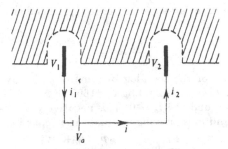

FIG. 145. Double probe immersed in a plasma.

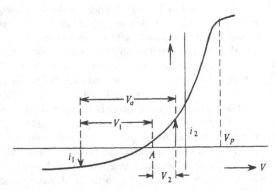

FIG. 146. Current i in a probe circuit as a function of the external applied voltage V_a.

Fig. 146 illustrates the distribution of potentials for a given current i. As V_a is increased from zero (point A) the current i rises first steeply, then less rapidly and would show saturation were it not for the expansion of the 'effective probe area' over which positive ions are collected from the plasma. The probe which assumes a negative potential against the plasma repels a larger fraction of electrons so that the increased probe current is due to an increase in a net positive flow of charges, whereas the other more positive probe allows a larger fraction of electrons to enter the probe. It is seen that this method operates with smaller probe currents than a single probe in that it uses only faster electrons in the tail of the energy distribution. Also as V_a is raised an increasing fraction of V_a drops off across the sheath which carries the ion saturation current, i.e. V_1 increases rapidly whereas V_2 becomes nearly constant.

If Δ is the width (in V) of the curve $i = f(V)$ plotted linearly (Fig. 147) the electron temperature in °K is

$$T_e = \tfrac{1}{4}\Delta . 11{,}700. \qquad (13.8)$$

This result is obtained thus: assume i^+ = constant. This is a good approximation since at high p the ions by charge transfer or other collisions lose readily any excess energy which they may have acquired, whereas at low p, v^+ depends

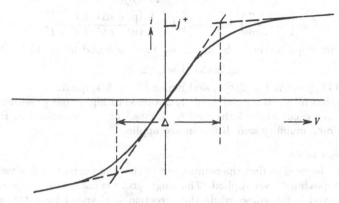

FIG. 147. Current-voltage 'characteristic' of a double probe. j^+ corresponds to the ion saturation current density, $\tfrac{1}{4}\Delta = kT_e$.

somehow on T_e which, however, remains constant. Thus the change in electron numbers (per cm² and sec) arriving at 1 and 2 is $\Delta F_1 = \Delta F_2 = \Delta F$. The probe current density $j = e . \Delta F$ and thus

$$j_1 = e \int_{v_{10}}^{v_1} v\, dN = j_2 = e \int_{v_2}^{v_{20} = v_{10}} v\, dN = j. \qquad (13.9)$$

The integration limits have been chosen so as to give the values of j_1 and j_2 the same sign.

The electron concentrations N_1, N_2 in the plasma and the average velocities are the same at 1 and 2, the latter defined by

$$(\tfrac{1}{2}m)\bar{v}^2 = kT_e. \qquad (13.10)$$

The upper and lower limits of integration are given by

$$v_1^2 = \frac{2e}{m}(V_p - \Delta V_1); \quad v_{10}^2 = \frac{2e}{m}V_p; \quad v_2^2 = \frac{2e}{m}(V_p + \Delta V_2),$$

V_p being the plasma potential.

With a Maxwellian distribution of the velocity components in the direction normal to the electrode surface

$$dN/N = (1/\sqrt{\pi})e^{-x^2}\, dx, \qquad (13.11)$$

where $x = v/\bar{v}$, and with (13.11) integration of (13.9) gives

$$j_1 = c[\exp\{-e(V_p - \Delta V_1)/kT\} - \exp(-eV_p/kT)],$$
$$j_2 = c[\exp(-eV_p/kT) - \exp\{-e(V_p + \Delta V_2)/kT\}], \qquad (13.12)$$

where $c = eN(kT/2\pi m)^{\frac{1}{2}}$. The probe voltage

$$\Delta V = \Delta V_1 + \Delta V_2. \qquad (13.13)$$

Since $j^+ = c\exp(-\Delta V_p/kT)$, where j^+ is the constant ion current to the probes, the total current densities to 1 and 2 are

$$j_1 = j^+\{\exp(-e\Delta V_1/kT)-1\} \quad \text{and} \quad j_2 = j^+\{1-\exp(-e\Delta V_2/kT)\}.$$

(13.14)

From (13.14) and the preceding expressions we find

$$\Delta V = -\frac{kT_e}{e}\ln\left[\frac{1+j/j^+}{1-j/j^+}\right]$$

and

$$j = \left[eN\left(\frac{kT_r}{2\pi m}\right)^{\frac{1}{2}}\exp(-eV_p/kT)\right]\frac{\exp(-e\Delta V/kT)-1}{\exp(-e\Delta V/kT)+1}.$$

(13.15)

The maximum slope is seen to be at $\Delta V = 0$ and is found from (13.15) to be of value

$$|dj/d(e\Delta V)| = j^+/2kT.$$

(13.16)

From Fig. 147 $dj/d(e\,\Delta V) = 2j^+/\Delta$, and hence $kT = \Delta/4$, q.e.d.

It can be shown that the same result is obtained at higher gas pressures where many collisions occur around the probe. Instead of the equations of free fall, those containing mobility and diffusion are applied.

(c) Directional probe

Probes can be used to find the component of electron current in a given direction. Two types have been applied. The single probe is insulated on one side and current received is measured while the direction is changed by rotation. The double probe consists of two metal electrodes in contact with but separated by a thin insulator and the difference of the currents to the electrodes measured as a function of angle. For details see pp. 331 ff. in (30).

SOME USEFUL INTEGRALS AND FUNCTIONS

$$\int_0^y e^{-x^2}\,dx = y - \frac{y^3}{3} + \frac{y^5}{5.2!} - \frac{y^7}{7.3!}$$

$y =$	0·1	0·2	0·5	1	2·0	$\geqslant 2{\cdot}5$
integral $\doteq$	0·1	0·2	0·46	0·75	0·88	see below

$$\int_0^\infty e^{-x^2}\,dx = \tfrac{1}{2}\sqrt{\pi} = 0{\cdot}887$$

$$\int_y^\infty e^{-x^2}\,dx = e^{-v^2}/2y \quad \text{for } y \geqslant 2{\cdot}5$$

$$\int xe^{-x^2}\,dx = -\tfrac{1}{2}e^{-x^2}$$

$$\int x^2 e^{-x^2}\,dx = \tfrac{1}{2}\left[\int e^{-x^2}\,dx - xe^{-x^2}\right].$$

Let n be a positive integer, then

$$\int x^n e^{-x^2}\,dx \ (\text{for } n \text{ even}) = \frac{(n-1)(n-3)\ldots1}{2^{(\frac{1}{2}n)}}\int e^{-x^2}\,dx -$$

$$-\tfrac{1}{2}e^{-x^2}\left[x^{n-1} + \frac{n-1}{2}x^{n-3} + \frac{(n-1)}{2}\frac{(n-3)}{2}x^{n-5} + \right.$$

$$\left. + \ldots + \frac{n-1}{2}\frac{n-3}{2}\ldots\frac{3}{2}x\right],$$

$$\int x^n e^{-x^2}\,dx \ (\text{for } n \text{ odd}) = -\tfrac{1}{2}e^{-x^2}\left[x^{n-1} + \frac{n-1}{2}x^{n-3} + \right.$$

$$+ \frac{n-1}{2}\frac{n-3}{2}x^{n-5} + \ldots + \frac{n-1}{2}\frac{n-3}{2}\ldots\frac{4}{2}x^2 +$$

$$\left. + \frac{(n-1)(n-3)\ldots2}{2^{\frac{1}{2}(n-1)}}\right].$$

$$\int_0^\infty x^{\frac{1}{2}}e^{-x^2}\,dx \quad \text{is related to} \quad \int_0^\infty x^{n-1}e^{-x}\,dx = \Gamma(n), \quad \text{see (207).}$$

$$\int_0^y x^n e^{-x}\,dx = (n,y)! \quad \text{see (207); } (n,\infty)! = n!$$

$$\int xe^{-x}\,dx = -e^{-x}[1+x]$$

$$\int x^2 e^{-x}\,dx = -e^{-x}[x^2 + 2x + 2]$$

$$\int \frac{e^x\,dx}{x} = \ln x + x + \frac{x^2}{2.2!} + \ldots.$$

For small values of y: $\displaystyle\int_0^y \frac{dx}{\ln x} = -\frac{y}{\ln(1/y)}$ and for

$y =$	1	1·4	2	10	25	100	500
integral $=$	$-\infty$	0	1·05	~ 6	18	30	102

$$J_0(x) = 1 - \frac{x^2}{2^2} + \frac{x^4}{2^2 . 4^2} - \dots .$$

CONVERSION FACTORS AND PHYSICAL CONSTANTS

1 e.s.u. of potential	$= 300$ V	
current	$= 3 \cdot 33 . 10^{-10}$ A	
power	$= 1 . 10^{-7}$ W	
charge	$= 3 \cdot 33 . 10^{-10}$ C	
resistance	$= 9 . 10^{11}\ \Omega$	
conductance	$= 1 \cdot 11 . 10^{-12}\ \Omega^{-1}$	
capacitance	$= 1 \cdot 11 . 10^{-12}$ F	
1 eV	$= 1 \cdot 6 . 10^{-12}$ erg/particle	
1 eV	$= 8086$ cm^{-1}	
1 eV	$= 2 \cdot 305 . 10^4$ cal/mole	

Charge of the electron $= 4 \cdot 8 . 10^{-10}$ e.s.u.

Mass of the electron $= 9 \cdot 1 . 10^{-28}$ g $= 5 \cdot 5 . 10^{-4}$ mass units (at. wt.)

Mass of the H atom $= 1 \cdot 673 . 10^{-24}$ g $= 1 \cdot 0076$ m.u.

Mass of the neutron $= 1 \cdot 675 . 10^{-24}$ g $= 1 \cdot 0090$ m.u.

Planck's constant h $= 6 \cdot 62 . 10^{-27}$ erg sec

Boltzmann's constant k $= 1 \cdot 38 . 10^{-16}$ erg/°K

Stefan's constant σ $= 1 \cdot 38 . 10^{-12}$ cal/cm^2 sec (°K)^4

Velocity of light *in vacuo* c $= 2 \cdot 998 . 10^{10}$ cm/sec

Avogadro's number $= 6 \cdot 025 . 10^{23}$ molecules/mole

Loschmidt's number $= 2 \cdot 69 . 10^{19}$ molecules/cm^3 (N.T.P.)

Radius of the first Bohr orbit $= 0 \cdot 53 . 10^{-8}$ cm

Area of the first Bohr orbit $= 8 \cdot 8 . 10^{-17}$ cm^2

$e/m = 1 \cdot 76 . 10^7$ e.m.u./gr $= 5 \cdot 27 . 10^{17}$ e.s.u./g

Gas constant R $= 82$ cm^3 atm/°K mole

PERIODIC TABLE

TABLE A. Periodic table, atomic numbers, and mean atomic weights

Group / Period	Ia	IIa	IIIa	IVa	Va	VIa	VIIa	VIII	VIII	VIII	Ib	IIb	IIIb	IVb	Vb	VIb	VIIb	0
1, 1s	(1) (H) (1)																1 H 1	2 He 4
2, 2s 2p	3 Li 7	4 Be 9											5 B 11	6 C 12	7 N 14	8 O 16	9 F 19	10 Ne 20
3, 3s 3p	11 Na 23	12 Mg 24											13 Al 27	14 Si 28	15 P 31	16 S 32	17 Cl 35	18 A 40
4, 4s 3d 4p	19 K 39	20 Ca 40	21 Sc 45	22 Ti 48	23 V 51	24 Cr 52	25 Mn 55	26 Fe 56	27 Co 59	28 Ni 59	29 Cu 64	30 Zn 65	31 Ga 69	32 Ge 73	33 As 75	34 Se 79	35 Br 80	36 Kr 83
5, 5s 4d 5p	37 Rb 85	38 Sr 88	39 Y 89	40 Zr 90	41 Nb 94	42 Mo 96	43 Tc ?	44 Ru 102	45 Rh 103	46 Pd 104	47 Ag 108	48 Cd 112	49 In 115	50 Sn 119	51 Sb 122	52 Te 128	53 I 127	54 Xe 130
6, 6s (4f) 5d 6p	55 Cs 133	56 Ba 137	57* La 139	72 Hf 179	73 Ta 182	74 W 184	75 Re 186	76 Os 191	77 Ir 193	78 Pt 195	79 Au 197	80 Hg 201	81 Tl 204	82 Pb 207	83 Bi 209	84 Po 210	85 At ?	86 Rn 222
7, 7s (5f) 6d	87 Fr ?	88 Ra 226	89† Ac 227															

*Lanthanide Series 4f	58 Ce 140	59 Pr 141	60 Nd 144	61	62 Sm 150	63 Eu 152	64 Gd 157	65 Tb 159	66 Dy 162	67 Ho 164	68 Er 167	69 Tm 169	70 Yb 173	71 Lu 175
†Actinide Series 5f	90 Th 232	91 Pa 231	92 U 238	93 Np	94 Pu	95 Am	96 Cm	97 Bk	98 Cf					

TABLE B. Electronic configuration of the elements

Groups: Inert gases (column **0**); Representative elements — alkalis etc. **s** (≤ 2) and **p** (≤ 6), with the last p‑column being the *halides*; Transition elements **d** (≤ 10); Inner transition elements **f** (≤ 14).

Number and type of inner electrons	n =	0	s 1	s 2	p 1	p 2	p 3	p 4	p 5 (halides)	d 1	d 2	d 3	d 4	d 5	d 6	d 7	d 8	d 9	f 1	f 2	f 3	f 4	f 5	f 6	f 7	f 8	f 9	f 10	f 11	f 12	f 13
$1s^2$	1	He	H																												
$2s^2\,2p^6$	2	Ne	Li	Be	B	C	N	O	F																						
$3s^2\,3p^6$	3	A	Na	Mg	Al	Si	P	S	Cl																						
(3)	(3)		K	Ca						Sc	Ti	V	Cr	Mn	Fe	Co	Ni														
$3s^2\,3p^6\,3d^{10}$	4	Kr	Cu	Zn	Ga	Ge	As	Se	Br																						
$4s^2\,4p^6$	5		Rb	Sr																											
(4)	(4)									Y	Zr	Nb	Mo	Tc	Ru	Rh	Pd														
$4s^2\,4p^6\,4d^{10}$	5	Xe	Ag	Cd	In	Sn	Sb	Te	I																						
$5s^2\,5p^6$	6		Cs	Ba																											
(5)	(5)									La																					
(5, 4)	(5, 4)																		Ce	Pr	Nd	----	Sm	Eu	Gd	Tb	Dy	Ho	Er	Tm	Yb
$4f^{14}\,5s^2\,5p^6$	(5)									Lu	Hf	Ta	W	Re	Os	Ir	Pt														
$5s^2\,5p^6\,5d^{10}$	6	Rn	Au	Hg	Tl	Pb	Bi	Po	At																						
$6s^2\,6p^6$	7		Fr	Ra																											
(6)	(6)									Ac																					
(6, 5)	(6, 5)																		Th	Pa	U	Np	Pu	Am	Cm	Bk	Cf	----			

Example: Fe is marked 4 (3) d 6. This means that 4 shells contain electrons and that $n = 3$ contains some d-electrons. The first column shows that its first 3 shells are filled. The configuration is: $1s^2$; $2s^2\,2p^6$; $3s^2\,3p^6$; $4s^2$; $3d^6$. This means there are 2 s-electrons ($l = 0$) each in $n = 1, 2, 3, 4$; further, 6 p-electrons each in $n = 2$ and $n = 3$, and 6 d-electrons in $n = 3$; altogether 26 electrons (see Table A). The atomic weight of the main isotope is ≈ 56 and there are 30 neutrons in the nucleus. For nomenclature see Chapter 3.

ANSWERS TO QUESTIONS

CHAPTER 2

(1) $i = 2\pi e N(k^+ + k^-)V/\ln r_2/r_1$; $i = (r_2^2 - r_1^2)\pi e\, dN/dt$;
   i (in e.s.u.) $= 0.39\,V$ (in e.s.u.); $i = 1.5 . 10^{-8}$ A.

(2) i_s (in A) $= 2.9 . 10^{-8}\, p$ (in mm Hg).

(3) $V = (32\pi\alpha/9)^{\frac{1}{3}}x^{\frac{4}{3}}$;   where $\alpha = (j^+/k^+ - j^-/k^-)$,
   $X = (8\pi\alpha x)^{\frac{1}{2}}$, $\rho = (\alpha/8\pi)^{\frac{1}{2}}x^{-\frac{1}{2}}$,
   $\alpha = (9/32\pi)V_0^2/d^3$.

(4) $(dV/dx)^2 = 8\pi j^- \sqrt{(2m/e)}\{\sqrt{V} - \sqrt{V_0} + \alpha\sqrt{(V_0 - V)}\}$; $\alpha = (j^+/j^-)\sqrt{(M/m)}$.

(5) (a) $rd^2V/dr^2 + dV/dr = j^-/\sqrt{(2eV/m)} - j^+/\sqrt{\{2e(V_0 - V)/M\}}$;   note that here j
is a current per cm length.

   (b) $V = \sqrt{2}\alpha . R_1\{\sec^{-1}r/R_1 - \sqrt{2} - \sqrt{(1 + R_1^2/r^2)}\}$,
      $\sin\alpha = j^+/k^+ - j^-/k^-$.

   (c) $j^+ = V_0^2 k^+/2R_1\{\sec^{-1}(R_2/R_1) - \sqrt{2} - \sqrt{(1 + R_1^2/R_2^2)}\}$.

(6) $f = 50/56$.

(7) $i_t = (qr_1 r_2/\Delta r)v/(\Delta r + r_1)^2$,
   $i_t \approx 3 . 10^{-11}$ A, $\Delta t = \Delta r/v = 3.5 . 10^{-8}$ sec.

(9) $\dfrac{Ze}{M-m} = \dfrac{l\sqrt{(2gh)}}{QR}$.

CHAPTER 3

(1) $i = 2 . 10^{-10}$ A; $\phi = 4.1$ V.

(4) $N^* = 7 . 10^6/\text{cm}^3$.

(5) $V_i = 23.7$ V; $V_i' = 54.4$ V.

(8) $f = 0.1$ and 0.5 respectively.

(10) $Z_i = \dfrac{K}{V_i + 5}$ ion pairs.

(11) $x_{\max} = 100$ km, $N_{e\max} = 10^5/\text{cm}^3$, $L_0 = 0.16$ ergs/sec.

(12) $j = 1 . 10^{-14}$ A/cm^2 at $10\,\mu$; $N^* = 5 . 10^6/\text{cm}^3$.

(15) $q_{ee} = 1.5 . 10^{-14}$ cm^2/(eV)$^2 = q_{ii}$.

(17) $V = 1500$ V.

(18) $i = 1.6 . 10^{-7}$ A.

(19) $r_{cr} \approx \left(\dfrac{e}{v}\right)^{\frac{1}{2}}\left(\dfrac{\alpha_0}{m}\right)^{\frac{1}{4}}$.

(21) $r = n^2\left(\dfrac{h}{2\pi e}\right)^2\dfrac{1}{mZ}$; $q = r^2\pi$.

CHAPTER 4

(5) $E(x) = \dfrac{e^{2\sqrt{(2\kappa)}(x/\lambda)}-1}{e^{2\sqrt{(2\kappa)}(x/\lambda)}+1}\left[\dfrac{eX\lambda}{\sqrt{(2\kappa)}}\right]$; $E_\infty = 2.10^{-9}$ ergs; $x = 14\cdot7$ cm.

(9) $x \approx 2\cdot5$ cm; positives/neutrals $= 1/18$.

(11) $p.k_e \approx \dfrac{(e/m)\lambda_1}{\sqrt{\{(e/m)(2\lambda_1/\sqrt{\kappa})(X/p)\}+v_g}}$; for N_2 at 1 atm $k_e = 1\cdot8.10^4$ cm^2/sec V;

v_g is the random velocity of the molecules; for the exact solution see (13).

CHAPTER 5

(4) $d-d_0 = x^2\left[\dfrac{e}{m}\dfrac{2\pi i}{v^3}\right]$.

(5) $D = \left(\dfrac{kT}{4\pi e^2 N}\right)^{\frac{1}{2}}$; $D = 7.10^{-3}$ cm.

(7) $X = 0\cdot09$ V/cm.

(8) $\dfrac{\Delta p}{p} \approx 30$; $\dfrac{D_H}{D_0} \approx \frac{1}{4}$.

(9) $\dfrac{1}{\tau} = \dfrac{1}{\tau_v}+\dfrac{1}{\tau_w} \approx \dfrac{p\bar{v}}{\lambda_{1v}^*}+\dfrac{\frac{2}{3}\bar{v}\lambda_1^*}{pd^2} = C_1 p+\dfrac{C_2}{pd^2}$,

λ_1^* being the mean free path associated with quenching, the index v referring to that in the gas; wall quenching efficiency $= 1$. $p = (1/d)\sqrt{(C_2/C_1)}$ for τ_{max}.

CHAPTER 6

(2) $p_{O_2} \approx 10^{-7}$ mm Hg.

(3) $\tau_H = \frac{1}{3}$ sec, $\tau_{H_2} = 5.10^{-10}$ sec.

(6) $\dfrac{N}{N_0} = \dfrac{1}{\{1+x\sqrt{(\rho_e N_0/6D_a)}\}^2}$; $x = 290$ cm.

(7) Symmetry of $N(x)$ for positive and negative values suggests
$$N = N_0+a_2 x^2+a_4 x^4+...;$$
hence from the differential equation one obtains
$$\dfrac{N}{N_0} = 1-\left(\dfrac{c-\rho N_0}{2D_a}\right)x^2+\dfrac{(c-2\rho N_0)(c-\rho N_0)}{24D_a^2}x^4-....$$

CHAPTER 7

(1) $\dfrac{i_0^+ - i_x^+}{i_0^-} = (e^{\alpha x}-1)\dfrac{v^-}{v^+}$.

(2) Solution is of the form $\ln(i/i_0) \propto pe^{-p}$.

(5) $j/j_0 = \dfrac{1}{\alpha\Delta}[e^{\alpha(d-a)}-e^{\alpha(d-a-\Delta)}]$; $j_0 = eN_0\Delta$.

(8) $V_s = (Bpd)/\ln[Apd/\ln\{1+p^2/aX^2\}]$.

(11) $d = 2\cdot8$ and $0\cdot35$ cm; $\gamma = 10^{-3}$ and 10^{-5}; $X_s = 36\cdot6$ kV/cm.

(12) 112 V; 5 and $8\cdot3$; $1\cdot0015$ and $1\cdot03$; $3\cdot4\cdot10^{-3}$ watts/cm^2.

(13) 24 cm assuming a line source; 41 m assuming a point source.

(14) $(r,p)_{\min} = \ln(1+1/\gamma)\ln n/[(n-1)A\{n^{1/(1-n)}-n^{n/(1-n)}\}]$.

(15) $p_{\max} \approx 3\cdot10^{-3}\dfrac{V}{d}$; $\alpha_{\max} \approx 1\cdot3\cdot10^{-2}\dfrac{V}{d}$.

CHAPTER 8

(5) From (8.5) using zero field mobility data: $j_c = 1\cdot4\cdot10^{-3}$ A/cm^2; from (8.10) with v_0^+ from charge transfer: $j_c = 1\cdot2\cdot10^{-4}$ A/cm^2; for $p = 760$ mm Hg and $800°$ C: $d = 1\cdot9\cdot10^{-3}$ cm; from the value found from (8.10) $j_c = 5$ A/cm^2.

(8) Find $\int_0^D \alpha\,dx$ using (7.11) and express analytically $\gamma_i = f(X/p)$.

(11) $\sigma \approx 2\cdot7\cdot10^{-3}$ e.s.u. (lower value).

(12) $cpR \geqslant 3\cdot10^{-4}$.

CHAPTER 9

(10) $X \approx 15$ V/cm; $L = 1\cdot5\cdot10^9$ ergs/cm sec.

REFERENCES

REVIEW ARTICLES, TABLES

1. K. T. COMPTON and I. LANGMUIR, Rev. Mod. Phys. 2 (1930), 123, and 3 (1931), 191.
2. M. J. DRUYVESTEYN and F. M. PENNING, ibid. 12 (1940), 87.
3. V. J. FRANCIS and H. G. JENKINS, Rep. Progr. Phys. 7 (1940), 230.
4. R. W. LUNT, A. VON ENGEL, and J. M. MEEK, ibid. 8 (1941), 338.
5. F. LL. JONES, ibid. 16 (1953), 216.
5a. A. VON ENGEL, Nuovo cim. 8 (1951), 42.
6. LANDOLT–BÖRNSTEIN, Tabellen, Springer, Berlin.
6a. L. B. LOEB, Brit. J. Appl. Phys. 3 (1952), 341.
6b. S. C. BROWN, Basic Data of Plasma Physics, Wiley, New York, 1959.
6c. L. SPITZER, Physics of Fully-ionized Gases, Interscience Publishers, New York, 1962.
6d. J. J. DELCROIX, Theory of Ionized Gases, Interscience Publ., New York, 1960.

GENERAL REFERENCES

7. J. S. TOWNSEND, Electricity in Gases, Clarendon Press, Oxford, 1915.
8. —— Handb. Radiol. 1, Akad. Verlag, Leipzig, 1920.
9. —— Motion of Electrons in Gases, Clarendon Press, Oxford, 1925.
10. —— Electrons in Gases, Hutchinson, London, 1947.
11. J. J. and G. P. THOMSON, Conduction of Electricity through Gases, Cambridge Univ. Press, 1933.
12. K. K. DARROW, Electrical Phenomena in Gases, Williams, Baltimore, 1932.
13. L. B. LOEB, Fundamental Processes, Wiley, New York, 1939.
14. K. G. EMELEUS, Conduction of Electricity through Gases, Methuen, London, 1951.
15. M. LAPORTE, Décharge électrique dans le gaz, Colin, Paris, 1948.
15a. M. BAYET, Physique électronique, Masson, Paris, 1958.
15b. R. PAPOULAR, Phénomènes électriques, Dunod, Paris, 1963.
16. R. SEELIGER, Physik der Gasentladungen, Barth, Leipzig, 1934.
17. A. VON ENGEL and M. STEENBECK, Elektrische Gasentladungen, vols. 1, 2, Springer, Berlin, 1932–4.
18. H. S. W. MASSEY and E. H. S. BURHOP, Electronic and Ionic Impact Phenomena, Clarendon Press, Oxford, 1952.
19. F. L. ARNOT, Collision Processes in Gases, Methuen, London, 1950.
20. J. M. MEEK and J. D. CRAGGS, Electrical Breakdown of Gases, Clarendon Press, Oxford, 1953.
21. J. D. COBINE, Gaseous Conductors, McGraw-Hill, New York, 2nd ed., 1941.
22. W. FINKELNBURG, Atomic Physics, McGraw-Hill, New York, 1950.
23. S. DUSHMAN in Taylor and Glasstone, Physical Chemistry, 1/2, van Nostrand, New York, 1946.
24. W. DE GROOT and F. M. PENNING, Handb. Phys. 23/1, Springer, Berlin, 1933.
25. R. BÄR, A. HAGENBACH, H. STÜCKLEN, and E. WARBURG, ibid. 14, Springer, Berlin, 1927.
25a. K. PRZIBRAM, ibid. 22/4, Springer, Berlin, 1926.

26. R. H. HEALEY and J. W. REED, *Slow Electrons in Gases*, Amalg. Wireless Australasia, Sidney, 1941.
26a. R. I. REED, *Ion Production*, Acad. Press, New York, 1962.
26b. F. LL. JONES, *Ionization and Breakdown*, Methuen, London, 1957.
27. R. SEELIGER and G. MIERDEL, *Handb. exp. Phys.* 13/3, Akad. Verlag, Leipzig, 1929.
28. E. RUTHERFORD, J. CHADWICK, and C. D. ELLIS, *Radioactive Substances*, Cambridge, 1930.
28a. F. M. PENNING, *Electrical Discharges*, Philips Technical Library, 1957.
29. L. B. LOEB, *Basic Processes*, University of California Press, 1955.
30. W. L. GRANOWSKI, *Der elektrische Strom im Gas*, I, Akad. Verlag, Berlin, 1955.
30a. C. W. ALLEN, *Astrophysical Quantities*, Athlone Press, London, 1955.
30b. G. FRANCIS, *Ionization Phenomena in Gases*, Butterworth, London, 1960.
30c. E. BADARAU and I. POPESCU, *Gaze ionizate*, 1, Edit. Tehnica, Bucharest, 1963.
30d. N. A. KAPZOW, *Elektr. Vorgänge in Gasen*, Verl. d. Wissensch. Berlin, 1955.

CHAPTER 1

31. C.-A. DE COULOMB, *Mém. Acad. Sci. Paris* (1785), 612.
32. O. VON GUERICKE, *Experimenta nova Magdeburgica*, Amsterdam, 1672; see *Handb. Phys.* 1, Springer, Berlin, 1926.
33. B. FRANKLIN, *Phil. Trans.* 47 (1751), 289; see J. A. CHALMERS, *Atmospheric Electricity*, Clarendon Press, Oxford, 1949.
34. G. C. LICHTENBERG, *Novi comment. Götting.* 8 (1777), 168.
35. H. DAVY, *Elements of Chemical Philosophy*, 1812; also *Phil. Trans.* 2 (1821), 487.
36. W. PETROFF, 'Galvano-Volta experiments with a giant battery', *Acad. med. chir.*, Petersburg, 1803.
37. H. AYRTON, *The Electric Arc*, The Electrician Printing Co., London, 1910.
38. M. FARADAY, *Researches in Electricity*, London, 1844.
39. J. PLÜCKER, *Pogg. Ann.* 103 (1858), 88.
40. W. HITTORF, ibid. 136 (1869), 1 and 197.
41. E. GOLDSTEIN, *Berl. Ber.* (1876), 271.
42. H. HERTZ, *Wied. Ann.* 9 (1883), 782.
43. W. CROOKES, *Phil. Trans.* 1 (1879), 135.
44. See G. C. DALLA NOCE and G. VALLE, *Scelta di scritti di A. Righi*, Zanichelli, Bologna, 1950; also A. VON ENGEL, *Nature, Lond.* 167 (1951), 174.
45. H. HERTZ, *Wied. Ann.* 31 (1887), 983.
46. W. HALLWACHS, ibid. 33 (1888), 301.
47. G. J. STONEY, *Proc. Dublin Soc.* 4 (1891), 563, and *Phil. Mag.* 38 (1894), 418.
48. A. PERRIN, *C.R.* 121 (1895), 1130.
49. J. J. THOMSON, *Phil. Mag.* 44 (1897), 293.
50. W. KAUFMANN, *Wied. Ann.* 61 (1897), 544.
51. E. WIECHERT, *Sitzber. Königsberg.* Jan. 1897.
52. P. LENARD, *Wied. Ann.* 52 (1894), 23, and *Quantitatives über Kathodenstrahlen*, Winter, Heidelberg, 1925.
53. E. GOLDSTEIN, *Berl. Ber.* (1886), 691.
54. J. J. THOMSON, *Rays of Positive Electricity*, Longmans, London, 1913.
55. F. W. ASTON, *Mass Spectra and Isotopes*, Arnold, London, 1933.
56. W. WIEN, *Wied. Ann.* 65 (1898), 440; see *Handb. Radiol.* 4/1, Akad. Verlag, Leipzig, 1923.
57. W. GERLACH, *Handb. Phys.* 22, Springer, Berlin, 1926.

CHAPTER 2

61. H. A. WILSON, *Rev. Mod. Phys.* **3** (1931), 156.
62. D. H. WILKINSON, *Ionization Chambers*, Cambridge Monogr. 1950.
63. B. F. J. SCHONLAND, *Atmospheric Electricity*, Methuen, London, 1953.
64. J. A. CHALMERS, *Atmospheric Electricity*, Clarendon Press, Oxford, 1949.
65. R. E. WOOLSEY, see (13), p. 315.
66. R. SEELIGER, *Atomphysik*, p. 139, Springer, Berlin, 1938.
67. C. W. RICE, *Phys. Rev.* **70** (1946), 228.
67a. J. K. THEOBALD, *J. Appl. Phys.* **24** (1953), 123.
67b. I. LANGMUIR, *Phys. Rev.* **38** (1931), 1656.
68. C. A. MEEK and R. W. LUNT, *Trans. Faraday Soc.* **32** (1936), 1273, 1284.
69. D. TAYLOR, *Radio Isotopes*, Methuen, London, 1951.
70. H. J. LOWE and D. H. LUCAS, *Brit. J. Appl. Phys.*, Suppl. 2 (1953), 40.
70a. A. HORNBECK and P. MOLNAR, *Phys. Rev.* **84** (1951), 615.
70b. A. VAN DER ZIEL, *Noise*, Wiley, New York, 1959.
70c. A. VON ENGEL and J. R. COZENS, *Adv. Electronics* **20** (1964).

CHAPTER 3 A

Section 1

71. K. F. HERZFELD in Taylor and Glasstone, *Physical Chemistry*, van Nostrand, New York, 1951.
72. L. B. LOEB, *Kinetic Theory*, p. 132, McGraw-Hill, New York, 1934.
72a. R. FÜRTH, *Handb. Phys.* **4**, Springer, Berlin, 1929.
73. CHR. FÜCHTBAUER et al., *Z. Phys.* **90** (1934), 403; **95** (1935) 1.
73a. S. CHAPMAN and T. G. COWLING, *Non-uniform Gases*, Cambridge University Press, 1939.
73b. M. H. CHOUDHURY, B.Sc. Thesis, Oxford, 1959.
73c. B. KIVEL, *Phys. Rev.* **116** (1959), 1484.

Section 2

74. R. FRAZER, *Molecular Beams*, Methuen, London, 1937.
75. I. ESTERMANN, *Rev. Mod. Phys.* **18** (1946), 300.
76. J. H. SIMONS et al., *J. Chem. Phys.* **11** (1943), 312.
76a. I. ESTERMANN et al., *Phys. Rev.* **71** (1947), 250.
76b. A. DALGARNO and R. A. BUCKINGHAM, *Proc. Roy. Soc.* A, **213** (1952), 506.
76c. A. G. GAYDON, *Dissociation Energies*, Hutchinson, London, 1947.
76d. J. B. HASTED, *J. Appl. Phys.* **30** (1959), 22.

Section 3

77. A. G. SHENSTONE, *Rep. Progr. Phys.* **5** (1938), 210; *Phil. Trans.* A, **241** (1948), 297.
77a. D. C. FROST and C. A. McDOWELL, *Canad. J. Chem.* **38** (1960), 407.
78. D. R. BATES et al., ibid. **243** (1950), 117.
78a. M. GOODRICH, *Phys. Rev.* **52** (1937), 259.
79. E. J. BOWEN, *Chemical Aspects of Light*, Clarendon Press, Oxford, 1946.
80. C. KENTY, *J. Appl. Phys.* **21** (1950), 1309.
81. E. H. S. BURHOP, *Auger-effect*, Cambridge Monogr. 1952.
81a. J. M. VALENTINE and S. C. CURRAN, *Rep. Progr. Phys.* **21** (1958), 1.
82. W. HANLE, *Phys. Z.* **33** (1932), 245.

82a. D. RIEDE, Z. Phys. 137 (1954), 313.
82b. H. RAUSCH VON TRAUBENBERG et al., Nature, Lond. 18 (1930), 417.
83. G. HERZBERG, Atomic Spectra, Dover, New York, 1944.
84. —— Molecular Spectra, Prentice-Hall, New York, 1950.
84a. K. T. DOLDER, M. F. A. HARRISON, and P. C. THONEMANN, Proc. Roy. Soc.
    A, 264 (1961), 367.
85. A. C. G. MITCHELL and M. W. ZEMANSKY, Resonance Radiation, Cambridge
    University Press, 1934.
85a. M. FABRE DE LA RIPELLE, J. Phys. Rad. 10 (1949), 319.
85b. M. GRYZINSKI, Phys. Rev. 115 (1959), 374.
86. P. C. THONEMANN, Progr. Nucl. Phys. 3 (1953), 219.
86a. L. S. FROST and A. V. PHELPS, Phys. Rev. 127 (1962), 1621.
86b. N.B.S. Monograph 50, Bibliography on Atomic Transition Probabilities, 1962.
86c. W. L. FITE et al., Phys. Rev. 116 (1959), 356.
86d. L. VRIENS, Phys. Letters, 8 (1964), 260.

Section 4

87. H. W. BERRY et al., Phys. Rev. 61 (1942), 63, and 62 (1942), 378.
87a. F. HORTON and D. MILLEST, Proc. Roy. Soc. A, 185 (1946), 381.
88. F. MANDL, A.E.R.E. Rep. T/R 1006, H.M. Stat. Off., 1952.
88a. P. SWAN, Proc. Phys. Soc. Lond. 68 B (1955), 1157.
89. A. ROSTAGNI, Nuovo Cim. 15/2 (1938), 117; 11 (1934), 34, 99.
90. G. L. WEISSLER et al., J. Opt. Soc. Amer. 42 (1952), 84; Proc. Roy. Soc. A, 220
    (1953), 71.
90a. J. P. CURTIS, Phys. Rev. 94 (1954), 908.
90b. E. C. BAUGHAN, Trans. Faraday Soc. 57 (1961), 1863; 59 (1963), 635.
91. W. WIEN, Handb. exp. Phys. 14, Akad. Verlag Leipzig, 1927.
92. J. T. TATE and P. T. SMITH, Phys. Rev. 46 (1934), 773.
93. M. BETHE, Handb. Phys. 21/1, Springer, Berlin, 1933.
93a. R. SCHADE, Z. Phys. 105 (1937), 595; M. Biondi, Phys. Rev. 88 (1952), 660.
93b. L. A. EDELSTEIN, Nature, 182 (1958), 932.
94. R. W. DITCHBURN et al., Proc. Roy. Soc. A, 181 (1943), 386; 219 (1953), 89.
94a. F. W. COOKE, Phys. Rev. 38 (1931), 1351.
95. F. A. LINDEMANN, Phil. Mag. 38 (1919), 669.
96. M. N. SAHA, ibid. 40 (1920), 472, 809.
97. A. UNSÖLD, Sternatmospheren, Springer, Berlin, 1938; see Rosseland, Astro-
    physics, Oxford, 1936; H. ALFVÉN, Cosmical Electrodynamics, Oxford,
    1953.
97a. E. J. B. WILLEY, Collisions of the Second Kind, Arnold, London, 1937.
97b. G. L. WEISSLER, Handb. Phys. 21, 304, Springer, Berlin, 1956.
97c. B. A. TOZER and J. D. CRAGGS, J. Electr. 8 (1960), 103.
97d. K. SOMMERMEYER and H. DRESEL, Z. Phys. 144 (1956), 388.
97e. N. G. UTTERBACK, Phys. Rev. 129 (1963), 219.
98. F. C. CHALKLIN and B. L. WORSNOP, X-Rays, Methuen, London, 1950.
98a. T. L. JOHN, M.N. Astr. 121 (1960), 41; L. M. BRANSCOMB and S. J. SMITH,
    Phys. Rev. 98 (1955), 1127.
98b. R.H.FOWLER and E.A.GUGGENHEIM, Statistical Thermodynamics, Cambridge
    Univ. Press, 1949.
98c. W. LICHTEN, J. Chem. Phys. 26 (1957), 306.
98d. W. CHRISTOPH, Ann. Phys. 23 (1935), 51.
99. W. LOCHTE-HOLTGREVEN, Naturw. 38 (1951), 258; Observatory, 72 (1952), 145.
100. H. S. W. MASSEY, Negative Ions, Cambridge Monogr. 1950.

100a. G. F. ROUSE and G. W. GIDDINGS, *Proc. Nat. Acad. Amer.* **11** (1925), 514; **12** (1926), 447.
100b. B. M. YAVORSKY, *Zh. eksper. teor. fiz.* 1/7 (1952), 126.
100c. M. PAHL, *Ergebn. exakt. Naturw.* **34** (1962), 182.
100d. J. B. HASTED and R. A. SMITH, *Proc. Roy. Soc.* A, **235** (1956), 349.
100e. J. D. CRAGGS, R. THORBURN, and B. A. TOZER, *Proc. Roy. Soc.* A, **240** (1957), 473; G. J. SCHULZ, *Phys. Rev.* **113** (1959), 816.
100f. W. STEUBING, *Phys. Z.* **10** (1909), 787.
100g. D. R. BATES (ed.), *Atomic Processes*, Acad. Press, New York, 1962.
100h. G. O. BRINCK, *Phys. Rev.* **134** (1964), A 345.
100j. D. L. EDERER and D. H. TOMBOULIAN, *Phys. Rev.* **133** (1964), A 1525.

CHAPTER 3B

Sections 5 and 6

101. T. J. JONES, *Thermionic Emission*, Methuen, London, 1936.
102. D. A. WRIGHT, *Proc. Inst. Elect. Engrs.* **100** (1953), part 3, 125.
102a. C. HERRING and M. H. NICHOLS, *Rev. Mod. Phys.* **21** (1949), 185.
103. J. H. DE BOER, *Electron Emission*, Cambridge Univ. Press, 1935.
103a. R. H. SLOANE and R. M. HOBSON, *Proc. Phys. Soc. Lond.* A, **66** (1953), 663.
103b. G. COUCHET, *Ann. Chim. (Phys.)*, **9** (1954), 731.
103c. O. HACHENBERG and W. BRAUER, *Adv. Electronics*, **11** (1959), 413.
103d. J. D. CRAGGS and H. S. W. MASSEY, *Hdb. Phys.* **37** (1959), 314.

Section 7

104. K. G. MCKAY, *Adv. in Electronics*, **1** (1948), 65.
105. H. BRUINING, *Secondary Electron Emission*, Pergamon Press, London, 1954.
106. H. FRIEDENSTEIN et al., *Rep. Progr. Phys.* **11** (1948), 335.
107. C. W. MUELLER, *J. Appl. Phys.* **16** (1945), 453.
107a. J. WOODS, *Proc. Phys. Soc. Lond.* B, **67** (1954), 843.
107b. O. HACHENBERG and W. BRAUER, *Adv. Electronics*, **11** (1959), 413.

Section 8

108. R. DORRESTEIN, *Physica*, **9** (1942), 447.
109. D. GREENE, *Proc. Phys. Soc. Lond.* **63** (1950), 876.
109a. H. G. HAGSTRUM, *Phys. Rev.* **89** (1953), 244; **96** (1954), 325, 336.
110. A. ROSTAGNI, *Ric. scient.* II/9, vol. **1**, 1938.
111. M. HEALEA and C. HOUTERMANS, *Phys. Rev.* **58** (1940), 608.
112. A. G. HILL et al., ibid. **55** (1939), 463.
113. A. KRUITHOF and F. M. PENNING, *Physica*, **4** (1937), 430.
114. F. LL. JONES, *Proc. Phys. Soc. Lond.* B, **64** (1951), 397, 519.
114a. M. SCHWARZ and P. L. COPELAND, *Phys. Rev.* **96** (1954), 1466.

Section 9

114b. R. F. STEBBINGS, *Proc. Roy. Soc.* A, **241** (1957), 270.
115. A. L. HUGHES and L. A. DU BRIDGE, *Photoelectricity*, McGraw-Hill, New York, 1932; H. SIMON and R. SUHRMANN, *Lichtelektrische Zellen*, Springer, Berlin, 1958.
116. A. SOMMER, *Photoelectric Tubes*, Methuen, London, 1951.

117. C. Kenty, *Phys. Rev.* **44** (1933), 891.
118. N. Wainfan *et al.*, *J. Appl. Phys.* **24** (1953), 1318.
118*a*. P. Gorlich, *Ann. Phys.* **13** (1932), 831.
118*b*. V. K. Rohatgi, *J. Appl. Phys.* **28** (1957), 951.

Sections 10 *and* 11

119. R. O. Jenkins, *Rep. Progr. Phys.* **9** (1943), 177.
120. J. E. Henderson and R. K. Dahlstrom, *Phys. Rev.* **55** (1939), 473.
121. E. W. Müller, *Z. Phys.* **102** (1936), 734.
121*a*. E. Y. Zandberg and N. I. Ionov, *Soviet Phys.* USPEHKI 67 (1959), 255.
121*b*. E. W. Müller, *Phys. Rev.* **102** (1956), 618.
121*c*. P. Görlich, *Adv. Electronics*, **11** (1959), 1.

CHAPTER 4

Section 1

122. A. M. Tyndall, *Mobility of Positive Ions*, Cambridge Tracts, 1939.
123. W. H. Bennet and L. H. Thomas, *Phys. Rev.* **62** (1942), 41.
124. K. H. Kingdon and E. J. Lawton, ibid. **56** (1939), 215.
124*a*. P. C. Hutton, *Proc. Phys. Soc. Lond.* **82** (1963), 526.
124*b*. L. Goldstein, *Adv. Electronics*, **7**, (1955), 401.
125. R. Meyerott, ibid. **70** (1946), 671.
126. J. A. Hornbeck, ibid. **83** (1951), 374.
127. R. L. F. Boyd, *Proc. Phys. Soc. Lond.* A, **63** (1950), 543.
128. R. J. Munson and A. M. Tyndall, *Proc. Roy. Soc.* A, **177** (1941), 187.
128*a*. H. Biondi and I. M. Chanin, *Phys. Rev.* **96** (1954), 831.
128*b*. R. N. Varney, ibid. **88** (1952), 362; **89** (1953), 708; *Proc. Conf. Ioniz., Phen., Paris*, 1963.
128*c*. T. Popescu, *Rev. Phys.* **4** (1959), 199.
128*d*. J. A. Dahlquist, *J. Chem. Phys.* **39** (1963), 1203.

Section 2

129. N. E. Bradbury and R. A. Nielsen, *Phys. Rev.* **49** (1936), 338, and **51** (1937), 69.
130. R. A. Nielsen, ibid. **50** (1936), 950.
131. J. A. Smit, *Physica*, **3** (1936), 543.
132. P. M. Davidson, *Proc. Phys. Soc. Lond.* B, **67** (1954), 159.
132*a*. E. W. B. Gill and A. von Engel, *Proc. Roy. Soc.* A, **197** (1949), 107.
132*b*. A. V. Phelps, *et al.*, *Phys. Rev.* **117** (1960), 470.
132*c*. J. C. Bowe, *Phys. Rev.* **117** (1960), 1411.

Section 3

133. J. B. Hasted, *Proc. Roy. Soc.* A, **205** (1951), 421; **227** (1955), 466.
134. B. M. Palynkh and L. A. Jena, *J. exp. theor. Phys. U.S.S.R.* **20** (1950), 481.
134*a*. R. F. Potter, *J. Chem. Phys.* **22** (1954), 974.
134*b*. A. C. Whittier, *Canad. J. Phys.* **32** (1954), 275.
134*c*. M. Sakuntala and A. von Engel, *J. Elect.* **9** (1960), 31.
134*d*. S. K. Allison, *Rev. Mod. Phys.* **30** (1958), 1137.
134*e*. E. W. McDaniel *et al.*, *Rev. Sci. Inst.* **33** (1962), 2.

CHAPTER 5

Section 1

135. W. Jost, *Diffusion*, Academic Press, New York, 1952.

Section 3

135a. W. P. Allis and D. J. Rose, *Phys. Rev.* 93 (1954), 84.
135b. K. B. Persson and S. C. Brown, *Phys. Rev.* 100 (1955), 729; H. H. Bromer, *Z. Phys.* 158 (1960), 133.
135c. M. A. Biondi and S. C. Brown, *Phys. Rev.* 75 (1949), 1700.
136. L. G. H. Huxley and A. A. Zaazou, *Proc. Roy. Soc.* A, 196 (1949), 402.
137. R. W. Crompton and D. J. Sutton, ibid. 215 (1952), 467.

Section 5

138. V. A. Bailey, *Phil. Mag.* 9 (1930), 560.
138a. J. S. Townsend and E. W. B. Gill, *Phil. Mag.* 26 (1938), 290.
138b. R. J. Bickerton and A. von Engel, *Proc. Phys. Soc. Lond.* B, 69 (1956), 468.

Section 6

139. E. E. Watson, *Phil. Mag.* 3 (1927), 849.
139a. O. Klemperer, *Electron Physics*, Butterworth, London, 1959.
139b. J. Dutton and E. M. Williams, *Proc. Phys. Soc.* 84 (1964), 171.
139c. T. C. Marshall and L. Goldstein, *Phys. Rev.* 122 (1961), 367.

CHAPTER 6

Section 3

140. J. Sayers, *Proc. Roy. Soc.* A, 169 (1938), 83.

Section 4

141. H. S. W. Massey, *Phil. Mag.*, Suppl. 1 (1952), 395.
142. F. L. Mohler, *Phys. Rev.* 31 (1928), 187; *Bur. Stand. J. Res.* 10 (1933), 771.
143. M. A. Biondi and S. C. Brown, *Phys. Rev.* 76 (1949), 1697.

Section 5

144. D. R. Bates, *Phys. Rev.* 78 (1950), 492.
144a. C. Kenty, *Phys. Rev.* 32 (1928), 624.
144b. D. R. Bates and A. E. Kingston, *Nature, Lond.* 189 (1961), 652.
144c. H. J. Oskam and V. R. Mittelstadt, *Phys. Rev.* 132 (1963), 1445.
144d. W. R. S. Garton, *J. Spectr.* 2 (1962), 335, and *Proc. Conf. Ioniz. Phen.*, Munich, 1961.

CHAPTER 7

Section 1

145. A. A. Kruithof and F. M. Penning, *Physica*, 3 (1936), 515.
145a. S. J. B. Corrigan and A. von Engel, *Proc. Phys. Soc. Lond.* 72 (1958), 786.
145b. —— —— *Proc. Roy. Soc.* A, 245 (1958), 335 and private communication.

146. K. G. EMELEUS, R. W. LUNT, and C. A. MEEK, *Proc. Roy. Soc.* A, **156** (1936), 394.
147. J. DUTTON, S. C. HAYDON, and F. LL. JONES, ibid. **213** (1953), 203; see *Nature*, **200** (1963), 58.
148. M. A. HARRISON and R. GEBALLE, *Phys. Rev.* **91** (1953), 1.
148a. J. P. MOLNAR, *Phys. Rev.* **83** (1951), 940.
148b. A. VON ENGEL, *Encycl. Phys.* **21**, Springer, Berlin, 1956.
148c. W. LEGLER, *Z. Phys.* **173** (1963), 169.

Section 2

149. P. MORTON, *Phys. Rev.* **10** (1946), 358.
149a. E. BADAREU and G. G. BRATESCU, *Brit. Soc. Rom. Fiz.* **45** (1944), 9.
149b. T. M. SHAW, *J. Chem. Phys.* **30** (1959), 1366.

Section 3

150. F. LL. JONES, and W. R. GALLOWAY, *Proc. Phys. Soc. Lond.* **50** (1938), 207.
151. R. GRIGOROVICI, *Z. Phys.* **111** (1939), 596.
152. J. H. BRUCE, *Phil. Mag.* **10** (1930), 476.
153. L. G. H. HUXLEY, ibid. **10** (1930), 185.
154. G. A. KACHICKAS and L. M. FISHER, *Phys. Rev.* **91** (1953), 775.
155. —— —— ibid. **88** (1952), 878.
155a. G. G. BRATESCU, *Ac. Rep. DIN*, **2** (1950), 12 Iunie.
155b. R. N. FRANKLIN, *Conf. Ioniz. Phen.*, *Uppsala*, **1** (1960), 164.
156. W. ROGOWSKI, *Arch. Elektrotech.* **20** (1928), 101.
157. J. ZELENY, *J. Appl. Phys.* **13** (1942), 444.
158. H. RAETHER, *Z. Phys.* **117** (1941), 375; *Ergebn. exakt. Naturw.* **22** (1949), 73.
158a. —— ibid. **33** (1961), 218.
158b. T. ITOH and T. MUSHA, *J. Appl. Phys.* **31** (1960), 744.
158c. J. D. CLARKE and P. J. HUTTON, *Proc. Conf. Ioniz. Phen. Paris*, 1963.
158d. R. W. MINCK, *J. Appl. Phys.* **35** (1964), 252.
158e. P. M. DAVIDSON, *Proc. Phys. Soc.* **80** (1962), 143.

CHAPTER 8

Section 1

159. T. SMITH *et al.*, *Proc. Leeds Phil. Soc.* **5/3** (1949), 207.
160. A. VON ENGEL, *Phil. Mag.* **32** (1941), 417.
160a. C. J. F. CHAUNDY, *Brit. J. Appl. Phys.* **5** (1954), 255.
160b. A. L. WARD and E. JONES, *Phys. Rev.* **122** (1961), 376.

Section 3

161. P. F. LITTLE and A. VON ENGEL, *Proc. Roy. Soc.* A, **224** (1954), 209.
162. K. WOLF, *Z. Phys.* **112** (1939), 96.
163. W. BRAUNBECK, ibid. **21** (1924), 204.
164. F. W. ASTON, *Proc. Roy. Soc.* A, **84** (1911), 526.
164a. W. STEUBING, *Ann. Phys. Lpz.* **10** (1931), 296.
164b. T. I. CAMPAN, *Z. Phys.* **91** (1934), 111; R. M. CHAUDRI and M. OLIPHANT, *Proc. Roy. Soc.* **137** (1932), 662.
164c. E. T. KUCHERENKO and A. G. FEDORUS, *Radiot. i elektronika*, **4** (1959), 1233.

164d. W. D. Davis and T. A. Vanderslice, *Phys. Rev.* **131** (1963), 21.
164e. L. Rothardt, *Proc. Conf. Ioniz. Phen., Munich*, 1961.
164f. R. Warren, *Phys. Rev.* **98** (1955), 1650.

Section 4

165. A. Guntherschulze, *Z. Phys.* **59** (1930), 433.
166. —— ibid. **61** (1930), 1.
167. —— *Z. tech. Phys.* **11** (1930), 49.
168. —— and H. Fricke, *Z. Phys.* **86** (1933), 451, 778, 821.
169. A. von Engel, R. Seeliger, and M. Steenbeck, ibid. **85** (1933), 144.
169a. E. Hantzsche, *Beitr. Plasmaphys.* **1** (1961), 179, 203.

Section 5

170. T. J. Killian, *Phys. Rev.* **35** (1930), 1238.
171. M. J. Druyvesteyn, *Z. Phys.* **81** (1933), 571.
172. G. D. Yarnold and S. Holmes, *Phil. Mag.* **22** (1936), 988.
173. G. Mierdel, *Wiss. Veröff. Siemens-Konz.* **17** (1938), 71.
174. L. Beckman, *Proc. Phys. Soc. Lond.* **101** (1948), 515.
175. P. C. Thonemann and W. T. Cowhig, *Proc. Phys. Soc. Lond.* B, **64** (1951), 345.
176. A. Lompe and R. Seeliger, *Ann. Phys.* **15** (1932), 300.
177. V. Elenbaas, *Z. Phys.* **78** (1932), 603.
178. R. Holm, *Wiss. Veröff. Siemens-Konz.* **3** (1932), 159.
179. B. Klarfeld, *Tech. Phys. U.S.S.R.* **4** (1937), 44.
180. H. D. Deas and K. G. Emeleus, *Phil. Mag.* **40** (1949), 460.
181. W. H. Bennett, *Phys. Rev.* **45** (1934), 890; **91** (1953), 1562.
181a. R. G. Fowler, *Proc. Phys. Soc. Lond.* **68** (1955), 130; **80** (1962), 620.
181b. C. Kenty, *Phys. Rev.* **126** (1962), 1235.

Section 7

182. M. J. Druyvesteyn, *Physica*, **2** (1935), 255.
183. R. Seeliger and H. Wulfhekel, *Phys. Z.* **34** (1933), 57.
184. R. Riesz and G. H. Diecke, *J. appl. Phys.* **25** (1954), 196.
184a. M. Steenbeck, *Wiss. Veröff. Siemens-Konz.* **15/2** (1936), 1.
184b. R. B. Cairns and K. G. Emeleus, *Proc. Phys. Soc. Lond.* **71** (1958), 694.
184c. L. B. Loeb, *J. Appl. Phys.* **29** (1958), 1369.

CHAPTER 9

Section 1

185. R. Seeliger, *Handb. exp. Phys.* **13/3**, Akad. Verlag, Leipzig, 1929.
186. A. Hagenbach, *Handb. Phys.* **14** (1927), 324; *Handb. Radiol.* **4/2**, Akad. Verlag. Leipzig, 1924.
186a. W. Rieder, *Z. Phys.* **146** (1956), 629.
186b. G. Ecker, *Ergebn. exakt. Naturw.* **33** (1961), 1; with 650 references.

Section 2

187. H. Maecker, *Ergebn. exakt. Naturw.* **25** (1951), 293; with W. Finkelnburg *Handb. Phys.* **22**, Springer, Berlin, 1956.
188. W. Grotrian, *Ann. Phys.* **47** (1915), 141.
189. A. von Engel, *Z. tech. Phys.* **10** (1929), 505.
190. H. Thoma and L. Heer, ibid. **13** (1932), 464.

Section 3

191. A. von Engel and M. Steenbeck. *Wiss. Veröff. Siemens-Konz.* **10** (1931), 155.
192. ———— ——— ibid. **12** (1933), 74.
193. H. Poritsky and C. G. Suits, *J. Exp. Phys.* **6** (1935), 190, 315.
194. W. Lochte-Holtgreven and H. Maecker, *Z. Phys.* **105** (1937), 1.
194a. H. Edels, *Proc. Inst. Elect. Engrs.* **108**A (1961), 55.
194b. L. A. King, *Sixth Intern. Spectrosc. Coll.*, Pergamon Press, London, 1956.

Section 4

195. M. Steenbeck, *Phys. Z.* **38** (1937), 1019.
196. C. Kenty, *J. Appl. Phys.* **10** (1939), 714.
197. W. Elenbaas, *Mercury Vapour Discharge*, North Holland Publ., Amsterdam, 1951.

Section 5

197a. J. D. Cobine and E. E. Burger, *Phys. Rev.* **93** (1954), 653.
197b. A. von Engel and K. W. Arnold, *Phys. Rev.* **125** (1962), 803.
198. K. D. Froome, *Proc. Phys. Soc. Lond.* **60** (1948), 424; ibid. B, **62** (1949), 805.
199. J. M. Somerville *et al.*, ibid. B, **65** (1952), 963; *The Electric Arc*, Methuen, London, 1959.

Section 6

199a. A. E. Robson and A. von Engel, *Proc. Roy. Soc.* A, **242** (1957), 217.

Section 7

199b. W. Mitkewicz, *J. Russ. Phys. Soc.* **35** (1903), 307, 675.
199c. A. E. Robson and P. C. Thonemann, *Proc. Phys. Soc. Lond.* **73** (1959), 508.
199d. A. von Engel and K. W. Arnold, *Nature, Lond.* **187** (1960), 1101.
199e. ———— ——— *Proc. Phys. Soc. Lond.* **79** (1962), 1098.
199f. W. Bez and K. H. Hocker, *Z. Naturf.* **11**a (1956), 118.

Section 8

200. L. Malter *et al.*, R.C.A. *Review*, **12** (1951), 415.
201. H. Kniepkamp, *Z. tech. Phys.* **17** (1936), 398.

APPENDIX 1

202. M. Steenbeck, *Wiss. Veröff. Siemens-Konz.* **11**/2 (1932), 36
203. F. Ll. Jones and G. D. Morgan, *Proc. Phys. Soc. Lond.* B, **64** (1951), 560.
203a. G. Francis, *Encycl. Phys.* **22**, Springer, Berlin, 1956.
203b. P. F. Little, ibid. **21**, Springer, Berlin, 1956.

APPENDIX 2

204. B. Yavorsky, *J. Phys. U.S.S.R.* **10** (1946), 476.
205. C. Kenty, *J. Appl. Phys.* **21** (1950), 1309.
205a. G. D. Latyschew and A. S. Leipunski, *Z. Phys.* **65** (1930), 111.

APPENDIX 3

206. W. P. ALLIS, *Encycl. Phys.* **21,** Springer, Berlin, 1956.

APPENDIX 5

206a. E. O. JOHNSON and L. MALTER, *Phys. Rev.* **80** (1950), 58.

APPENDIX 6

207. E. JAHNKE and F. EMDE, *Tables of Functions,* Dover Publ., New York, 1945.
208. P. O. PEIRCE, *Tables of Integrals,* Ginn, Boston, 1910.
209. S. J. B. CORRIGAN (private communication); see also D. T. STUART and
     E. GABATHULER, *Proc. Phys. Soc.* **72** (1958), 287.
210. M. KARMINSKY, *Impact Phenomena on Metal Surfaces,* Springer, Berlin, 1964.
211. *Collected works of I. Langmuir,* vol. 3, Pergamon Press, Oxford, 1961.

GLOSSARY

(list of terms, γλωσσα = *glossa* = tongue)

affinity: *af* = *ad; affinis* = neighbouring, connected with ; *finis* = end, boundary.

ambipolar: *ambi* = on both sides; πολος = *polos* = axis.

anode: ανα = *ana* = up; οδος = *hodos* = way.

cathode: κατα = *cata* = down.

centrifugal (-petal): *centrum* = centre; *fugere* = flee (*petere* = seek).

cluster: clot.

diffuse: dif = *dis* = apart; *fundere* = pour, spread.

dissociation: *socius* = comrade.

electron = ηλεκτρον: amber, substance electrically charged by rubbing, smallest negative charge.

electrolysis = . . . λυσις = . . . *lysis* = setting free.

electrophoresis: φορεω = *phoreo* = carry about.

emission: *e* = *ex* = out; *mittere* = send.

excitation: *citare* = *ciere* = set in rapid motion from (ex) rest, excite.

ion = ιον: that which goes, carrier of charge.

ionize: to make ions.

isotropic: ισο = *iso* = equal; τροπος = *tropos* = turn.

metastable: μετα = *meta*, indicating change: liable to change in stability.

microscopic: μικρος = *micros* = small; σκοπεω = *scopeo* = look at.

mutual: *mutuus* = borrowed, lent; reciprocal.

oscillogram: *oscillare* = swing ; γραμμα = *gramma* = writing.

photon: φως = *phos* = light.

plasma: πλασσω = *plasso*: moulded thing = electrons, ions, gas molecules, photons brought together.

polarize (*see* **ambipolar**): arrange in definite direction.

precipitate: *prae* = before ; *caput* = head: send headlong, make tumble.

quantum: *quantus* = as much as: quantity.

quasi: as if.

resonance: *re* = again; *sonare* = sound.

spontaneous: *sponte* = of one's own accord.

virtual: *virtualis* = for practical purposes but not strictly.

INDEX

(T) *refers to table,* (F) *to figure*

abnormal cathode fall, 23 (F).
absorption coefficient, 36, 78 (F) (T).
— of quanta, 50.
— cross-sections, 78 (F) (T), 295.
affinity energy, 86.
age of ions, 115, 158, 168.
allowed transitions, 42.
ambipolar diffusion, 145 (F).
angular distribution of electrons, 39 (F).
— momentum, 42.
anode, arc, 280.
— dark space, 218 (F).
— fall, glow, 221.
— spots, glow, 221.
anomalous terms, 46.
arc anode, 280.
— cathode fall, 275 (T).
— cathode; hot, cold, 273, 276.
— discovery of, 1.
— externally heated cathode, 287 (F).
— electric field, 262 (F).
— gas temperature, 265 (F).
— spectrum, 259.
— transitions, 280 (F).
— voltage, 260 (F).
Aston, dark space, 218 (F).
atom-atom collisions, 71 (F).
atomic beam, 108.
attachment, coefficient for electrons, 188 (F).
— cross-sections, 87 (F).
— energy, 86.
Auger effect, 79.
auto-electronic emission, 103.
— ionization, 49 (F).
avalanches, 178, 209.
average energy, 29.

back-scattering, 21.
band-spectra, arc, 267.
beam measurements, 108.
Bessel function, 241, 301.
black-body radiation, 294.
Bohr theory, 58.
Boltzmann distribution, 83.
boundary, glow, 218.
breakdown, 178, 193.
Brownian motion, 112.
build-up time, 205.

canal rays, 3, 127.
capture of electrons, 86, 87 (F).

cataphoresis, 256.
cathode fall, arc, 275 (T).
— — glow, 223 (F), 229 (T).
cathode rays, 2.
cavity measurements, 169.
characteristic of ionized gas 8, 9 (F).
charge distribution, positive column, 240.
charge transfer, 130 (F), 131 (T).
chemical reactions, arc, 270.
cloud chamber, 79.
clusters, of ions, 116.
cold arc, 276.
collision, frequency, 36.
— second kind, 77, 291 (F).
Compton effect, 81.
conservation of spin, 290.
constricted glow, 234.
contraction, positive column, 253, 281.
corona discharge, 18, 254.
cosmic radiation, 9.
Coulomb force, 37.
critical potentials, 43 (T), 59 (T).
Crookes dark space, 218 (F).
cross-sections, definition, 35.
— elastic, 33 (F).
— excitation, 44 (T) ff.
— ionization, 63 (F).
— for ions, 70 (F).
— for atoms, 71 (F).
current density, arc, 275 (T).
— — glow, 230 (T).
current pulse theorem, 25.

Dalton's law, 82, 257.
dark discharge, 203.
dark space glow, 227 (F), 230 (T).
de Broglie wave, 31, 40, 104.
detachment, 86.
detailed balancing, 290.
diffusion, 140 (T).
— ambipolar, 145 (F).
— plasma, 145 (F).
— in electric field, 142.
— in magnetic field, 146, 148 (F).
— and repulsion, 151.
dipole effect, mobility, 116.
discharges, discovery, 1.
— types, 2.
dissociation, cross-section, 53 (F).
— by electrons, 52 (F), 53 (F).
— in fields, 214 (F).